Impact of Artificial Reefs on the Environment and Communities

Jorge H. P. Ramos
CinTurs, University of Algarve, Portugal

IGI Global
PUBLISHER of TIMELY KNOWLEDGE

A volume in the Practice, Progress, and Proficiency in Sustainability (PPPS) Book Series

Published in the United States of America by
IGI Global
Engineering Science Reference (an imprint of IGI Global)
701 E. Chocolate Avenue
Hershey PA, USA 17033
Tel: 717-533-8845
Fax: 717-533-8661
E-mail: cust@igi-global.com
Web site: http://www.igi-global.com

Library of Congress Cataloging-in-Publication Data

Names: Ramos, Jorge, 1972- editor.
Title: Impact of artificial reefs on the environment and communities / Jorge Ramos, editor.
Description: Hershey, PA : Engineering Science Reference, [2022] | Includes bibliographical references and index. | Summary: "This book makes the connection between research in the natural or social sciences and the tourism industry - as a major engine of the world economy - based on the use of artificial reefs, often inaccessible and forgotten by the public"-- Provided by publisher.
Identifiers: LCCN 2021059218 (print) | LCCN 2021059219 (ebook) | ISBN 9781668423448 (hardcover) | ISBN 9781668423455 (paperback) | ISBN 9781668423462 (ebook)
Subjects: LCSH: Artificial reefs--Social aspects. | Artificial reefs--Economic aspects. | Marine ecotourism. | Sustainable tourism.
Classification: LCC SH157.85.A7 I47 2022 (print) | LCC SH157.85.A7 (ebook) | DDC 639.9/2--dc23/eng/20211209
LC record available at https://lccn.loc.gov/2021059218
LC ebook record available at https://lccn.loc.gov/2021059219

This book is published in the IGI Global book series Practice, Progress, and Proficiency in Sustainability (PPPS) (ISSN: 2330-3271; eISSN: 2330-328X)

British Cataloguing in Publication Data
A Cataloguing in Publication record for this book is available from the British Library.

Practice, Progress, and Proficiency in Sustainability (PPPS) Book Series

Ayman Batisha
International Sustainability Institute, Egypt

ISSN:2330-3271
EISSN:2330-328X

Mission

In a world where traditional business practices are reconsidered and economic activity is performed in a global context, new areas of economic developments are recognized as the key enablers of wealth and income production. This knowledge of information technologies provides infrastructures, systems, and services towards sustainable development.

The **Practices, Progress, and Proficiency in Sustainability (PPPS) Book Series** focuses on the local and global challenges, business opportunities, and societal needs surrounding international collaboration and sustainable development of technology. This series brings together academics, researchers, entrepreneurs, policy makers and government officers aiming to contribute to the progress and proficiency in sustainability.

Coverage

- Sustainable Development
- Socio-Economic
- Knowledge clusters
- Outsourcing
- Environmental informatics
- ICT and knowledge for development
- E-Development
- Green Technology
- Global Content and Knowledge Repositories
- Eco-Innovation

IGI Global is currently accepting manuscripts for publication within this series. To submit a proposal for a volume in this series, please contact our Acquisition Editors at Acquisitions@igi-global.com or visit: http://www.igi-global.com/publish/.

The Practice, Progress, and Proficiency in Sustainability (PPPS) Book Series (ISSN 2330-3271) is published by IGI Global, 701 E. Chocolate Avenue, Hershey, PA 17033-1240, USA, www.igi-global.com. This series is composed of titles available for purchase individually; each title is edited to be contextually exclusive from any other title within the series. For pricing and ordering information please visit http://www.igi-global.com/book-series/practice-progress-proficiency-sustainability/73810. Postmaster: Send all address changes to above address.

Titles in this Series

For a list of additional titles in this series, please visit: http://www.igi-global.com/book-series/practice-progress-proficiency-sustainability/73810

Wetland Biodiversity, Ecosystem Services, and the Impact of Climate Change
Abdelkrim Ben Salem (Mohammed V University, Morocco) Laila Rhazi (Mohammed V University, Morocco) and Ahmed Karmaoui (University of Moulay Ismail, Morocco & Southern Center of Culture and Science,Morocco)
Engineering Science Reference • © 2022 • 310pp • H/C (ISBN: 9781799892892) • US $225.00

Handbook of Research on Organizational Sustainability in Turbulent Economies
Rafael Ignacio Perez-Uribe (Santo Tomas University, Bogota, Colombia) Carlos Salcedo-Perez (Politecnico GranColombiano University Institución, Colombia) and Andres Carvajal-Contreras (EAN University, Colombia)
Business Science Reference • © 2022 • 428pp • H/C (ISBN: 9781799893011) • US $295.00

Examining Algae as a Sustainable Solution for Food, Energy, and the Environment
Mostafa M. El-Sheekh (Tanta University, Egypt) Norhayati Abdullah (Universiti Teknologi Malaysia, Malaysia) and Imran Ahmad (Universiti Teknologi Malaysia, Malaysia)
Engineering Science Reference • © 2022 • 330pp • H/C (ISBN: 9781668424384) • US $195.00

Implications for Entrepreneurship and Enterprise Development in the Blue Economy
Lukman Raimi (Universiti Brunei Darussalam, Brunei) and Jainaba M. L. Kah (Office of the President, Gambia)
Business Science Reference • © 2022 • 302pp • H/C (ISBN: 9781668433935) • US $245.00

Technological Development and Impact on Economic and Environmental Sustainability
Yilmaz Bayar (Bandirma Onyedi Eylul University, Turkey) Mahmut Unsal Sasmaz (Usak University, Turkey) and Omer Faruk Ozturk (Usak University, Turkey)
Engineering Science Reference • © 2022 • 352pp • H/C (ISBN: 9781799896487) • US $215.00

Handbook of Research on Monitoring and Evaluating the Ecological Health of Wetlands
Ashok K. Rathoure (Akone Services, India)
Engineering Science Reference • © 2022 • 370pp • H/C (ISBN: 9781799894988) • US $295.00

Blockchain Technologies for Sustainable Development in Smart Cities
P. Swarnalatha (Vellore Institute of Technology, India) and S. Prabu (Pondicherry University, India)
Engineering Science Reference • © 2022 • 285pp • H/C (ISBN: 9781799892748) • US $215.00

IGI Global
PUBLISHER of TIMELY KNOWLEDGE

701 East Chocolate Avenue, Hershey, PA 17033, USA
Tel: 717-533-8845 x100 • Fax: 717-533-8661
E-Mail: cust@igi-global.com • www.igi-global.com

Table of Contents

Section 4
Sustainable Options

Detailed Table of Contents

Artificial reefs (ARs) have gained considerable attention in Turkey, due to their positive biological and socioeconomic implications. The most recent and comprehensive AR deployment was completed in Edremit Bay. The authors investigated the stakeholders' perception and satisfaction along with social, economic, and management concerns about the AR deployment for the period October-December 2014. They first defined user and non-user AR stakeholders and their interests in the project. They represented multiple interactions among stakeholders about the AR deployment by focusing on user types, attitudes, or behaviors towards the AR deployment. They queried existing and potential problems about the project and the deployment area to produce consistent management solutions. They suggest that public participation and multiple collaboration would enhance communication and deal among the stakeholders towards finding solutions for the existing problems. Future studies should reveal social, economic, and management aspects of the stakeholders by the medium and long-term monitoring studies.

For 59 million people in the world, fishing represents their main source of income. In addition, for 3.2 billion people, fishery resources provide 20% of the protein of animal origin in their diet. One tool that could contribute to recovering stocks in a sustainable way is artificial reefs. In Mexico, 750 structures were deployed in Barra de Pichi and 500 in Caleta de Campos municipality of Lázaro Cárdenas, Michoacán, between 2016 and 2018. The objective of this work is to evaluate the level of sustainability of the artisanal fishing system between Barra de Pichi and Caleta de Campos, in relation to the artificial reefs (ARs) and with the purpose of assessing its usefulness as a public policy tool. The framework for the evaluation of natural resource management systems incorporating sustainability indicators (MESMIS) was used as a methodological guide, comparing two groups: AAusers (users ARs) and non-AAusers (non-users of ARs). The results showed that the artisanal fishing activity carried out by ARs users is more sustainable than non-ARs users.

Diving tourism has been very popular in the world and Turkey. In addition to reasons such as seeking hobbies and socializing, marine biodiversity, natural reefs, wrecks, and corals cause more and more people to dive every day. In many regions in Turkey, which has many underwater riches, diving is prohibited for reasons such as archaeological and military zones and natural reserves. Therefore, alternative diving points have been created for diving tourism by using artificial reefs in recent years. Within the scope of 35 artificial wreck reefs, in which diving tourism has been carried out in Turkey, 3000 amphoras, many thematic sculptures, 22 ships, 9 planes, and 1 tank were launched. While decommissioned vehicles were used in the first years, underwater parks with different themes have been created in recent years. In this study, artificial wreck reefs in Turkey were examined, and the permit process, the structures, and the interests of divers were evaluated. In addition, mistakes made in material and location selection and interactions with other sectors are also discussed.

Artificial reefs are structures that can be natural or artificial materials that are intentionally placed on the seabed, placed at one time or over a long period of time, and ideally will work as part of a natural ecosystem. In artificial reefs deployment, it is necessary to pay attention to several factors so that the artificial reef itself functions optimally. Therefore, competent human resources are needed through the correct technical understanding of artificial reef deployment. The less-than-optimal deployment of artificial reefs in Damas waters is caused by a lack of understanding of human resources, so that when monitoring was carried out, it was found that 22 of the 25 artificial reefs found were mostly in poor condition.

This chapter describes the innovative design and fabrication of a complex reef system for Sydney Harbour, Australia. The Bio-Shelters project is an ongoing collaborative investigation into the application of computational design and Industry 4.0 hybrid technologies to create site-specific artificial reefs. The location of the artificial reefs is Blackwattle Bay in Sydney Harbour. The Bay has been used for heavy industry for over 100 years, resulting in polluted water and severely damaged natural marine ecosystems. The chapter discusses the design approach to the Bio-Shelters, the marine species targeted, the iterative computational design, and sustainable material investigations that took place during the development. The authors then detail the hybrid fabrication processes using laser-cut steel waffle structures and shotcrete surface treatments. The chapter concludes with the current status of the ongoing project and defines the next steps using robotic fabrication.

The artificial reef is a vital submerged structure used to improve marine biological conditions and fishery resources. An artificial reef, with its different designs and structures, reshapes the local flow field by creating upwelling flow, decelerating flow, and back eddy. The shift in flow might result in nutrient exchange in the water and an increase in local primary production. Artificial reefs have been built all around the world, but there is no standardised technique for determining the effect range of an artificial reef.

In recent times, decommissioning of offshore platforms has become an even more discussed topic, for its relevant environmental, social, and economic repercussions. In particular, by carrying out economic considerations, all the divestiture possibilities applicable to an offshore platform and the relative sustainable business models (SBMs) will be analyzed in a wide framework of the circular economy and sustainable principles. In this scenario, sustainable decommissioning (SD) of offshore platforms process refers to multidimensional and interdisciplinary challenges, which requires a deep understanding of technical, legal, economic, financial, social, and environmental variables. The decommissioning of these structures is an issue that has gained a great deal of international attention and will require in the next years an open dialogue and exchange between institutions, oil and gas companies, enterprises, and the environment.

Preface

In recent decades there has been a collective awareness across the world of the important need to preserve marine resources for future generations. One of the ways, seen as useful for the preservation of marine resources, has been the use of artificial reefs in depths lacking in marine life aggregating structures (Pickering et al., 1999).

The concept of artificial reefs has been of civilizational knowledge for hundreds of years (Spanier et al., 1985). Our ancestors, through experience and empirical knowledge, accidentally observed that structures such as stones and tree logs – placed in fresh, brackish or sea water bottoms – allowed the attraction of fish and other biological resources that sought shelter and refuge in these structures. This knowledge and learning allowed to supply an additional value for the fishers, who thus managed to be more certain about their catch, spending less effort and being more certain about the place of capture. In a certain manner, the development of this concept of artificial reefs facilitated the livelihoods of these people. After reaching that level of learning, it seemed quite simple to put into practice and was apparently a very efficient technique from the point of view of the results obtained.

Since then, the concept of artificial reefs has evolved much. Artificial reefs can be categorized in diverse ways. According to Ramm et al. (2021), in a simplistic way, artificial reefs can be classified into three categories. A number of artificial reefs are accidentally sunk relics (shipwrecks). Others that were intentionally sunk serve for coastal protection or for the exploitation of fossil fuels or for other forms of energy production. There are still others, which, having been intentionally sunk, are found at low depths and serve for the protection of living organisms.

Among the artificial structures that are intentionally sunk – such as the platforms for oil and gas production – there are ecological implications when their useful life ends (Coolen et al., 2020). While artificial reefs that are intentionally sunk having objectively a reef function, can present an infinity of forms, and exhibit an immense panoply of complexities to positively maximize the ecological implications derived from their deployment (Na, 2021).

The usefulness of artificial reefs has been seen from different perspectives. For example, a few countries have developed national plans for the sinking of modular or other structures to achieve various aims (MacDonald, 1994). As more generalized objectives, they have been, inter alia, the creation of habitats for the attraction and/or production of marine resources, or to serve as a mitigating measure for fishing practices that may be more destructive and inflict damage to the seabed (Bortone et al., 2011). Modular forms are of great interest for the protection of living marine organisms and allow for more sustainable fishing if their management is properly safeguarded. In this sense, it is particularly important to conduct experimental fisheries to assess the potential usefulness of artificial reefs and propose appropriate management measures (Whitmarsh et al., 2008).

In another aspect, the creation of artificial reefs is also related to the creation of habitats that are attractive for human recreational activities, namely scuba diving. In this category of artificial reefs, there is a wide range of structures that are used – in addition to the modular forms – and that are of more interest for underwater recreational activities. In this category are a series of structures that have a second life, namely a whole panoply of obsolete vehicles – which after being properly treated – are later converted into artificial reefs (Yozzo et al., 2004). Not all of these structures trigger a strong interest on the part of enthusiasts of leisure and tourist activities, as is the case of vessels with an associated historical legacy (Leeworthy et al., 2006). There are, however, other vehicles or structures that do not arouse interest, namely because they are devoid of relevant cultural or historical facts to which they are associated. Another way that has been developed is the creation of natural coral reefs by human intervention methods (Carr & Hixon, 1997).

The materials used to build artificial reefs have also been a constant challenge. Few decades ago, it was thought that virtually all opportunity materials could make good artificial reefs (Carral-Couce et al., 2017). It was thought that artificial reefs could be a desirable alternative for dumping unwanted materials and substances, once incorporated into reef structures, or serving as artificial reefs themselves. This did not happen, and there were many errors that needed to be resolved and corrected. Currently, in terms of materials, there are great developments in the sense of incorporating recycled materials together with inorganic materials derived from organisms. More recently, there is also the manufacture of reef structures using industry 4.0 tools and techniques (Mustafa et al., 2019).

Reef structures are often deployed, and after their maturation they are used commonly by the primary stakeholders who take advantage of the resources that develop from the existence and maturation of the reef. However, over time, there are several places with reef structures that lack proper management (Fabi et al., 2015). Sometimes, the creation of artificial reefs only implies a great confluence of expensive human and material resources, in a brief period pre- and post-reef deployment phase. In the long term, human and financial resources are often lacking to conduct an adequate follow-up, either in terms of monitoring or in terms of managing the structures by their users. By carrying out stakeholder analysis – which identifies relevant actors, their expectations concerning the structures put in place, and their behavior – it is added a valuable contribution towards the understanding of artificial reefs systems as a whole. This is a necessary step in understanding the reef arrangements and its users (Milon et al., 2000).

Among the most important reasons towards why artificial reefs have been created, are the protection of the seabed with macroalgae, the possibility of improving fishing, and meeting the expectations and demands of tourists. Coastal managers have realized that it is important that artificial reefs are useful to people, especially local communities, and that it is therefore important that they are promoted by government entities for various professional and leisure activity use. Coastal managers must take this into consideration when reefs are designed. They must also define the ecosystem services promoted as well as the main uses for which the reefs are created.

CHALLENGES

It is particularly important to be able to gather enough resources to implement artificial reefs. The planning must be well studied and include fundamental aspects such as the materials used in the construction, as well as the location of the deployed structures. In a way, planning must involve as many stakeholders as possible from an early stage of a reef project. However, the interest and degree of involvement of the different stakeholders varies over time, as the different stages of the reef project develop.

From the beginning of the 1970s, science related to artificial reefs began to be produced on a more regular basis. Several areas of study were developed, with a great predominance of some natural sciences (e.g., biology and ecology). There has been a bunch of research on artificial reefs related with management and fisheries aspects, and a little less to tourism. In general, the social sciences have been less represented when it comes to scientific research with artificial reefs. Authors such as Seaman Jr. (2002) have been pinpointing and reviewing artificial reef challenges. Regarding the usefulness of structures for human activities, there are some challenges to ponder:

- The challenge of adjusting sunken artificial structures to serve the purposes for which they were created.
- The challenge of knowing if there are users, whether they extract living resources or not, and if they are satisfied with the structures.
- The challenge of an adequate management of structures, in a sustainable way in terms of biological resources and users in the long term.
- The challenge of identifying when there are enough structures supplied or when they are still lacking.
- The challenge of being able to resolve any conflicts that may arise from post-deployment use between users.

Researchers and other artificial reef stakeholders have previously identified all these challenges. However, it is important never to neglect these challenges having in mind to provide a better approach and an adequate mode of treatment. The aim is artificial reefs must be effectively useful from the point of view of the people who use them.

ORGANIZATION OF THE BOOK

The book *Impact of Artificial Reefs on the Environment and Communities* is organized into twelve chapters distributed by four sections. The chapters focus on various aspects of knowledge, from natural sciences, social sciences and engineering. They are case studies conducted in various parts of the planet. Below is a brief description of each section and its chapters:

Section 1, "Stakeholder Analyses and Other Socioeconomic Aspects," includes three chapters, all related to the use of artificial reefs by different stakeholders. Real or perceived interactions by different users and other parties involved are investigated, from an economic and social point of view.

Chapter 1 addresses the issue of reef efficiency after the maturation of structures, and additionally relate them to the biological components. The perceptions of artificial reefs by human activities, namely primary users, are identified as fundamental. An aspect of how one can indirectly monitor artificial reefs through statistical series related to the primary sector in municipalities where the coastal sea is explored.

Chapter 2 explores a stakeholder analysis by modeling their social networks in relation to artificial reefs. There are used indicators to compare the evolution of stakeholders at various stages of artificial reef projects – inception, implementation and management. With these indicators, the positions of stakeholders within a network are compared to identify whether artificial reef projects succeed or fail.

Chapter 3 reports the stakeholder analysis of a reef implemented in 2014. This analysis is done in relation to the multiple interactions between the stakeholders. Questions are raised about potential problems related with the project, and management solutions are suggested. To this end, multiple public participation and collaboration is proposed, to improve communication and so that stakeholders can deal with and contribute to a better problem-solving approach.

Section 2, "Management of Fisheries in Local Communities," covers three chapters focusing on aspects that relate artificial reefs to local fishing communities and determine the level of engagement with users. It includes the need to find appropriate management strategies in order to achieve artificial reef projects that can be seen as successful.

Chapter 4 focuses on issues that relate to the question how reef deployment and the resources generated can cause conflicts between fishers. Matters between entities entering the reef deployment process and the way in which cooperatives and individual fishers are involved have to do with sustainable management incentives. The control of fishing pressure – which is exerted on fishery resources – is related with their (over)exploitation.

Chapter 5 explores the information artisanal fishers have about artificial reef deployment. Immense importance is given to the localization of modules in mitigating conflicts between users. The importance of reef structures in attracting species of commercial value is mentioned. The degree of interaction between fishing communities and reefs is addressed, as well as how appropriate plans can be tailor-made to suit management.

Chapter 6 mentions that management issues are especially important to successfully implement artificial reef deployment for local fishing communities. It is also mentioned that, although conservation and fisheries management can be included in the purposes of artificial reef projects, their fullness is difficult to achieve. The question is raised that human activities on reefs without adequate management plans can cause unwanted effects such as illegal fishing and conflicts between users. Alternative management measures are proposed to overcome this deficiency.

Section 3, "Dive Tourism Use," comprises three chapters dedicated to the tourism potential of certain artificial reef structures and which environmental, social and economic contributions they can make to local communities. Some management possibilities are suggested, envisioning to provide a better usefulness of artificial reefs.

Chapter 7 focuses on the question of the usefulness of reef modules for human purposes – in addition to fishing –, namely in terms of leisure and tourism. Even when they are found at shallow depths, they may be of limited use for scuba diving. The modules are useful for diversifying dive sites, thus imposing less human pressure on natural reefs. However, the modules are often of low attractiveness, due to the monotony of the structures.

Chapter 8 reports the decommissioning of civil and military vehicles for the creation of artificial reefs. It is a practice used in a number of countries, which gives a second use to vehicles, namely in the creation of biodiversity and new fishing and leisure places. The creation of this type of reef is in line with the growth of tourist diving activity, particularly in places close to military, archaeological or protected areas where diving is not allowed. However, it is referred that there is a lack of studies to corroborate the usefulness of this type of reef for human use.

Chapter 9 refers to the use of artificial reefs sunk in 2017 as an alternative to coral reef beaches that are in poor condition. This chapter mentions the importance of having tourism awareness groups to participate in improving the management of tourist villages where artificial reefs are placed. It is mentioned that coordination between stakeholders and the government is important for the smooth running of a reef program.

Finally, Section 4, “Sustainable Options,” covers three chapters related to the sustainability of materials used in the manufacture of artificial reefs, aspects of reef functionality related to physical factors, and sustainable practices for decommissioning structures giving them a second life.

In Chapter 10, ways of making bio shelters with hybrid manufacturing techniques are presented. Different forms of reef geometries achieved through industry 4.0 tools are presented. Different structures, stacking modes, robotic tools and how to operate them are evaluated. Adjustments are suggested for future work on a larger scale, namely in processes of regeneration of lost marine habitats.

Chapter 11 explores the fluid dynamics of reef modules in computational mode. An experiment is conducted with a reef set, exploring the usable volume based on geometry. A forecast is made to help decision-makers in the best location of artificial reefs, to cause a greater biological impact, namely in attracting species and restoring ecosystems.

Chapter 12 discusses the decommissioning of offshore structures used in oil and gas production. The wide variety of reconversion projects of these structures is mentioned. Alternative uses or conversion of multipurpose platforms at sea are expected to have a positive impact on several marine activities, such as: tourism, culture, transport, oil and gas, and leisure.

Impact of Artificial Reefs on the Environment and Communities presents insights on the connection between artificial reefs and the tourism and fishery industries in order to encourage the sharing of experiences and development of new innovative possibilities that create value for society. It presents the experiences, testimonies, and innovations of those who work closely with artificial reefs. Covering topics such as computational fluid dynamics, dive tourism, and Industry 4.0, this book is an essential resource for entrepreneurs, researchers, students and educators of higher education, coastal managers, government officials, leaders in tourism, engineers, and academicians.

Jorge H. P. Ramos
CinTurs, University of Algarve, Portugal

REFERENCES

Bortone, S. A., Brandini, F. P., Fabi, G., & Otake, S. (Eds.). (2011). *Artificial reefs in fisheries management* (pp. 1–332). CRC Press. doi:10.1201/b10910

Carr, M. H., & Hixon, M. A. (1997). Artificial reefs: The importance of comparisons with natural reefs. *Fisheries (Bethesda, Md.)*, *22*(4), 28–33. doi:10.1577/1548-8446(1997)022<0028:ARTIOC>2.0.CO;2

Carral Couce, L. M., Guerreiro, M. J. R., Fraguela Formoso, J. A., Alvarez-Feal, J. C., & Vizoso, A. F. (2017, October). Green artificial reef PROARR: Repopulation of coastal ecosystems and waste recycler of the maritime industries. In *Pan-American Conference of Naval Engineering* (pp. 363–373). Springer.

Coolen, J. W., Bittner, O., Driessen, F. M., van Dongen, U., Siahaya, M. S., de Groot, W., Mavraki, N., Bolam, S. G., & van der Weide, B. (2020). Ecological implications of removing a concrete gas platform in the North Sea. *Journal of Sea Research*, *166*, 101968. doi:10.1016/j.seares.2020.101968

Fabi, G., Scarcella, G., Spagnolo, A., Bortone, S. A., Charbonnel, E., Goutayer, J. J., ... Trommelen, M. (2015). Practical guidelines for the use of artificial reefs in the Mediterranean and the Black Sea. *General Fisheries Commission for the Mediterranean. Studies and Reviews (General Fisheries Commission for the Mediterranean)*, (96), I.

Leeworthy, V. R., Maher, T., & Stone, E. A. (2006). Can artificial reefs alter user pressure on adjacent natural reefs? *Bulletin of Marine Science*, *78*(1), 29–38.

MacDonald, J. M. (1994). Artificial reef debate: Habitat enhancement or waste disposal? *Ocean Development and International Law*, *25*(1), 87–118. doi:10.1080/00908329409546027

Milon, J. W., Holland, S. M., & Whitmarsh, D. J. (2000). Social and economic evaluation methods. *Artificial reef evaluation: with application to natural marine habitats*, 165-194.

Mustafa, S., Estim, A., Tuzan, A. D., Ann, C. C., Seng, L. L., & Shaleh, S. R. M. (2019). Nature-based and technology-based solutions for sustainable blue growth and climate change mitigation in marine biodiversity hotspots. *Environmental Biotechnology*, *15*(1), 15. doi:10.14799/ebms302

Na, W. B. (2021). Structural Complexity of Artificial Reefs. *Computational Structural Engineering*, *34*(3), 19–26.

Pickering, H., Whitmarsh, D., & Jensen, A. (1999). Artificial reefs as a tool to aid rehabilitation of coastal ecosystems: Investigating the potential. *Marine Pollution Bulletin*, *37*(8-12), 505–514. doi:10.1016/S0025-326X(98)00121-0

Ramm, L. A., Florisson, J. H., Watts, S. L., Becker, A., & Tweedley, J. R. (2021). Artificial reefs in the Anthropocene: A review of geographical and historical trends in their design, purpose, and monitoring. *Bulletin of Marine Science*, *97*(4), 699–728. doi:10.5343/bms.2020.0046

Seaman, W. Jr. (2002). Unifying trends and opportunities in global artificial reef research, including evaluation. *ICES Journal of Marine Science*, *59*(suppl), S14–S16. doi:10.1006/jmsc.2002.1277

Spanier, E., Tom, M., & Pisanty, S. (1985). Enhancement of fish recruitment by artificial enrichment of man-made reefs in the southeastern Mediterranean. *Bulletin of Marine Science*, *37*(1), 356–363.

Whitmarsh, D., Santos, M. N., Ramos, J., & Monteiro, C. C. (2008). Marine habitat modification through artificial reefs off the Algarve (southern Portugal): An economic analysis of the fisheries and the prospects for management. *Ocean and Coastal Management*, *51*(6), 463–468. doi:10.1016/j.ocecoaman.2008.04.004

Yozzo, D. J., Wilber, P., & Will, R. J. (2004). Beneficial use of dredged material for habitat creation, enhancement, and restoration in New York–New Jersey Harbor. *Journal of Environmental Management*, *73*(1), 39–52. doi:10.1016/j.jenvman.2004.05.008 PMID:15327845

Acknowledgment

The editor would like to acknowledge the help of all the people involved in this project. Without their support, this book would not have become a reality.

First, the editor would like to thank each one of the authors for their contributions. A sincere gratitude goes to the chapters' authors who contributed their time and expertise to this book.

Second, the editor wishes to acknowledge the valuable contributions of the reviewers regarding their underlying role in improving the quality of the manuscripts. Most authors also served as referees. It is highly appreciated their double task.

Third, some of the researchers and scholars contacted, despite not having contributed as authors/coauthors have an experience in the work with artificial reefs on diversified areas of knowledge and provided their time and expertise in revising and gave some advice on the chapters (Editorial Advisory Board).

Fourth, the editor communicates a sincere gratitude also to the IGI Global team – particularly the Assistant Development Editors of the Book Division – who were always available to provide all the technical support needed to make this book a palpable reality.

Fifth, the editor would also express gratefulness to the Portuguese Foundation for Science and Technology (FCT), through project UIDB/04020/2020, and the Research Centre for Tourism, Sustainability and Wellbeing (CinTurs) (UID/SOC/04020/2019).

Jorge H. P. Ramos
CinTurs, University of Algarve, Portugal

Section 1

Stakeholder Analyses and Other Socioeconomic Aspects

Chapter 1
Efficiency of Artificial Reefs for Local Communities Over Time

Jorge H. P. Ramos
https://orcid.org/0000-0002-5115-7919
CinTurs, University of Algarve, Portugal

Miguel Neves dos Santos
International Commission for the Conservation of Atlantic Tunas, Spain

ABSTRACT

Half a century ago it was thought that artificial reefs (AR) would be the solution to numerous problems. After the disappearance of this panacea, people started to think about AR more from the point of view of their efficiency and usefulness. However, some issues persist. Will it be possible some time after the reef deployment to reach the goals initially outlined by the proponents? It is very important to get answers to this and other questions to know about the efficiency of reef deployment. However, answers to these questions are not always easy to obtain. The AR are submerged, often difficult to access, and the resources available for research are generally scarce and very well delimited in time. In this chapter, the authors approach the case study of artificial reefs deployed off the Algarve coast in the south of Portugal over time. The focus will be on the contribution of reef structures to human activities at economic and social levels, which in some ways have benefited from these improvements on the seabed.

INTRODUCTION

From the point of view of efficiency, when designing an artificial reef (AR), several objectives to be achieved after the implementation of the structures must be anticipated (Seaman & Jensen, 2000; Ramm et al. 2021). To achieve efficiency in the project, it is equally important to consider from the beginning that a good development of the entire process will have to include the set of stakeholders interested and involved in the initiative, the monetary funds available, the geographic areas with potential for use and the selection criteria to be used (Claudet & Pelletier, 2004). The selection criteria cover several aspects such as: not only construction materials, size, or performance of the structures (Kim et al., 2021) and

DOI: 10.4018/978-1-6684-2344-8.ch001

their typology (e.g., modules or ex-vessels), but also what purpose will the structures serve, what type of users will be covered, and which are the rules to follow to an adequate management of the areas where the structures were sunk (Bohnsack & Sutherland, 1985; Tunca et al., 2019). After all these initial assumptions have been considered, the modular construction or reuse of the structures is done, and the reef deployment is consummated (Martin, 2004).

Sometime after the reef deployment – that is, during the structure maturation process – it is convenient to verify to what extent the initially proposed objectives have been achieved. This long process makes it possible to determine whether the reef deployment has been efficient, either from a structural, ecological, social, or economic point of view, or another that has been defined (Seaman, 2000).

For example, as the deployment of reefs essentially consists in the creation of three-dimensional spaces that can mimic the habitats of certain organisms, what is initially done to verify the success of the deployment of reefs is to carry out scientific research on the progressive achievement of the colonization of living marine resources (Holland et al., 2021). It is of great importance to get answers to questions related to the types of living marine organisms that colonize the reefs in each geography and depth, or what type of ecological arrangements are formed on the different sides of the reef, regardless of if they are modules or another type of structures (Choi et al., 2002). This monitoring is essential, as if there is not an adequate colonization, this fact can be an indicator of a possible problem (Kolian et al., 2017). In such a case, reef efficiency is called into question.

After verifying reef colonization and evaluating its eventual biological and ecological success, it is important to understand how AR are accepted by potential human users (Adams et al., 2006). In turn, these potential users will empirically assess the usefulness of the structures relating them to the satisfaction achieved in their own activity, whether professional or recreational (Layman & Allgeier, 2020). The income derived from businesses that may eventually operate taking advantage of the reef's surplus value, enter in this process of assessing the efficiency of AR (Fletcher et al., 2011). For instance, it is possible to find the average wage that each worker in the primary sector earns (i.e., small-scale fisheries). The diversification of tourist activities of companies that promote sport fishing and leisure diving can also be evaluated. The efficiency of reef structures derives from the surplus that these firms will take from having more choice options to offer their clients (Monteiro, 2016).

The existence of high biodiversity in these reef structures is a fundamental factor to assess and consider the issue of ecological efficiency (Huddart, 2019). There may also be factors with a more negative predisposition that may influence a good performance in the efficiency of AR. These types of negative factors can often be related to siltation caused by undercurrents on some structures, or fishing gear that has been stranded or lost on some modules (Barber et al., 2009). In both cases there is a latent loss of reef efficiency. In these cases, it is essential to have an adequate identification of the problem and steps should be taken for a reef maintenance (Roseman et al., 2017).

The aim of this chapter is primarily to relate AR to socioeconomic efficiency. In the first section it is introduced the theme of AR efficiency and achievement of deployment objectives. Second, the literature on reef deployment in Portugal since its inception is reviewed. In the third section, the issue of AR efficiency is described, namely in the amount of scientific production over time, its trend, and which areas of knowledge have been covered by research on artificial reefs in Portugal, structured in a simple conceptual model to elucidate how the transfer of knowledge from science to users (firms) can be done, and what inputs and outputs are to be considered in terms of the efficiency of AR. Subsequently, in the fourth section some of the expectations of coastal communities in relation to reef efficiency are summarized, relating it to the proxy of the average income of the primary sector worker by municipality. Reef efficiency objectives in the Portuguese context based on the literature gap and using a simple indicator, are then briefly discussed in the fifth section. Finally, some conclusions are drawn.

LITERATURE REVIEW

The Deployment of Artificial Reefs in Portugal

In Portugal, the first experiments with artificial reefs date back to the early 1980s (Table 1). At that time, some experiments were made with cars, tires, and wooden boats in the Madeira archipelago, with the aim of improving local fisheries by getting rid of obsolete structures and vehicles, giving them a new utility (Lipcius & Burke, 2006).

In 1990, concrete modules relating to the first two pilot reefs (Faro and Olhão) were sunk for the first time in mainland Portugal (Santos et al., 1996). The project was initiated by the national Fisheries Research Institute. Sometime later, due to the successful pilot program, there was an incentive to implement modules on a larger scale, along the entire southern coast of mainland Portugal (Monteiro & Santos, 2000). There was also the deliberate sinking of an ex-fishing trawler and the accidental sinking of a barge (Ramos, 2021).

In the early 2000s, there were some reef deployments in the archipelagos of Madeira (Wirtz et al., 2020) and the Azores (Gomes-Pereira et al., 2012). The structures used ranged from reef modules (Madeira) to a barge pontoon (Faial – Azores).

In 2003 the largest artificial reef in Portugal was completed (Seaman et al., 2011). Shortly thereafter, there was an interest on the part of a municipality in deploying an AR on the west coast (Ramos et al., 2019). During the following years, more reef deployments were carried out in the Madeira archipelago. In 2006 a project was signed to develop reefs associated with surfing. Studies were also carried out with multifunctional reefs, namely for the protection of the coast and for surfing, at Leirosa beach (Mendonça et al., 2012). Studies with identical AR were carried out for a beach in Estoril (2008) (Bicudo et al., 2008). In 2010 the Nazaré AR was implemented. Other structures have been deployed since then in other parts of the country and islands (Ribeiro & Neves, 2020). The Ocean Revival project consisted of the sinking of four former navy ships for the practice of diving and education (2012-2013) (Monteiro, 2016). More recently, some studies have been carried out with experimental reefs using 3D printing technology (industry 4.0) (Ly et al., 2021).

Table 1. The AR supply in Portugal. Source: Adapted from Ramos (2007)

Year	Region	Contribute	AR type	Volume/unit	Number	Area
1983	Madeira archipelago	car bodies, tires, and wooden boats (Ponta S. Lourenço)	various	various	various	sparse
1990	Algarve	first pilot artificial reefs off Faro and Olhão	modules	2.7m³ ; 174m³	2 x (735 ; 20)	2 x 60ha
1996	Algarve	iron hull trawler "Eng António Nunes" wreck off Faro	ship	40m (l) x 8m (w)	1	~300m²
1998	Algarve	large scale deployment was initiated	modules	2.7m³ ; 174m³	>20,000	~43Km²
1999	Algarve	the barge "Titan" carrying a derrick sunk accidentally off Cacela	ship	100m (l) x 30m (w)	1	3,000m²
2000	Madeira archipelago	the cargo ship "Madeirense" was sunk off Porto Santo Island	ship	70m (l) x 10m (w)	1	~700m²
2000	Madeira archipelago	two AR sets deployed (Paúl do Mar/Jardim do Mar)	modules	2.7m³	100 + 450	~50ha
2000	Madeira archipelago	prismatic cubes (Paúl do Mar/Jardim do Mar)	modules	45m³	16	< 1ha
2003	Azores archipelago	the barge "Pontoon 16" was sunk off Faial Island	platform	40m (l) x 17m (w)	1	680m²
2003	Algarve	the largest artificial reef was completed	modules	2.7m³ ; 174m³	~5,800; 72	12.2 Km²
2004	Central coast	proposal for an artificial reef of Nazaré, first studies	modules	2.7m³		
2004	Madeira archipelago	two AR sets deployed (Ponta da Galé)	modules	2.7m³	100	~10ha
2006	Madeira archipelago	cubic blocks deployed near Ponta Pequena	modules	2.7m³	2,500	~80ha
2006	Madeira archipelago	project proposal for a surf reef in Lugar de Baixo	multipurpose		1	
2006	Lisbon & Tagus valley	signed a project for two AR (surf and protection reefs)	multipurpose		2	
2006	Central coast	experiments with a surfing protection reef at Leirosa beach	multipurpose		1	
2010	Central coast	AR deployed off Nazaré coast (aiming fisheries and diving)	modules	2.7m³	1,400	~100ha
2013	Algarve	4 ex-navy vessels were intentionally sunk (Ocean Revival)	ships	44-102m length	4	~5,000m²
2016	Madeira archipelago	wreck of the "NRP General Pereira d'Eca", intentionally sunk	ship	85m (l) x 12.5m (w)	1	~1,100m²
2018	Madeira archipelago	wreck of the "NRP Afonso Cerqueira" was intentionally sunk	ship	85m (l) x 10m (w)	1	~850m²
2020	North coast	3D printed AR modules (off Leixões cruise terminal)	modules	~1m³	9	< 100m²

Pre-Deployment, Deployment and Expected Efficiency in Portuguese AR

The literature recommends that in the conceptual phase of artificial reefs, it is important to investigate to what extent there is a need to establish sites with these structures (demand) (Bohnsack & Sutherland, 1985; Fabi et al., 2011). There is a whole panoply of stakeholders who want or should be involved in the decision-making process of the structures (Sutton & Bushnell, 2007). There are a few factors that are important and fundamental to consider, such as the location, the type of materials to be used, the depths where the structures will be sunk, for what purposes they will serve, etc. (Sayer & Wilding, 2002).

During and right after the reef deployment, there is usually a "moratorium" period for any fishing or leisure activities that may take place in those places (Leeworthy et al., 2006). This period is extremely important so that there can be an adequate colonization of structures and that a whole biological and ecological chain can be developed in the new habitat.

As mentioned above, there are several aspects of the Portuguese reef deployment. The main region of deployment is located on the south coast in the Algarve region. Throughout this area of deployment, the sunk structures were essentially concrete blocks of two types, with various purposes, but mainly to serve as shelter for species that make the transition from inland waters to the sea, and as production zones to enhance catches by small-scale fishing vessels (Santos & Monteiro, 1998).

The expected efficiency for this type of reef structures was then to collect information indicating that small scale fishermen, and/or possibly other activities related to tourism (onboard recreational / charter boat fishing and diving activities) could generate additional income in their activities due to the presence of the AR (Ramos et al., 2007). For economic activities to capitalize on the reef's presence, it is important to note that there is an adequate colonization of all structures, allowing for the existence of a good level of biodiversity. Literature on AR states that it is important that the reef modules are functional, that is, without being silted up or covered with fishing gear, so that they can serve the purposes for which they were implemented (Paredes et al., 2019).

In the case of a Portuguese study the literature indicates that there are also some other multifunctional reef structures (do Carmo et al., 2011). In these, the expected efficiency is essentially focused on the coastal protection part, and eventually on the sports practice of surfing.

During and after reef maturation, it is important to carry out studies in different areas of knowledge, to understand how modules are colonized and their colonization progression over time (Moura et al., 2006; Leitao et al., 2008). It is also necessary to check whether the structures are stable from a physical or chemical point of view and fit into the surrounding environment (Vicente et al., 2008; Falcão et al., 2009; Hiscock et al., 2010). The literature states that if all these nuances are verified beyond the period of maturation of the structures and its communities, the minimum conditions are created to demonstrate that the structures are being efficient in relation to the initially proposed designs (Encarnação & Calado, 2018).

The Efficiency of the AR

Trend of Scientific Production

As recommended in the literature, in the case of artificial reefs deployed in Portugal, several studies were carried out over time, in a way following the maturation of the AR assemblages/communities (Santos & Monteiro, 2007). Areas of research and trends in studies have varied somewhat over the past three decades, as shown in Figure 1. In the first years after the implementation of the pilot reefs, there was no production of indexed scientific articles. The scientific production of indexed articles only started when the second phase of reef deployment was initiated by the promoting entity, i.e., the national institute for fisheries research (IPIMAR at the time, IPMA at present). It was only from the mid-2000s that most scientific articles were produced. There was a sharp drop in the following years. There has been again a growth in scientific production in recent years, demonstrating, in a way, the resumption of interest in the subject of artificial reefs. The recent introduction of new technologies, such as 3D printing (from industry 4.0), has increased the scientific production related to AR (Dunn et al., 2019; Pioch et al., 2020, Ly et al., 2021).

Figure 1. Trend in the production of AR refereed articles in Portugal over time (period 1990-2021)
Source: Google Scholar

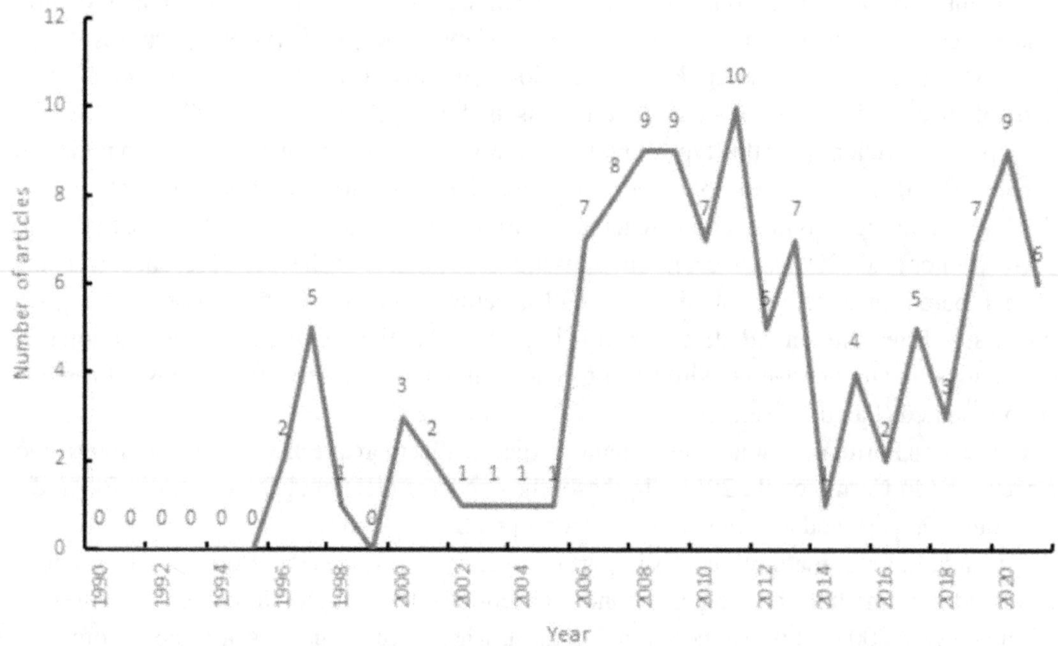

The Main Lines of Research in AR

Studies involving AR cover several areas of knowledge (Figure 2). However, in the disciplines related to biology (and ecology) there has been an incidence of most studies in the areas of benthic colonization. Some studies on physical and chemical oceanography were also published, with the aim of ascertaining the quality of water and sediment in the area where the reef structures are located. There are several studies on the biology of some species, namely fish, and their relationship with artisanal fisheries. Another aspect is that of some studies referring to the economic efficiency of the reef deployment and other socio-economic components, with a particular focus on fisheries management and on the analysis of stakeholders (Lima et al., 2019).

In the first decades of the reef deployment, biology and ecology studies were the ones that prevailed in terms of quantity of articles published. In the last decade, the study of the management of reef structures has gained some notoriety.

Figure 2. Sunburst plot showing scientific production in different areas of research in AR in Portugal from the early 1990s to the present
Source: Authors' own based on Google Scholar data

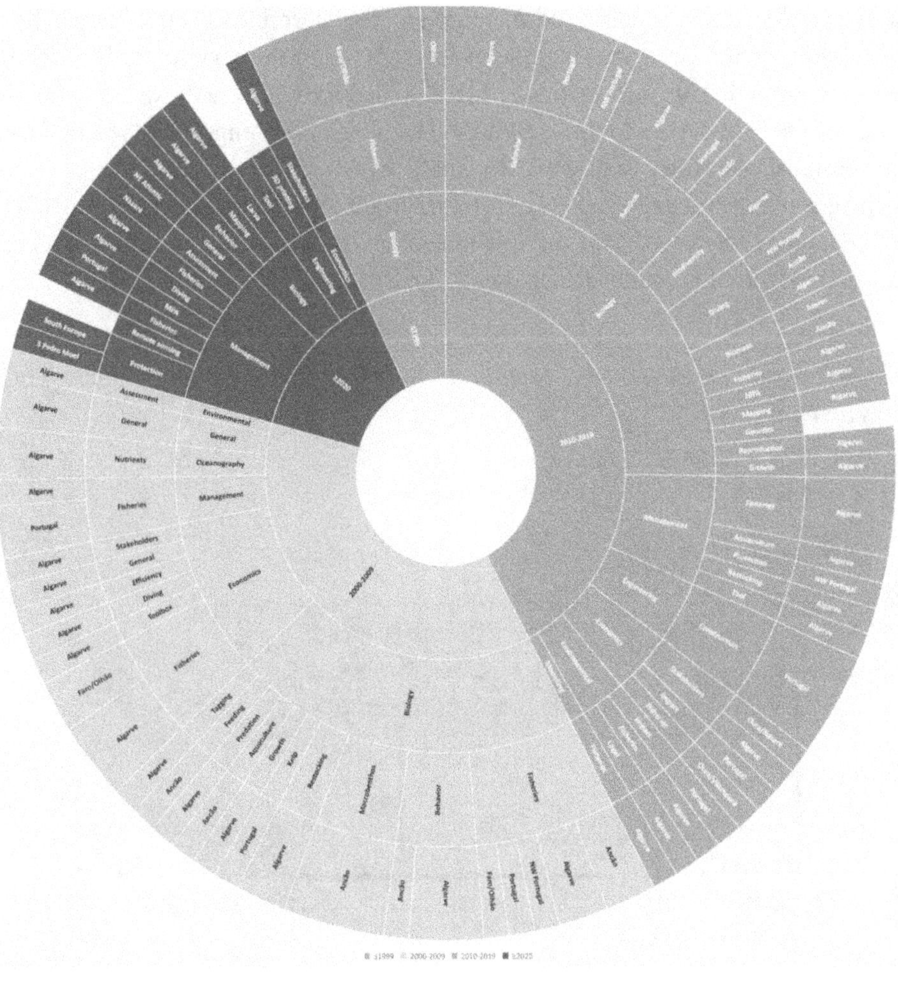

Simple Conceptual Model

After consulting the various key stakeholders, which culminates in the best decision for the deployment of AR by the entities that promote this type of structures (demand-supply), it is advisable to make efforts to verify the potential of the inputs produced by reef deployment (Figure 3). In this sense, several state and academic research institutions are usually involved. In addition, what has been verified is that scientific research should cover as many areas of knowledge as possible, as well as having adequate human resources and means.

It is important that there is a transfer of knowledge from research to firms (Quinn, 2009). This interaction between the different players is beneficial to the entire system. An adequate transfer of knowledge also enhances the efficiency of AR in relation to human activities that take place in the vicinity where the structures are placed.

Assessing the efficiency of AR allows us to ascertain to what extent local fishing communities, as well as tourism agents related to sport fishing and diving, manage to generate surplus production (Stolk et al., 2007). This surplus production at the firm level is a contribution towards greater sustainability of employment in these locations. A rather rudimentary way to determine the efficiency indicator is to use as a proxy time series of the economic income of workers in the primary sector for each municipality where the reefs are located. This information is useful to deduce what are the average salaries practiced by the activities of the primary sector, namely in the local small fishing. This type of data is available from Portuguese statistics entities (e.g., INE, Pordata).

On the issue of activities related to leisure and tourism (e.g., recreational fishing and SCUBA diving), the aggregated data referring to tourism are only indicators of the potential number of tourists that may eventually enjoy this type of activities (Shani et al., 2012).

Figure 3. Simple conceptual model to ascertain on AR efficiency on coastal communities (research areas and firm activities)

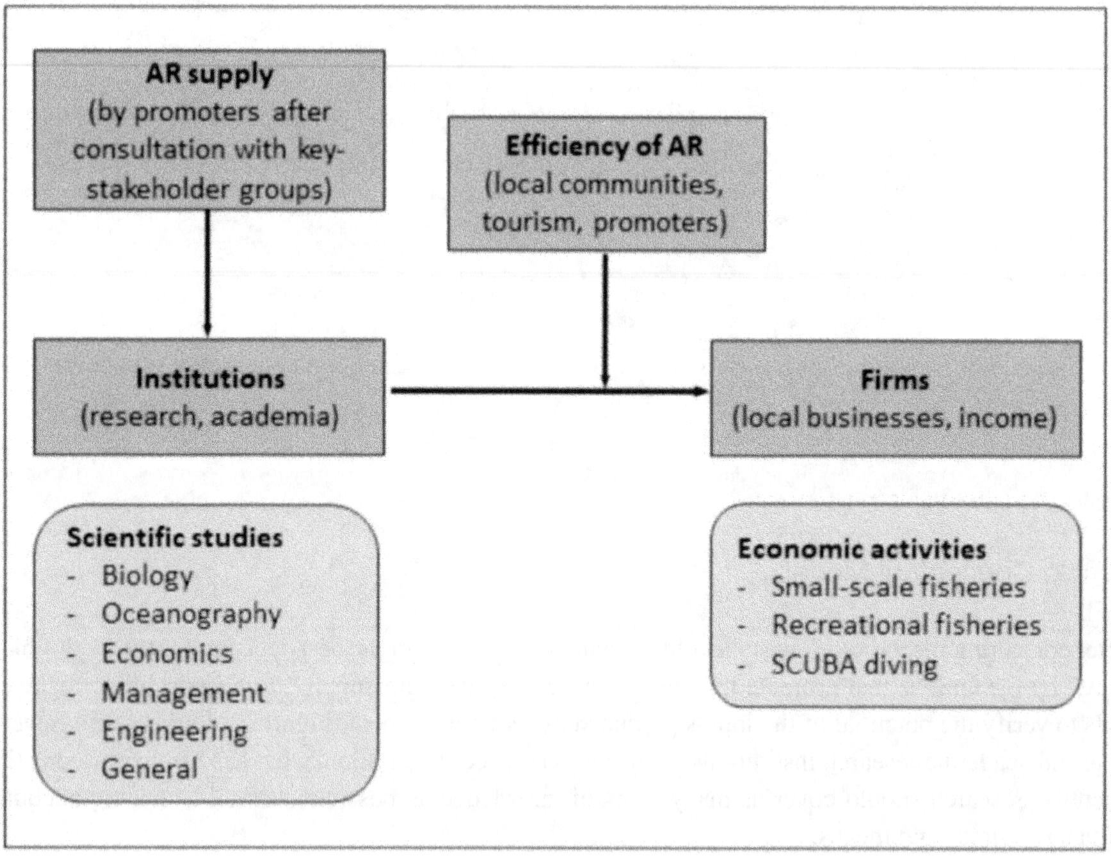

Efficiency Inputs and Outputs

Table 2 briefly describes the input and output criteria considered important for assessing the efficiency of AR. These in terms of inputs essentially focus on criteria related to the areas of biology, ecology and living resources (in terms of protection and release of fish), policy and management (in terms of monitoring, funding, implementation), engineering and technology (regarding cooperation with universities, institutional research equipment).

The criteria for evaluating the efficiency of the AR in the outputs part focus essentially on the economy (in terms of business income or salary paid to the primary sector worker, catches of the main fish species sold in the market), related aspects with ecology and biology (namely with biodiversity standards), and with possible losses of efficiency (derived from reported negative impacts of silting or loss of fishing gears).

Table 2. Criteria and items considered relevant to the efficiency of the deployment of artificial reefs

Criteria		Item	Units	Attribute	Reference(s)
INPUTS	*Biological and ecological resources*	Species protection	Good/bad	+	Santos et al., 2005; Santos & Monteiro, 2007; Ramos et al., 2011
		Releasing of fish (restocking)	Thousands	+	Santos et al., 2006; Lino et al., 2009; Abecasis et al., 2013
	Policy and management	Monitoring	Times/year	+	Ramos et al., 2006a ; Roa-Ureta et al., 2019
		Funding	M€	+	Santos & Monteiro, 2001; Santos et al., 2011
		Policies implemented	Number	+	Monteiro & Santos, 2000
	Engineering and technology	Cooperation with Universities	Number	+	Moura et al., 2007 ; Leitao et al., 2016
		Research equipment	Yes/no	+	Seaman et al., 2011; Leitão, 2013
OUTPUTS	*Economy*	Business revenue (income)	M€	+	Santos & Monteiro, 1998; Whitmarsh et al., 2008; Pordata (no date)
		Outputs of main species	Tons	+	Santos & Monteiro, 1997, 1998; Leitao et al., 2009
	Ecology and biology	Standard of biodiversity	Number	+	Boaventura et al., 2006; Moura et al., 2008
	Loss	Reported negative impacts	Number	-	Erzini et al., 1997 ; Ramos et al., 2007; Ramos et al., 2019

AR Economy in Coastal Communities in the Algarve

The objectives of the AR that were implemented along the Algarve coast are intrinsically related to the improvement of living marine resources and the facilitation of their management (Figure 4). The goals are mainly linked to biological and environmental issues. However, it is important that AR are also useful for activities carried out by people (e.g., small-scale fishing, recreational fishing, scientific and tourist diving) (Fabi & Spagnolo, 2011). The reef's usefulness is associated with the goods and services provided by the existence of the structures and the surplus obtained from them (ecosystem services). This surplus can be monetized through the goods and services produced by the existence of the reef and its production/attraction potential (Pickering & Whitmarsh, 1997; Oh et al., 2008).

Figure 4. Map of Portugal and location of the artificial reef program in the Algarve, which occurred between 1990 and 2003, and was promoted by the national fisheries and marine research institute (currently IPMA); each red bullet point represents an AR set of thousands of concrete made modules
Source: Authors' own

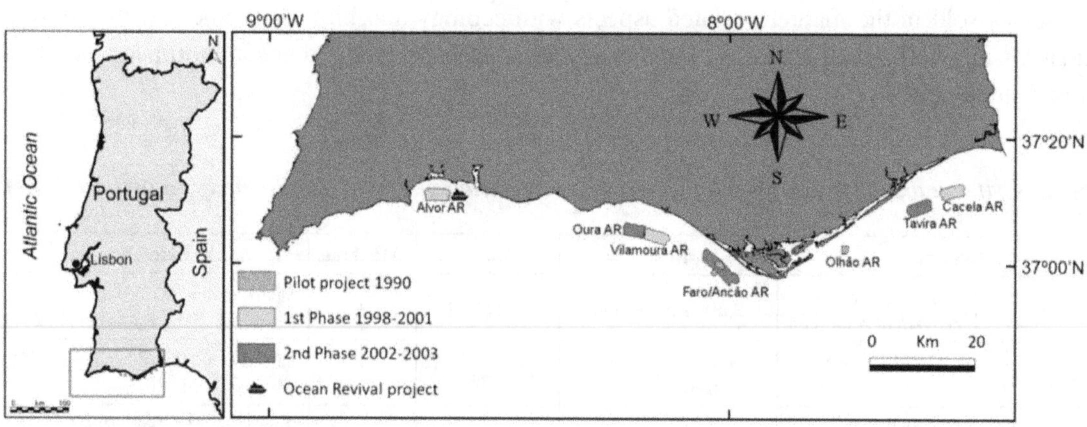

In the part of small-scale artisanal and commercial fishing, the surplus that is provided by the existence of AR, quantified by the quantities of fish caught in that area, allows us to perceive a surplus derived from the structures. If there is a surplus provided by AR, which remains indefinitely over time, it can be said that the structures have been an asset and that there is sustainability in the reef system.

However, if the structures provided only an increase in fishing effort or other externalities, which eventually led to the depletion of fisheries resources supposedly derived from the reef's existence, then the reef's long-term usefulness can be said to have disappeared. In this hypothetical case, it is important to apply management practices in line with the situation. In a utility approach, it is imperative that the existence of the reef brings added value to the local communities, which can be reflected in an economic benefit promoted by the structures. Figure 5 shows the time series relating to the average wages earned by workers in the primary sector for the different Algarve municipalities.

Similar situations can occur for other types of activities, such as recreational ones. However, recreational activities must be approached differently. Although, the surplus of the reef benefit in these cases is not derived from an extractive activity, but only from the consumption of a service. These benefits are usually expressed in terms of satisfaction.

All along the Algarve coast, the AR were deployed in areas devoid of rock, to mimic at some level the natural reefs. In these places there are, in most cases, fishing communities based on small local fishing. Many of these local fishermen provide a good number of fresh fish for nearby markets and catering activities. The presence of artificial reefs is, in principle, an asset to strengthen the safety of catches by this small fishery, which is more sustainable and more balanced from a social point of view, as it provides more employment and produces fresher fish.

Figure 5. Box and whisker plots showing monthly primary sector workers' income along time in the municipalities of the Algarve region (1991 to 2019); key: Municipality with no fishers or no AR (pale red background), AR far away (pale yellow background), AR nearby (pale green background)
Source: Pordata (no date)

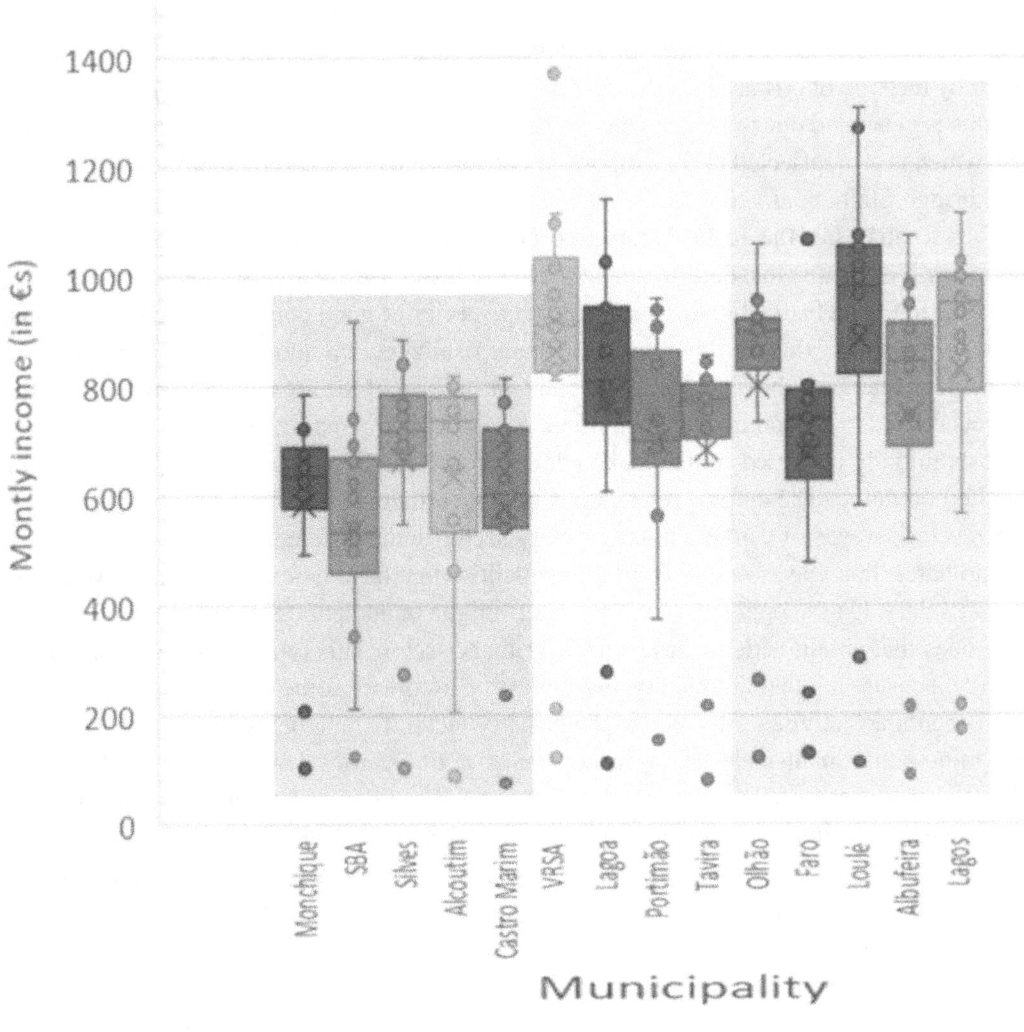

DISCUSSION

In the case of the Portuguese case study, in a social context, it can be said that the AR off the Algarve coast were designed in such a way that eligible fishermen – who are essentially from small scale artisanal fishing – could use the structures. Likewise, entrepreneurs of other activities – such as diving tourism and recreational fishing by charter boats – and their collaborators can also take satisfaction from the simple fact that there are AR available to widen their plunging spots and fishing grounds. Nevertheless, recreational and/or tourist diving activity are essentially limited by the depth of the modules and the interest generated by this type of structure. Sunken vessels generate higher satisfaction for recreational/tourist divers than usual concrete modules (Ramos et al., 2006b; Kirkbridge-Smith et al., 2013).

On the other hand, there is some scientific evidence that proves that the small fishing activity takes advantage of the man-made concrete structures (Whitmarsh et al., 2008). In the case of recreational fishing, there is also some evidence that the presence of AR generates some surplus in firms (e.g., charter boat companies advertise sea fishing trips 3-5 miles from the coast, between 20-30 meters deep). In the case of recreational and tourist diving, some additional income is generated, but it is very dependent and limited by the type of AR used. In the case of ships or other sunken vessels, there is some demand, and a surplus is generated due to the presence of these structures. In the case of modules, the demand is quite low, which is also reflected in the proportion of the surplus obtained by firms (Sutton & Bushnell, 2007; Kirkbridge-Smith et al., 2016).

Therefore, it can be said that in the Algarve most users are commercial fishermen eligible to fish in the AR bathymetry. The factors that determine the greater or lesser use are diverse. These factors are related to the distance to the reefs, the likelihood of catching species of good commercial size and obtaining a reasonable commercial value. In this regard, there are fishing communities that can benefit more than others, whether because of the distance to the reefs, or the size of the area covered by the structures, or other aspects related to coastal management (e.g., ease of access, area restrictions).

The goals initially proposed by the stakeholders of AR can be induced through indicator proxies and consulted literature. The outputs related to economic efficiency can be induced through the evolution of the average wages of workers in the primary sector in the different municipalities of the region (Pordata, no date). In a way, workers from municipalities that have no contact with a river, or the sea have a lower salary. On the opposite side, workers from the municipalities that have AR in their vicinity are the ones that obtain better salaries in the primary sector. This type of indicator can be used as an efficiency measure in the municipality, but it must be used with some caution, as each municipality has several economic activities allocated to the primary sector, fishing being just one of them. It is also important to note that small-scale fishing is not exclusive to fisheries generated by AR. Therefore, the income indicator is aggregated to other workers who exercise other professions in the primary sector.

There are several studies that show that in reef areas there are good colonization rates of important commercial species, and this item is important to evaluate positively regarding the efficiency of AR (e.g., Leitao 2008, 2009; Santos et al., 2011). The literature shows that in places where there is good colonization and good levels of biodiversity, the biomass created is an asset for the entire system, generating various ecosystem services (Schulze et al., 2020).

One of the major issues related to the efficiency of reef structures is associated with the longevity of ecosystem service delivery (Sato et al., 2021). Literature states that concrete structures (modules) have a great longevity (e.g., Vivier et al., 2021). Here, however, there are two strands of thought. One states that structures after a few decades – say 50 years – of existence begin to deteriorate and lose their efficiency (e.g., Grace, 2001). There is another line of thought that from a certain point onwards colonizing organisms that secrete calcium carbonate or other substances have already established themselves and created sufficiently solid structures that not only mimic natural rock habitats, but also give them unlimited longevity to the initial modular structures (Firth et al., 2014).

CONCLUSION AND FINAL CONSIDERATIONS

AR only become efficient after their maturation. If there is good colonization and biodiversity, ecological efficiency can be achieved. To have the perception of efficiency of AR for human activities, there are some difficulties. Likewise, it is important that there is human knowledge of the structures, mainly by the primary stakeholders who are the main users and who directly benefit from the ecosystem services provided by the man-made structures.

Monitoring artificial reefs over a long period of time requires a lot of human resources and financial means, which are not always available. This situation makes proper data collection difficult. The alternatives that are available – for example data from statistical series – are not the most suitable but allow us to infer about the efficiency of AR in local communities.

ACKNOWLEDGMENT

The authors are grateful for the valuable review insights provided by two anonymous reviewers in a previous version of the manuscript. J. Ramos was financed by National Funds made available by FCT Foundation for Science and Technology through project UIDB/04020/2020.

REFERENCES

Abecasis, D., Bentes, L., Lino, P. G., Santos, M. N., & Erzini, K. (2013). Residency, movements and habitat use of adult white seabream (Diplodus sargus) between natural and artificial reefs. *Estuarine, Coastal and Shelf Science, 118*, 80–85. doi:10.1016/j.ecss.2012.12.014

Adams, C., Lindberg, B., & Stevely, J. (2006). The economic benefits associated with Florida's artificial reefs. *EDIS, 2006*(14).

Barber, J. S., Chosid, D. M., Glenn, R. P., & Whitmore, K. A. (2009). A systematic model for artificial reef site selection. *New Zealand Journal of Marine and Freshwater Research, 43*(1), 283–297. doi:10.1080/00288330909510001

Bicudo, P., Custódio, M., Cardoso, N., Fortes, C., Neves, D. G., Mendes, L., ... Miguel, L. (2008, July). Viability Study of an Artificial Surf Reef in S. Pedro do Estoril, Cascais, Portugal. In *The Eighteenth International Offshore and Polar Engineering Conference*. OnePetro.

Boaventura, D., Moura, A., Leitao, F., Carvalho, S., Curdia, J., Pereira, P., & Monteiro, C. C. (2006). Macrobenthic colonisation of artificial reefs on the southern coast of Portugal (Ancão, Algarve). In *Marine Biodiversity* (pp. 335–343). Springer. doi:10.1007/1-4020-4697-9_29

Bohnsack, J. A., & Sutherland, D. L. (1985). Artificial reef research: A review with recommendations for future priorities. *Bulletin of Marine Science, 37*(1), 11–39.

Choi, C. G., Takeuchi, Y., Terawaki, T., Serisawa, Y., Ohno, M., & Sohn, C. H. (2002). Ecology of seaweed beds on two types of artificial reef. *Journal of Applied Phycology, 14*(5), 343–349. doi:10.1023/A:1022126007684

Claudet, J., & Pelletier, D. (2004). Marine protected areas and artificial reefs: A review of the interactions between management and scientific studies. *Aquatic Living Resources, 17*(2), 129–138. doi:10.1051/alr:2004017

Do Carmo, J. S. A., Neves, M. G., & Ten Voorde, M. (2011). Designing a multifunctional artificial reef: Studies on the influence of parameters with most influence in the vertical plane. *Journal of Coastal Conservation, 15*(1), 99–112. doi:10.100711852-010-0124-1

Dunn, K., Haeusler, M. H., Zavoleas, Y., Bishop, M., Dafforn, K., Sedano, F., Yu, D., & Schaefer, N. (2019). Recycled Sustainable 3D Printing Materials for Marine Environments. In Blucher Design Proceedings. Editora Blucher.

Encarnação, J., & Calado, G. (2018). Effects of recreational diving on early colonization stages of an artificial reef in North-East Atlantic. *Journal of Coastal Conservation, 22*(6), 1209–1216. doi:10.100711852-018-0630-0

Erzini, K., Monteiro, C. C., Ribeiro, J., Santos, M. N., Gaspar, M., Monteiro, P., & Borges, T. C. (1997). An experimental study of gill net and trammel net\'ghost fishing\'off the Algarve (southern Portugal). *Marine Ecology Progress Series, 158*, 257–265. doi:10.3354/meps158257

Fabi, G., & Spagnolo, A. (2011). *Artificial reefs in the management of Mediterranean Sea fisheries.* CRC Press.

Fabi, G., Spagnolo, A., Bellan-Santini, D., Charbonnel, E., Çiçek, B. A., García, J. J. G., ... Santos, M. N. D. (2011). Overview on artificial reefs in Europe. *Brazilian Journal of Oceanography, 59*(spe1), 155-166.

Falcão, M., Santos, M. N., Drago, T., Serpa, D., & Monteiro, C. (2009). Effect of artificial reefs (southern Portugal) on sediment–water transport of nutrients: Importance of the hydrodynamic regime. *Estuarine, Coastal and Shelf Science, 83*(4), 451–459. doi:10.1016/j.ecss.2009.04.028

Firth, L. B., Thompson, R. C., Bohn, K., Abbiati, M., Airoldi, L., Bouma, T. J., Bozzeda, F., Ceccherelli, V. U., Colangelo, M. A., Evans, A., Ferrario, F., Hanley, M. E., Hinz, H., Hoggart, S. P. G., Jackson, J. E., Moore, P., Morgan, E. H., Perkol-Finkel, S., Skov, M. W., ... Hawkins, S. J. (2014). Between a rock and a hard place: Environmental and engineering considerations when designing coastal defence structures. *Coastal Engineering, 87*, 122–135. doi:10.1016/j.coastaleng.2013.10.015

Fletcher, S., Bateman, P., & Emery, A. (2011). The governance of the Boscombe artificial surf reef, UK. *Land Use Policy, 28*(2), 395–401. doi:10.1016/j.landusepol.2010.08.002

Gomes-Pereira, J. N., Tempera, F., Ribeiro, P. A., & Porteiro, F. M. (2012). Notes on fauna associated with an opportunistic artificial reef near cold-water corals. *ARQUIPÉLAGO. Life and Marine Sciences.*

Grace, R. A. (2001). The factors and processes that influence artificial reef longevity. *Marine Technology Society Journal, 35*(2), 3–13. doi:10.4031/002533201788001910

Hiscock, K., Sharrock, S., Highfield, J., & Snelling, D. (2010). Colonization of an artificial reef in south-west England—ex-HMS 'Scylla'. *Journal of the Marine Biological Association of the United Kingdom, 90*(1), 69–94. doi:10.1017/S0025315409991457

Holland, M. M., Becker, A., Smith, J. A., Everett, J. D., & Suthers, I. M. (2021). Characterizing the three-dimensional distribution of schooling reef fish with a portable multibeam echosounder. *Limnology and Oceanography, Methods*, *19*(5), 340–355. doi:10.1002/lom3.10427

Huddart, D. (2019). Recreational SCUBA diving and snorkelling. In *Outdoor Recreation* (pp. 361–394). Palgrave Macmillan. doi:10.1007/978-3-319-97758-4_14

Kim, D., Jung, S., & Na, W. B. (2021). Evaluation of turbulence models for estimating the wake region of artificial reefs using particle image velocimetry and computational fluid dynamics. *Ocean Engineering*, *223*, 108673. doi:10.1016/j.oceaneng.2021.108673

Kirkbride-Smith, A. E., Wheeler, P. M., & Johnson, M. L. (2013). The relationship between diver experience levels and perceptions of attractiveness of artificial reefs-examination of a potential management tool. *PLoS One*, *8*(7), e68899. doi:10.1371/journal.pone.0068899 PMID:23894372

Kirkbride-Smith, A. E., Wheeler, P. M., & Johnson, M. L. (2016). Artificial reefs and marine protected areas: A study in willingness to pay to access Folkestone Marine Reserve, Barbados, West Indies. *PeerJ*, *4*, e2175. doi:10.7717/peerj.2175 PMID:27547521

Kolian, S. R., Sammarco, P. W., & Porter, S. A. (2017). Abundance of corals on offshore oil and gas platforms in the Gulf of Mexico. *Environmental Management*, *60*(2), 357–366. doi:10.100700267-017-0862-z PMID:28526900

Layman, C. A., & Allgeier, J. E. (2020). An ecosystem ecology perspective on artificial reef production. *Journal of Applied Ecology*, *57*(11), 2139–2148. doi:10.1111/1365-2664.13748

Leeworthy, V. R., Maher, T., & Stone, E. A. (2006). Can artificial reefs alter user pressure on adjacent natural reefs? *Bulletin of Marine Science*, *78*(1), 29–38.

Leitão, F. (2013). Artificial reefs: From ecological processes to fishing enhancement tools. *Brazilian Journal of Oceanography*, *61*(1), 77–81. doi:10.1590/S1679-87592013000100009

Leitao, F., Baptista, V., Teodósio, M. A., Hughes, S. J., Vieira, V., & Chícharo, L. (2016). The role of environmental and fisheries multi-controls in white seabream (Diplodus sargus) artisanal fisheries in Portuguese coast. *Regional Environmental Change*, *16*(1), 163–176. doi:10.100710113-014-0726-5

Leitao, F., Santos, M. N., Erzini, K., & Monteiro, C. C. (2008). Fish assemblages and rapid colonization after enlargement of an artificial reef off the Algarve coast (Southern Portugal). *Marine Ecology (Berlin)*, *29*(4), 435–448. doi:10.1111/j.1439-0485.2008.00253.x

Leitao, F., Santos, M. N., Erzini, K., & Monteiro, C. C. (2009). Diplodus spp. assemblages on artificial reefs: Importance for near shore fisheries. *Fisheries Management and Ecology*, *16*(2), 88–99. doi:10.1111/j.1365-2400.2008.00646.x

Lima, J. S., Zalmon, I. R., & Love, M. (2019). Overview and trends of ecological and socioeconomic research on artificial reefs. *Marine Environmental Research*, *145*, 81–96. doi:10.1016/j.marenvres.2019.01.010 PMID:30837123

Lino, P. G., Bentes, L., Abecasis, D., dos Santos, M. N., & Erzini, K. (2009). Comparative behavior of wild and hatchery reared white sea bream (Diplodus sargus) released on artificial reefs off the Algarve (Southern Portugal). In Tagging and tracking of marine animals with electronic devices (pp. 23-34). Springer.

Lipcius, R., & Burke, R. P. (2006). *Abundance, biomass and size structure of eastern oyster and hooked mussel on a modular artificial reef in the Rappahannock River*. Chesapeake Bay.

Ly, O., Yoris-Nobile, A. I., Sebaibi, N., Blanco-Fernandez, E., Boutouil, M., Castro-Fresno, D., Hall, A. E., Herbert, R. J. H., Deboucha, W., Reis, B., Franco, J. N., Teresa Borges, M., Sousa-Pinto, I., van der Linden, P., & Stafford, R. (2021). Optimisation of 3D printed concrete for artificial reefs: Biofouling and mechanical analysis. *Construction & Building Materials*, *272*, 121649. doi:10.1016/j.conbuildmat.2020.121649

Martin, K. S. (2004). GIS in marine fisheries science and decision-making. *Geographic information systems in fisheries*, 237-258.

Mendonça, A., Fortes, C. J., Capitão, R., da Graça Neves, M., Moura, T., & Do Carmo, J. S. A. (2012). Wave hydrodynamics around a multi-functional artificial reef at Leirosa. *Journal of Coastal Conservation*, *16*(4), 543–553. doi:10.100711852-012-0196-1

Monteiro, C. C., & Santos, M. N. (2000). Portuguese artificial reefs. In *Artificial Reefs in European Seas* (pp. 249–261). Springer. doi:10.1007/978-94-011-4215-1_15

Monteiro, P. V. (2016). The Role of knowledge-intensive service activities on inducing innovation in co-opetition strategies: Lessons from the maritime cluster of the Algarve region. *International Journal of Management and Enterprise Development*, *15*(1), 78–95. doi:10.1504/IJMED.2016.075876

Moura, A., Boaventura, D., Curdia, J., Carvalho, S., Da Fonseca, L. C., Leitão, F. M., & Monteiro, C. C. (2007). Effect of depth and reef structure on early macrobenthic communities of the Algarve artificial reefs (southern Portugal). In *Biodiversity in Enclosed Seas and Artificial Marine Habitats* (pp. 173–180). Springer. doi:10.1007/978-1-4020-6156-1_15

Moura, A., Boaventura, D., Cúrdia, J., Santos, M. N., & Monteiro, C. C. (2006). Biomass production of early macrobenthic communities at the Faro/Ancão artificial reef (Portugal): Effect of depth and reef layer. *Bulletin of Marine Science*, *78*(1), 83–92.

Moura, A., Da Fonseca, L. C., Cúrdia, J., Carvalho, S., Boaventura, D., Cerqueira, M., Leitão, F., Santos, M. N., & Monteiro, C. C. (2008). Is surface orientation a determinant for colonisation patterns of vagile and sessile macrobenthos on artificial reefs? *Biofouling*, *24*(5), 381–391. doi:10.1080/08927010802256414 PMID:18612908

Oh, C. O., Ditton, R. B., & Stoll, J. R. (2008). The economic value of scuba-diving use of natural and artificial reef habitats. *Society & Natural Resources*, *21*(6), 455–468. doi:10.1080/08941920701681953

Paredes, M. G., Padilla-Rivera, A., & Güereca, L. P. (2019). Life cycle assessment of ocean energy technologies: A systematic review. *Journal of Marine Science and Engineering*, *7*(9), 322. doi:10.3390/jmse7090322

Pickering, H., & Whitmarsh, D. (1997). Artificial reefs and fisheries exploitation: A review of the 'attraction versus production'debate, the influence of design and its significance for policy. *Fisheries Research, 31*(1-2), 39–59. doi:10.1016/S0165-7836(97)00019-2

Pioch, S., de Monbrison, D., & Simard, F. (2020). Artificial Reefs in France: Current State-of-the-Art and Recent Innovative Projects. In Modern Fisheries Engineering (pp. 31-40). CRC Press.

Pordata. (n.d.). *PORDATA: Base de dados Portugal contemporâneo*. Available online from: pordata.pt

Quinn, T. P. (2009). *The influence of artificial reef associated fish assemblages and varying substrates on coral recruitment*. Academic Press.

Ramm, L. A., Florisson, J. H., Watts, S. L., Becker, A., & Tweedley, J. R. (2021). Artificial reefs in the Anthropocene: A review of geographical and historical trends in their design, purpose, and monitoring. *Bulletin of Marine Science, 97*(4), 699–728. doi:10.5343/bms.2020.0046

Ramos, J. (2021). Online Perception of Artificial Reef Risk and Safety by Stakeholders (Including Residents and Tourists) Via the Analytic Hierarchy Process. *Journal of Spatial and Organizational Dynamics, 9*(3), 189–204.

Ramos, J., Lino, P. G., Himes-Cornell, A., & Santos, M. N. (2019). Local fishermen's perceptions of the usefulness of artificial reef ecosystem services in Portugal. *PeerJ, 6*, e6206. doi:10.7717/peerj.6206 PMID:30647998

Ramos, J., Santos, M. N., Whitmarsh, D., & Monteiro, C. C. (2006a). Patterns of use in an artificial reef system: A case study in Portugal. *Bulletin of Marine Science, 78*(1), 203–211.

Ramos, J., Santos, M. N., Whitmarsh, D., & Monteiro, C. C. (2006b). The usefulness of the analytic hierarchy process for understanding reef diving choices: A case study. *Bulletin of Marine Science, 78*(1), 213–219.

Ramos, J., Santos, M. N., Whitmarsh, D., & Monteiro, C. C. (2007). Stakeholder perceptions regarding the environmental and socio-economic impacts of the Algarve artificial reefs. In *Biodiversity in Enclosed Seas and Artificial Marine Habitats* (pp. 181–191). Springer. doi:10.1007/978-1-4020-6156-1_16

Ramos, J., Santos, M. N., Whitmarsh, D., & Monteiro, C. C. (2011). Stakeholder analysis in the Portuguese artificial reef context: winners and losers. *Brazilian Journal of Oceanography, 59*(SPE1), 133-143.

Ramos, J. H. P. (2007). *Socio-economic implications of artificial reef deployment: a Portuguese case study* (Doctoral dissertation). University of Portsmouth.

Ribeiro, C., & Neves, P. (2020). Habitat mapping of Cabo Girão Marine Park (Madeira island): A tool for conservation and management. *Journal of Coastal Conservation, 24*(2), 1–16. doi:10.100711852-019-00724-9

Roa-Ureta, R. H., Santos, M. N., & Leitão, F. (2019). Modelling long-term fisheries data to resolve the attraction versus production dilemma of artificial reefs. *Ecological Modelling, 407*, 108727. doi:10.1016/j.ecolmodel.2019.108727

Roseman, E. F., McLean, M., Pritt, J. J., Fischer, J. L., & Kennedy, G. (2017). Artificial reefs and reef restoration in the Laurentian Great Lakes. *Biodiversity, Conservation, and Environmental Management in the Great Lakes Basin*, 33-46.

Santos, M., Monteiro, C., & Lasserre, G. (1996). Faune ichtyologique comparée de deux récifs artificiels du littoral de la Ria Formosa (lagune Portugal): Résultats préliminaires. *Oceanologica Acta, 19*(1), 89–97.

Santos, M. N., Lino, P. G., Pousão-Ferreira, P., & Monteiro, C. C. (2006). Preliminary results of hatchery-reared seabreams released at artificial reefs off the Algarve coast (southern Portugal): A pilot study. *Bulletin of Marine Science, 78*(1), 177–184.

Santos, M. N., & Monteiro, C. C. (1997). The Olhao artificial reef system (south Portugal): Fish assemblages and fishing yield. *Fisheries Research, 30*(1-2), 33–41. doi:10.1016/S0165-7836(96)00567-X

Santos, M. N., & Monteiro, C. C. (1998). Comparison of the catch and fishing yield from an artificial reef system and neighbouring areas off Faro (Algarve, south Portugal). *Fisheries Research, 39*(1), 55–65. doi:10.1016/S0165-7836(98)00169-6

Santos, M. N., & Monteiro, C. C. (2001). Modern Aquaculture in the Coastal Zone 281. *Modern Aquaculture in the Coastal Zone: Lessons and Opportunities, 314*, 281.

Santos, M. N., & Monteiro, C. C. (2007). A fourteen-year overview of the fish assemblages and yield of the two oldest Algarve artificial reefs (southern Portugal). In *Biodiversity in Enclosed Seas and Artificial Marine Habitats* (pp. 225–231). Springer. doi:10.1007/978-1-4020-6156-1_19

Santos, M. N., Monteiro, C. C., & Lasserre, G. (2005). Observations and trends on the intra-annual variation of the fish assemblages on two artificial reefs in Algarve coastal waters (southern Portugal). *Scientia Marina, 69*(3), 415–426. doi:10.3989cimar.2005.69n3415

Santos, M. N., Monteiro, C. C., & Leitão, F. (2011). The role of artificial reefs in the sustainability of artisanal fisheries. In *Artificial reefs in Fisheries Management.* Taylor & Francis Group.

Sato, M., Inoue, N., Nambu, R., Furuichi, N., Imaizumi, T., & Ushio, M. (2021). Quantitative assessment of multiple fish species around artificial reefs combining environmental DNA metabarcoding and acoustic survey. *Scientific Reports, 11*(1), 1–14. doi:10.103841598-021-98926-5 PMID:34593907

Sayer, M. D. J., & Wilding, T. A. (2002). Planning, licensing, and stakeholder consultation in an artificial reef development: The Loch Linnhe reef, a case study. *ICES Journal of Marine Science, 59*(suppl), S178–S185. doi:10.1006/jmsc.2002.1270

Schulze, A., Erdner, D. L., Grimes, C. J., Holstein, D. M., & Miglietta, M. P. (2020). Artificial Reefs in the Northern Gulf of Mexico: Community Ecology Amid the "Ocean Sprawl". *Frontiers in Marine Science, 7*, 447. doi:10.3389/fmars.2020.00447

Seaman, W. (2000). *Artificial reef evaluation: with application to natural marine habitats.* CRC Press. doi:10.1201/9781420036633

Seaman, W. Jr, Grove, R., Whitmarsh, D., Santos, M. N., Fabi, G., Kim, C. G., & Pitcher, T. (2011). *Artificial reefs as unifying and energizing factors in future research and management of fisheries and ecosystems.* CRC Press. doi:10.1201/b10910-3

Seaman, W., & Jensen, A. C. (2000). *Purposes and practices of artificial reef evaluation*. CRC Press LLC. doi:10.1201/9781420036633

Shani, A., Polak, O., & Shashar, N. (2012). Artificial reefs and mass marine ecotourism. *Tourism Geographies*, *14*(3), 361–382. doi:10.1080/14616688.2011.610350

Stolk, P., Markwell, K., & Jenkins, J. M. (2007). Artificial reefs as recreational scuba diving resources: A critical review of research. *Journal of Sustainable Tourism*, *15*(4), 331–350. doi:10.2167/jost651.0

Sutton, S. G., & Bushnell, S. L. (2007). Socio-economic aspects of artificial reefs: Considerations for the Great Barrier Reef Marine Park. *Ocean and Coastal Management*, *50*(10), 829–846. doi:10.1016/j.ocecoaman.2007.01.003

Tunca, S., Ünal, V., & Lizaso, J. L. S. (2019). Stakeholder Analysis of the Artificial Reefs in Edremit Bay (Northern Aegean), Turkey: Contributions to management. *International Biodiverstiy & Ecology Sciences Symposium*.

Vicente, M., Falcão, M., Santos, M. N., Caetano, M., Serpa, D., Vale, C., & Monteiro, C. (2008). Environmental assessment of two artificial reef systems off southern Portugal (Faro and Olhão): A question of location. *Continental Shelf Research*, *28*(6), 839–847. doi:10.1016/j.csr.2007.12.009

Vivier, B., Dauvin, J. C., Navon, M., Rusig, A. M., Mussio, I., Orvain, F., Boutouil, M., & Claquin, P. (2021). Marine artificial reefs, a meta-analysis of their design, objectives and effectiveness. *Global Ecology and Conservation*, *27*, e01538. doi:10.1016/j.gecco.2021.e01538

Whitmarsh, D., Santos, M. N., Ramos, J., & Monteiro, C. C. (2008). Marine habitat modification through artificial reefs off the Algarve (southern Portugal): An economic analysis of the fisheries and the prospects for management. *Ocean and Coastal Management*, *51*(6), 463–468. doi:10.1016/j.ocecoaman.2008.04.004

Wirtz, P., Genner, M., & Relvas, C. (2020). *First record of Cobia (Rachycentron canadum) (Pisces: Rachycentridae) at the coast of Madeira Island*. NE Atlantic Ocean.

KEY TERMS AND DEFINITIONS

Algarve Coast: Is the southernmost region of Portugal mainland where most artificial reefs are deployed.

Artificial Reef Demand: When there are stakeholders (usually primary such as users) that seek artificial reefs as a complement or substitute for their professional (e.g., small-scale fishers) or recreational (e.g., scuba diving) activities.

Artificial Reef Program: Between 1990 and 2003 occurred In the Algarve (south region of Portugal mainland), deployment actions promoted by the national fisheries and marine research institute (currently IPMA).

Artificial Reef Stakeholders: All users or non-users that have interest in the structures either as primary, secondary, or external actors.

Artificial Reef Supply: When there are artificial reef deployment structures promoted by decision entities. Depending on the type of structures and the objective of the project, artificial reefs supplied are generally available for different purposes including biodiversity promotion, coastal protection, fisheries enhancement.

Coastal Communities: Usually villages or small towns that employ many people that are highly reliant on coastal goods and services, namely sea resources and related activities. For example, fishers, scuba divers and tourism operators may benefit of coastal areas where there are diversified and abundant fish resources, as well as low pollution levels can be found.

Efficiency: The efficiency of artificial reefs can be seen from the point of view of the usefulness of the structures for the benefit of human activities (e.g., fishing, diving), so that there is no unnecessary waste of resources – materials, energy, money, time – and that the benefits achieved by the stakeholders be the desired ones.

Chapter 2
Artificial Reef Stakeholders in France: A Social-Network Analysis

Jessica Salaün
Atlantique Landes Récifs, France

Sylvain Pioch
Paul-Valéry University, France

Jean-Claude Dauvin
Caen Normandie University, France

ABSTRACT

As coastal zones are subjected to cumulative human pressures, it is vital to protect these productive and sensitive marine ecosystems. Artificial Reefs (ARs) are recognized as useful tools to overcome certain challenges to these ecosystems, if managed and regulated appropriately. While ARs were originally used for fishery purposes, for almost two decades they have also been widely used in ecological restoration to recover specific habitat functions. Most assessments of ARs focus on their biological and physical results: few have investigated how stakeholders contribute to their outcome. This chapter seeks to shed light on this aspect, examining the role and characteristics of stakeholders in ARs using social network analysis.

INTRODUCTION

It has been accepted for decades that coastal areas provide various ecosystem services, defined as benefits coming from nature that contribute to human well-being. For 60 years this area has undergone many human developments and a diversification of its use (Trouillet, 2006; Lacroix *et al.*, 2021). Yet, traditional coastal activities such as fishing (that rely on this productive ecosystem) must now deal with emerging industrial activities (Borja *et al.,* 2016). Human activities such as tourism, renewable energy production, oil & gas drilling and mining, generate increasing pressure on the coastal ecosystem. Moreover, the land use shift with artificialization of this coastal area are identified as one of the major causes of

DOI: 10.4018/978-1-6684-2344-8.ch002

the biodiversity decrease, with also over-exploitation, climate change, pollution and species dissemination (IPBES, 2019). This resulting in the urgent need to better protect and manage this productive and sensitive ecosystem (Pinarbasi *et al.*, 2019).

To overcome certain challenges facing environmental management of this ecosystem, Artificial Reefs (ARs) had been identified as suitable tools. While originally ARs were used for fishery purposes, recently they had been deployed for ecological restoration to recover specific habitat functions: sheltering, spawning and feeding (Baine, 2001; Jensen, 2002; Lacroix *et al.*, 2002). As the ecological restoration and engineering turn into a major environmental answer against biodiversity erosion and climate change for the next decade (Assemblée Générale des Nations-Unies, 2019), ARs could participate to this end. In this context and to overcome challenges, it becomes necessary to better understand the role and the effect of this tool on the ecosystem of coastal area (Ramos *et al.*, 2019).

To enhance its effectiveness, ARs need to be managed with adapted regulation rules (Lacroix *et al.*, 2002). As multiple activities coexisted on the coastal area, several types of stakeholders might have an interest in using ARs (Milon *et al.*, 2000; Salaün *et al.*, 2020). In fact, by definition, ARs are human-made structures designed to be used directly by fishermen, divers, scientists or sailors and indirectly for tourist activities or environmental purposes. They can also be employed in certain human activities, e.g., for mooring or as a substitute for natural reefs for recreational diving (Pioch *et al.*, 2019). As a consequence, social interactions around ARs as to be taken into consideration to define their influence on ARs functioning and ensure their effectiveness.

To this end, this chapter investigates the types of stakeholders involved in AR projects, using network analysis to attempt to answer two main questions: (1) What roles do different stakeholders play during each stage of AR development? (2) What are the features of AR social networks?

This chapter adopts a multi-case approach. The investigation is based on an analysis of ten case studies from the metropolitan coastline of France, for each of which the social network was modelled to reveal the stakeholders' positions and roles. The ARs were deployed in France between 1968 and 2020 along the Mediterranean, Atlantic and English Channel coasts.

This chapter begins by describing the background that led to this choice of analysis, then presents the three stages of the network analysis and discusses solutions and recommendations. The findings indicate that this type of approach could help to better assess the social criteria that leads to the success of ARs.

BACKGROUND

Biological Focuses

ARs have been deployed to mimic natural reef ecological functionalities and/or to sustain artisanal fisheries. The definition of ARs induces a biological aim to increase fish biomass, species diversity and provide shelter against predators. These structures are dimensioned to mimic natural reef effect in providing supplementary surface area for primary and secondary producers (Bortone *et al.*, 2000). Despite of the ARs worldwide deployment and the increasing research on their design, performance and management, the understood of their efficiency still remain a point of interest (Lee at al., 2018; Lima *et al.*, 2020; Vivier *et al.*, 2021; Taormina *et al.*, 2022). As their main goal is to increase fish biomass, their studies preferably focus on the variation of certain ecological component such as fish assemblage, abundance, species richness and have proven the ARs usefulness to provide habitat and suitable shelter against predators (Véron *et al.*, 2008; Folpp *et al.*, 2011; Neves dos Santos and Zalmon, 2015; Becker *et al.*, 2018). More recently biological studies have investigated the production of ARs through functional approaches (Cresson 2013). The feeding relation has been explored using trophic approach with isotopic analysis or gut content and clearly demonstrate that ARs have a production as well as attraction effect (Fabi *et al.*, 2006; Smith *et al.*, 2016; Cresson *et al.*, 2019; Roa-Ureta *et al.*, 2019; Mavraki *et al.*, 2021; Salaün, 2022). Other studies also demonstrate their real contribution in the production of commercial fishes within their similitude to the natural reef fish assemblage (Simon *et al.*, 2013; Pereira *et al.*, 2016; Streich *et al.*; 2018, Wu *et al.*, 2019). The comparison with natural reefs also illustrated how ARs design providing shelters and new substrate with greater prey available influence the growth rates and productivity (Scarcella *et al.*, 2011; Rouse *et al.*, 2020)

However, one aspect of ARs that remains poorly studied is how stakeholders contribute to their outcome (Milon *et al.*, 2000; Baine, 2001; Ramos *et al.*, 2008; Véron *et al.*, 2008; Tessier, Dalias and Lenfant, 2015; FAO, 2015; Lee *et al.*, 2018).

Lack of Social Analysis

The end users of ARs are the direct and undirect beneficiaries of these structures and are commonly considered to be the benthic species, pelagic fish and also humans (Milon *et al.*, 2000). While several studies have been conducted to investigate links between ARs and human end-users, few have taken a holistic approach (Koeck *et al.*, 2011).

Of those social studies that have been conducted, most have investigated links between ARs and fishermen (Santos and Monteiro, 2007; Simard, 1997; Nakamura, 1985). These were based on interviews and questionnaires to reveal the perception and satisfaction of professional fishermen, but also new users such as divers (Ramos *et al.*, 2011a; Tessier, 2013; Kirkbride-Smith, 2016). Economic approaches had also been carried out, but once again only on fisheries or diving activities (Sutton and Bushnell, 2007; Fletcher *et al.*, 2011; Chen *et al.*, 2013; Keller *et al.*, 2016). Yet, many other undirected stakeholders are concerned by ARs such as political, environmental organization, scientific organization or contractor. Our study attempts to address this gap using an approach on all ARs stakeholders.

The principal aim of such an approach is to integrate all the stakeholders that have an interest in ARs deployment and reveal the links between them (Freeman, 1984; Meffe *et al.*, 2002; Sutton and Bushnell, 2007; Simpson *et al.*, 2016; De Juan *et al.*, 2017; Schumacher *et al.*, 2018; Francis *et al.*, 2019; Molino *et al.*, 2020). One stakeholder approach had concerned an AR project in Portugal, in which stakeholders' involvement was described based on socio-economic results as winner or losers (Ramos *et al.*, 2011b). Indeed, this type of approach could be applied to qualify and quantify the benefits for each end user of ARs (*Op. cit.*). Such an approach could also indicate how successful the management of the ARs sites and his resource was (Cárcamo *et al.*, 2014).

METHODOLOGY

Social Network Analysis

To supplement biological studies, this chapter focuses on stakeholders, using social network analysis to identify their role, their characteristics as well as their position within the social network. This network analysis was applied for the three main stages of the ARs project (Ramos *et al.,* 2011b) (Figure 1):

- Initiation stage: the identification of the need and planning of the ARs
- Implementation stage: the deployment of the ARs
- Management stage: monitoring the ARs and evaluating its benefits.

Figure 1. The three stages of ARs development

This approach allows understanding and modelling social processes (Wasseman and Faust, 1994). It defines a network as elements – which can be individuals or organizations – that have connections between them (Newman, 2003). The model uses mathematical concepts that allow complex networks to be simply represented (Lazega, 2014). This approach is used in various domains and applied to social sciences and in particular sociology of organizations during the 1970s with the development of the Social Network Analysis field (Dumoulin Kervran and Pepin-Lehalleur, 2012).

In this social network analysis applied to ARs, one social network was modelled for each stage in the development of an AR integrating all end-users defined as stakeholders. The data was analysed using social network modelling in order to understand the structure of social relationships around AR projects. Social network models were built with nodes that symbolize stakeholders (individuals or organizations), with links representing the connections between them.

In order to build the social network model, data was collected using several methods: open-ended interviews, archival research and informal conversations. For the open-ended interviews, guidelines were established to explore three areas at each stage of the project: (1) the stakeholders, (2) the relationships with other stakeholders, and (3) the links between them. Interviewees were selected using the snowball sampling method, which was often used in stakeholder network analysis (Reed *et al.,* 2009). This consisted of identifying the project initiator, and then asking him or her to identify other individuals involved in the ARs project in order to interview them. The interviewing process ended when interviewees were not able to identify new individuals who played a significant role in the project or identified only stakeholders already interviewed (Kaufmann, 2016). The open-ended interviews were conducted from July 2019 to December 2020. A total of 134 face-to-face or phone interviews were carried out with stakeholders involved in ARs projects in France, principally with a role of project initiator, project manager, project developer, funder or regulator, but also stakeholders who provided scientific and technical advice or were consulted for their opinion (e.g., marine users).

For the model, stakeholders were identified and classified according to several scales, from a local to national level (Grimble and Wellard, 1997). Stakeholders were identified through the organizations they represent, and these organizations were classified in categories and types (Figure 2).

Figure 2. Stakeholder classification levels

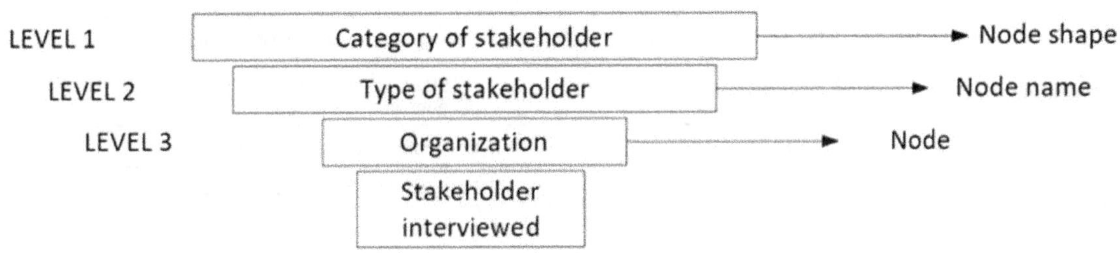

A wide variety of actors may be involved in ARs projects. To classify these actors, four main groups have been defined: Europe, national authorities, local authorities and civil society. And on the function of the stakeholder, 16 types of stakeholders are determined (Table 1).

Table 1. Node descriptions

CATEGORY OF STAKEHOLDER (NODE SHAPE)	TYPE OF STAKEHOLDER (NODE NAME)	DESCRIPTION
Civil society	Commercial organizations	Engineering, industrial, press
	Environmental organizations	NGOs, local and national non-profit organizations
	Leisure organizations	Recreational fishermen, scuba divers, boaters
	Fishermen's organizations	Professional fishermen
	Professional organizations	Non-profit organizations
	Unorganized individuals	Individuals not representing an organization
Local authorities	Township	/
	Interlinked municipalities	Group of municipalities
	County	/
	Region	/
National authorities	Decentralized service	Environment and marine departments
	Public institutions	Laboratories, research institutes, water agencies, chambers of commerce, marine parks
	Agencies	Environment and marine agencies
Europe	European institutions	European Structural and Investment Funds
	Other countries	Portugal
Mediterranean	Mediterranean institution	International institution focusing on the Mediterranean

Each group has a certain role and function on an AR project:

- National authorities: coordinate implementation of policy and manages public property. Also, they support research and development and can provide funding and expertise.
- Local authorities (region, county, township, etc.): involved in policy, management and funding; role depends on delegation of authority.
- Civil society (environmental and leisure organizations, professional and commercial organizations): mainly involved as users of the territory. Categories have been made to distinguish users with profit expectation as commercial organizations or professional fishermen and users with non-profit expectation so with scientific (Professional organizations), environment (Environmental organizations) or recreational interests (Leisure organizations).

The relationship between stakeholders was defined by exchanges that could involve technical materials, monetary resources, involvement, skills or information (Table 2). The links were weighted on the frequency of the relationship determined by the stakeholders (1: irregular; 2: at least once a year; 3: more than once a year).

Table 2. Links descriptions

Category of link	Type of link	Descriptions
Technical materials	Boat	/
	Harbour facilities	/
Monetary	Funding	One-off, for a project or part of a project
	Fees	Payment for a service
	Grant	Annual request
Involvement	Decision-making	/
	Cooperation	Discussion that can influence decisions at a certain level
	Consultation	Feedback collected
Skills	Technical	Implementation, construction
	Scientific	Monitoring, analysis
	Collaboration	Exchange of skills
	Activities	Cultural events, awareness-raising campaigns
Information	Technical	Reports
	Scientific	Papers
	Administrative	Reports, authorization, exclusion zones
	Media	Press, newscasts

The data analysis and graphic outputs were generated using NodeXL, a Microsoft Excel package.

In order to analyse the structural network properties, eight indicators were calculated for each social network:

- **Number of connected nodes:** the number of stakeholders involved in the network. This indicator provides information on the presence or absence of stakeholders (Wasserman and Faust, 1994).
- **Degree centrality:** the number of links for a given node. This indicator is used to assess the structural importance of a stakeholder (Freeman, 1979; Friedman *et al.,* 2020).
- **Betweenness centrality:** the number of times a stakeholder is on a path to join other nodes. This measure indicates the stakeholder's capacity to connect to other stakeholders through his/her presence (Freeman, 1979; Ingold, 2017; Friedman *et al.,* 2020).
- **Closeness centrality:** the shortest path between nodes (Delmas, 2018).
- **Eigenvector centrality:** the influence of a node in the network considering the degree centrality of the node and the nodes related to it (Ospina-Alvarez, 2020).
- **Average geodesic distance:** the average number of nodes that separate each node.
- **Density:** the number of links in the network out of the total number of links possible.
- **Clustering:** the grouping of nodes according to neighbouring links (Delmas, 2018).

The analysis centred on the project initiator (ego) and the relationships around it. For each ARs project, three networks were built: (1) the social network during the period of building the AR project (initiation stage), (2) a transitory social network (implementation stage) and (3) the current social network (management stage). The aim of the transitory network was to represent a different network state between the origin of the project and the current state. In most cases, there was a change after five years of deployment because the European funding contract ended and with it the necessity of conducting ecological surveys. To be able to compare the network properties of different projects, each network was built with a maximum geodesic path of four (representing the shortest path between two extreme nodes).

The social analysis consisted of three study parts: the nodes, the links and the overall network.

Case Studies

This study focused on ten over 49 ARs sites, spread over the three metropolitan French coasts and deployed at different periods with production and/or protection goals (Figure 3):

- **English Channel**: Cherbourg and Etretat
- **Atlantic:** Croisic, Oléron and Capbreton
- **Mediterranean Sea**: Gruissan, Agde, Carry-le-Rouet, Marseille and Vallauris.

While the sites represent around 25% of artificial reefs in France, as some ARs had the same project initiator and project manager, the studied social networks represent about 33% of French ARs.

Figure 3. Location of French ARs and case studies

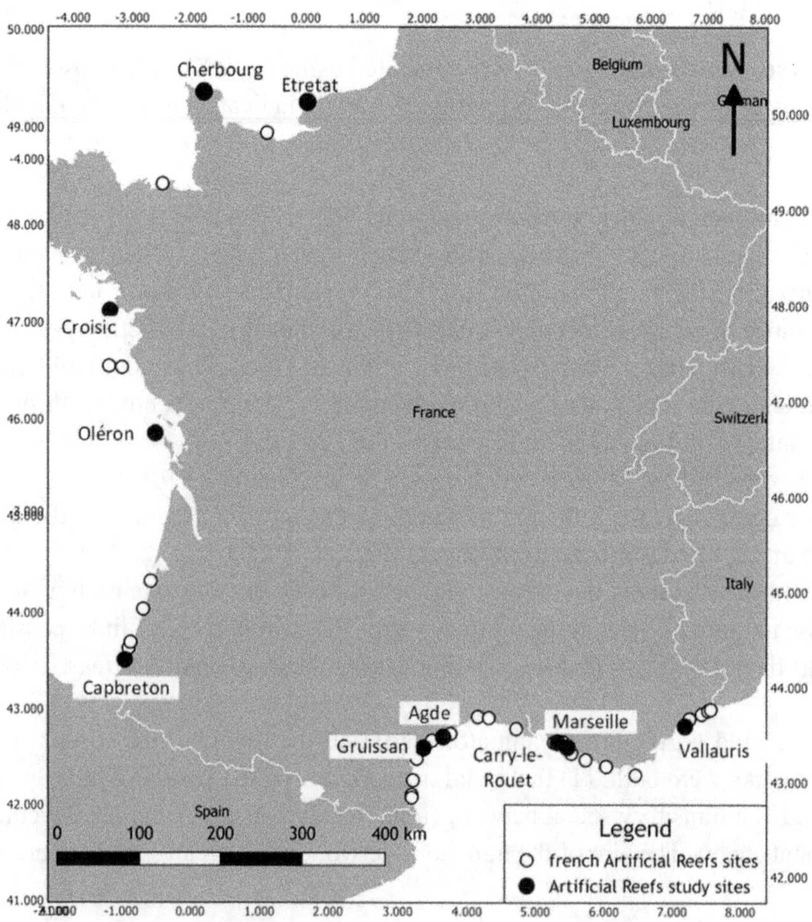

Two of the study sites emerged during the 1980: Carry-le-Rouet and Vallauris. They were initiated by fishermen's organizations, which asked for support to face a decrease in marine fauna. To address their issues, the fishermen relied on local government support. At Carry-le-Rouet the main objectives were production and protection from illegal trawling (Charbonnel and Bachet, 2010), whereas at Vallauris the objectives were to restore the ecological functionalities and support professional fishing by increasing the resource. These ARs deployments had been mainly financed by European funds, the Region and their respective departments and represented a total volume of 8 094 m^3 and 1 093m^3, respectively at Vallauris and Carry-le-Rouet (Table 3).

During the next decades (1990–2010), the fishing crisis had grown more acute. To face it, ARs projects were initiated at various location in France (Table 3):

- In the English Channel, the Chamber of Commerce of Fécamp had conducted an ARs deployment of 450 m^3 in order to support the fishing industry and improve the knowledge of the reef effect in Etretat (Castetz, 2008).
- In the Atlantic Ocean, the projects were carried out directly by the fishermen's organizations (professional or leisure). They led the ARs projects to test various shape and size of ARs and the influence attempt to solve their problem of marine resource management (Croisic, Capbreton).
- In the Mediterranean Sea, the fishermen's organizations asked their townships to be the project holder of ARs deployment (Gruissan, Agde). The main goal of these projects was to help fishermen to manage their fishing territory by creating a specific area for marine fauna production.

Some projects took almost 10 years to complete. This was the case of the Marseille and Oléron AR projects. The Marseille project had been initiated by the local authorities in the aim of establishing management of the coastal environment by creating a marine protected area (Bernard *et al.,* 1999). A total of 401 ARs for a total volume of 27 300 m^3 were deployed in 2008 (Table 3). The Oléron project began in the context of the 1990–2010 period, with the aim to find a solution for the fishing sector economic difficulties. But their deployment only occurs 2018 because of the difficulties to find sufficient funding (Table 3).

The younger ARs site studied was the Cherbourg site. It was part of the reflections and actions carried out for the improvement of the ecosystem of the Channel and for a better management of marine resources (Auzas and Maurouard, 2015). Deployed in 2015 in collaboration with scientists, the aim was to improve the biodiversity and production of the Channel marine ecosystem (Table 3).

Table 3. Characteristics of the ten case studies

Case study	Location	Project initiator	Period of deployment	Goal	Volume (m3)
Etretat	English Channel	Chamber of commerce Fécamp	2008	Production	450
Cherbourg	English Channel	Harbour institution	2015	Production	97,2
Croisic	Atlantic Ocean	Fishermen organization	2003	Production and protection	276
Oléron	Atlantic Ocean	Fishermen organization	2018	Production	5 742
Capbreton	Atlantic Ocean	Group of municipalities	1999	Production and protection	830
Gruissan	Mediterranean Sea	Township of Gruissan	2002	Production and protection	9 350
Agde	Mediterranean Sea	Township of Agde	1995	Production and protection	2 503
Carry-le-Rouet	Mediterranean Sea	Marine Park	1983	Production and protection	1 093
Marseille	Mediterranean Sea	Township of Marseille	2007	Production	27 300
Vallauris	Mediterranean Sea	County	1980	Production	8 094

ARTIFICIAL REEF SOCIAL NETWORKS

The Position of Stakeholders

As a similar method was used to construct networks for each studied site, it was possible to compare the value of calculated indicators. The indicators in this multi-site analysis highlighted the various positions stakeholders hold within the ARs networks.

National authorities such as agencies are the most frequent stakeholders within the studied networks. In France, in contrast to other countries such as the USA, China, Japan and Portugal, agencies do not lead AR projects, but they play a regulatory role. They are considered key members at each AR stage. At the initiation stage they must authorize the AR deployment, then they control the success of the project based on annual reports sent by the project initiator.

The stakeholders with the highest number of relationships within the AR networks are interlinked municipalities and region. They have at least four connections with other stakeholders. They are key stakeholders as they maintain a large number of relationships (Degree centrality).

Township, interlinked municipalities and environmental organizations make it possible to develop the network and ensure the participation of remote stakeholders (high betweenness centrality). They have the capacity to bring together stakeholders by maintaining the connection between them and to extend the network by creating new relationships. This role is linked with their functions. Environmental organizations are already connected to local authorities and institutions for funding. Interlinked municipalities are also intermediaries between counties and townships. They facilitate the spread of information, so having them in the AR network can help to develop it.

Leisure organizations and decentralized services are relay nodes that enable other stakeholders to reach the others (high closeness centrality). They have a physically central position within the network.

Region and decentralized services are also strategic stakeholders within the networks (high eigenvector centrality). They are in the vicinity of other central stakeholders with high degree centrality.

Using Principal Component Analysis (PCA), four main positions based on these indicators were distinguished (Figure 4):

Figure 4. PCA identifying four main positions

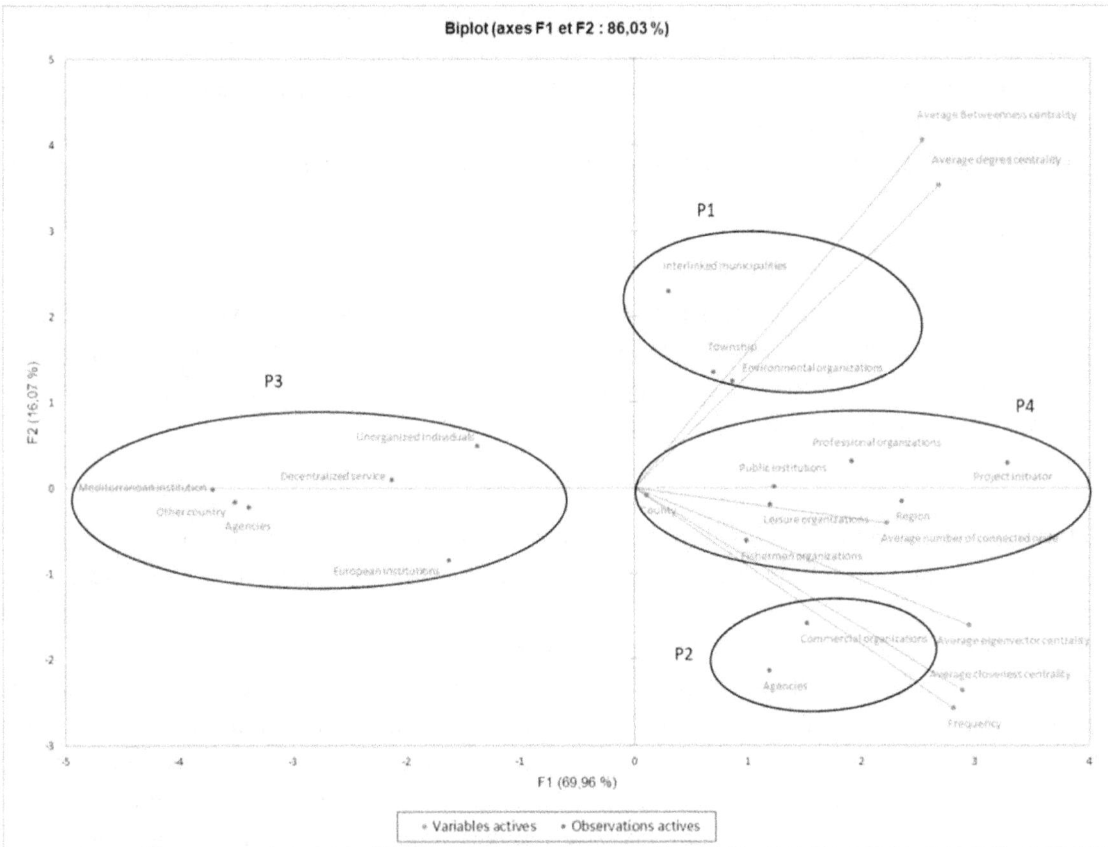

Position 1 (P1): mobilizing and relational organizations

This group represents the types of organisations with a high value of Degree and Betweenness centrality, strongly participating in mobilizing new players and creating a link between them. These are the townships, interlinked municipalities and environmental organizations.

Position 2 (P2): Organizations that are essential at each stage of AR projects

The frequency of the organization within networks defines which are needed at each stage of AR projects. These are the marine agencies and the commercial organizations.

Position 3 (P3): Organizations not involved at each stage of AR projects

These types of organizations are not required at each stage of AR projects so they are less involved in the project. These include the other country, other agencies, decentralized service, the European institutions and Mediterranean institutions.

Position 4 (P4): Central organizations

As defined above, several centrality indicators are used to define the position of organizations within networks. The role may be strategic, such as information disseminator, or connected to other types of organizations. The central stakeholders are the project initiator alongside local authorities.

Each stage of the ARs was characterized by a specific combination of stakeholders within the network:

- **Initiation stage**: dominated by the centrality of fishermen's and professional organizations, interlinked municipalities and region.
- **Implementation stage**: involves local authorities, public institutions, and environmental, leisure and commercial organizations.
- **Management stage**: centred on interlinked municipalities and region.

Connections

The links between two stakeholders is the second component of network analysis. The relationships identified allow characterizing how stakeholders are involved in the network. While their presence in the network described in the previous section shows what role they play in the network, the connections show how they maintain this role.

Involvement can take several forms:

- **Passive**: no commitment of any kind between stakeholders, just information flow.
- **Active**: stakeholders take part by providing economic, technical, human or skill resources.

Within AR networks, active forms of connection are preferred and represent 75% of total relationships versus 25% of passive relationships. However, apart from material connections, which are little represented, the proportion of other relationships (skills, people, money and information) is equivalent. Some organizations reveal a preference for one or two types of connections, whereas others show no preferences. For example, commercial businesses choose to exchange skills and information with other stakeholders. From a general point a view, skills exchanges are principally linked with public institutions or commercial businesses, human resources are linked with municipalities and public institutions, economic resources are deployed by municipalities and regional governments, and technical resources are provided by counties. The more remote from the local area an organization is, the fewer its relationships within the network.

The frequency of relationships allows distinguishing temporary from lasting connections and indicates the strength of a network. The most frequent connections are the economic links. At the implementation stage, the average distribution of funds (in 6 studied sites) was: 40% from the European Institution, 18% from the region, 12% from the county, and 30% from interlinked municipalities and townships. In only one case, public institution supported AR deployment, but they can provide other temporary funding. Other frequent connections are annual regulation and technical loans such as boats or diving infrastructure needed during the implementation and management stages, but this is one-time rather than lasting.

Irregular connections include skills exchange in order to give public presentations about the project. This type of connection is developed by the interlinked municipalities and project manager with the aim of raising awareness about biodiversity and the necessity to protect the environment. This type of link was present in six out of ten study cases and at each stage of the AR.

Some relationships are also essential in connecting two stakeholders who would not otherwise be in contact. These relationships are vital to maintain the network and the number of stakeholders involved and are identified by overlooking relationships that link stakeholders with high betweenness centrality and more generally central stakeholders. The connections favoured by the central stakeholders are collaboration (21%), funding (17%) and consultation (14%).

In the global overview, no specific type of link characterized each stage of an AR. However, the type of connection over the evolution of the project tended to favour collaborative relationships.

Network Structure

The last component of the network analysis was based on the structural characteristics and emergent properties of the network. The structural characteristics refer to the number of components and the shape. Across the ten AR projects studied, 208 organizations were involved in total, referring to all the ten projects. This involvement was fairly homogeneous through the stages of an AR, with significant involvement (141 organizations) during the initiation stage, a decrease in involvement during the implementation stage (134 organizations), and renewed participation (159 organizations) during the management stage. The decrease in involvement between the first two stages might suggest a decline in stakeholder interest; however, these stages are quite different and do not have the same needs. During the implementation stage, the authorization and acceptance of public authorities are already acquired as well as some or all of the necessary funds for deployment. The project leader is also already defined and the AR is in the process of deployment. The organizations required before the project is less necessary during this stage of reef colonization. At this stage, twenty-eight organizations left the network as they no longer had a role to play (decentralized service, commercial organization). Conversely, twenty-one new organizations joined the network during this period, responding to the need to carry out scientific studies (commercial organization and public institutions).

At the management stage, stakeholders were remobilized to meet the aims of monitoring the project results. Involvement increased with twenty-nine new organizations – particularly public institutions (34%) – solicited compared to only 4 organizations that left the network. Some organizations (6 stakeholders) present during the initiation period became reinvolved in the AR project at this time.

The types of stakeholders most present in the network structure are public institutions, commercial organizations and townships. The involvement of these organizations is related to the needs at different stages: technical support during implementation and scientific support during the management stage. This strong presence of these types of organizations can also be explained by their diversity. Public institutions alone accounted for 19% of the stakeholders involved, commercial organizations accounted for 14% and townships 11%.

In terms of the number of stakeholders involved during the different stages of the project, several changes were observed (Figure 5):

- A continuous increase in the number of stakeholders (Agde, Marseille, Vallauris, Carry-le-Rouet)

- A general increase or constant number of stakeholders between the initiation stage and the management stage (Capbreton, Oléron)
- A continuous decrease in the number of stakeholders (Etretat and Cherbourg)
- A general decrease between the initiation stage and the management stage (Croisic and Gruissan).

More generally, a dichotomy is observed between AR sites with more than fifty stakeholders involved in the network (6 sites) and those with fewer than 30 (4 sites).

Figure 5. Change in the number of stakeholders involved

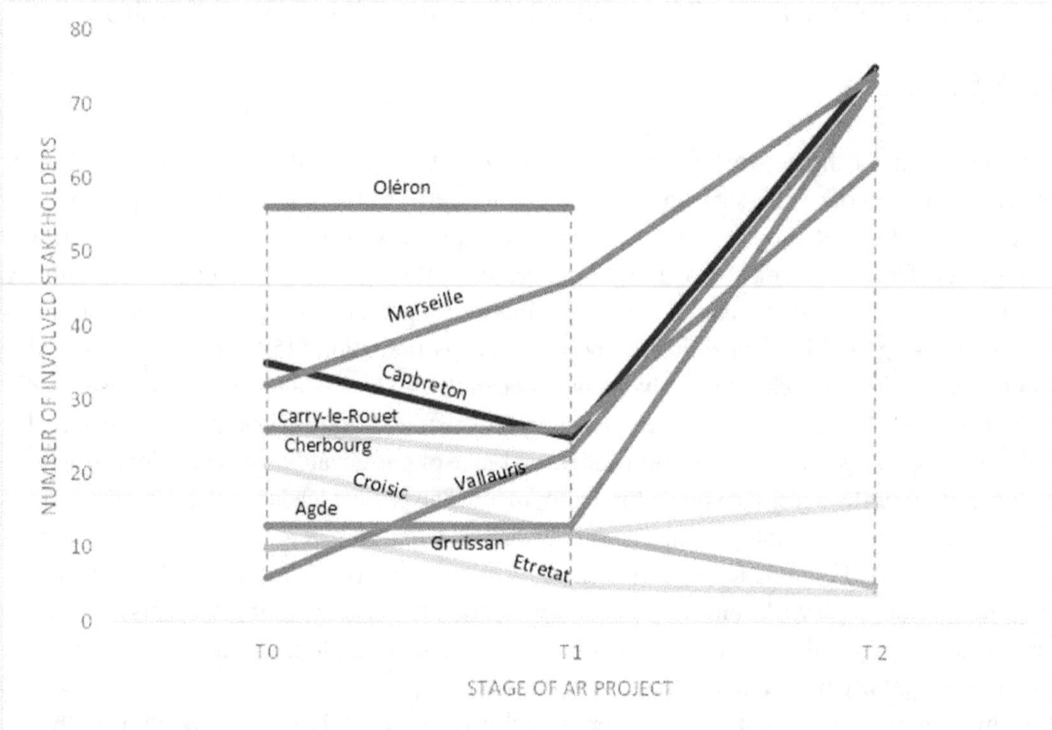

The shape of the network gives an indication of how relationships are established and evolving (Figure 6). A planar shape indicates a network whose connections do not intersect; in other words, it is possible to represent the network without overlapping links (Mueller, 2012). One type of planar network is star-shaped. In this case, the path from one stakeholder to another is unique. In this analysis, eighteen out of the 28 networks studied were planar, and three study sites had a planar network for the three stages of the ARs (Croisic, Etretat and Gruissan).

A complex shape indicates a network with a high number of nodes and relationships. Analyses of social media networks can include many thousands of nodes and relationships. In this study, networks with more than one hundred relationships are defined as complex networks.

In a global overview of the ten case studies, the structural characteristics of the network are identified for each stage of an AR:

- **Initiation stage**: dominated by a planar network
- **Implementation stage**: dominated by a planar network
- **Management stage**: dominated by a complex network.

This shows a trend towards network complexity over the life of an AR.

Figure 6. Network shapes

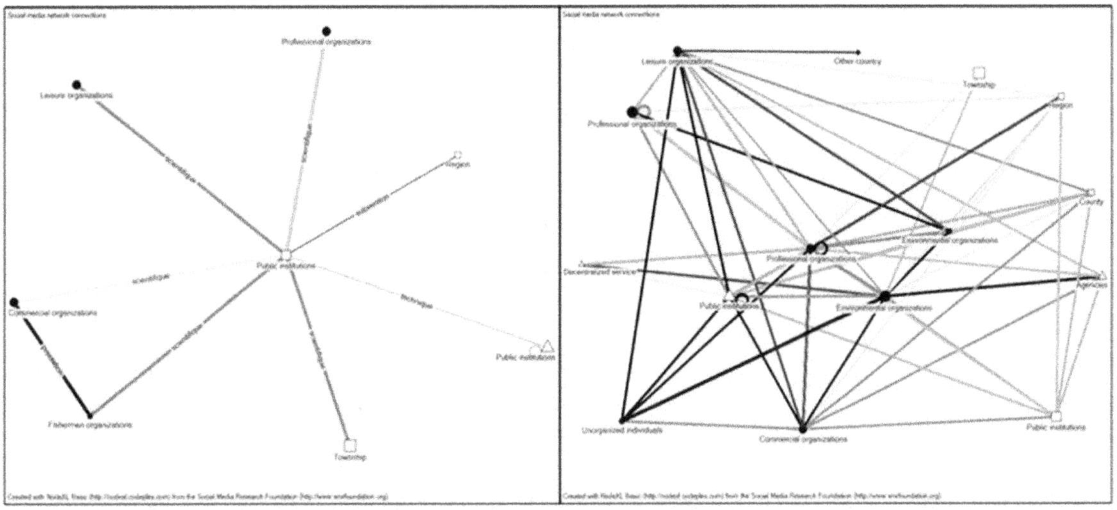

Planar network — Complex network

The two major network properties are connectivity and connexity. Network connectivity represents the real number of connections, whereas connexity represents the possibility of linking stakeholders with each other (Lagesse *et al.,* 2016).

To reveal the emergent properties of a network, four indicators (density, degree, geodesic average and cluster) are used in a PCA. The Pearson correlation describes 92% of the variables (Figure 7). The density and the geodesic average indicators are negatively correlated and uncorrelated to the cluster and degree indicators. Cluster and degree are positively correlated – if one increased, the other indicator evolved in the same way. This statistical analysis allows four types of networks with different properties to be distinguished.

Figure 7. PCA identifying four main types of networks

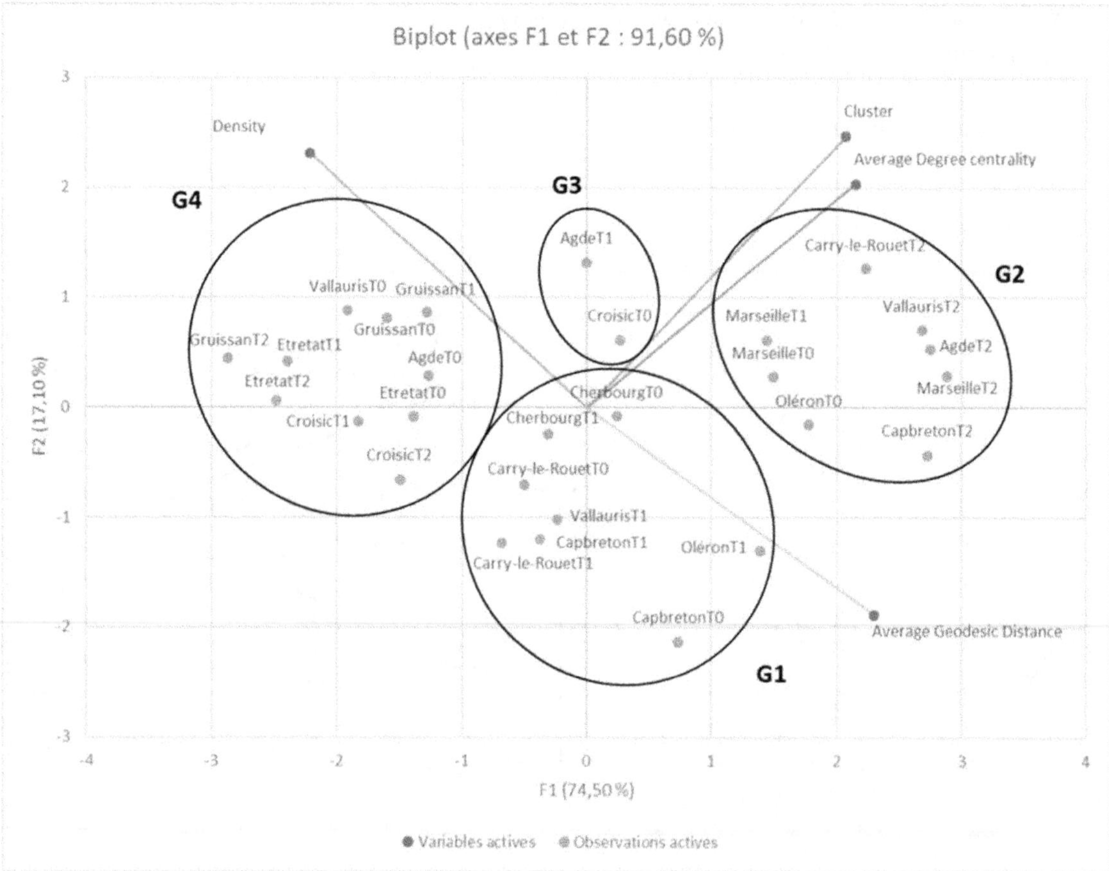

Group 1 (G1): Low connectivity network

This type of network has low density, with no cluster identified and a low geodesic average. These are characteristics of low connectivity.

Group 2 (G2): High connexity network

This type of network is complex, with a high number of stakeholders involved. The network is large, with a high geodesic average. While it is characterized by low connectivity, within the network a small group of stakeholders are totally connected and form a cluster of 4 or 5 nodes. Although the network connectivity is low, the capacity to connect a high number of stakeholders is higher than in other types of networks.

Group 3 (G3): Intermediary network

This type of network has medium values of connectivity and connexity compared to other network types. It has a cluster of two highly connected nodes. This may be a transitional stage towards a high connexity network.

Group 4 (G4): High connectivity network

This type of network has high density and a low geodesic average. The connexity is low and no connected cluster forms. These are characteristic of high connectivity. It has a low number of stakeholders, so the possibility of connections is low, explaining this high connectivity.

In terms of the evolution of network characteristics over the stages of an AR, three trends are distinguished:

- **Increase in connexity:** revealing an intensification of the stakeholders involved and the capacity of the network to extend relationships.
- **Decrease in connectivity:** may imply a decline of interest in the project or a transitory period between two stages. As each stage of an AR mobilizes different stakeholders, a decrease in stakeholder participation could be due to this rather than declining interest.
- **Stability of network properties:** indicating network solidity and sustainability.

SOLUTIONS AND RECOMMENDATIONS

A Central Role: A Source of Power or a Constraint?

The network analysis highlights key stakeholders who occupy a principal position within the networks based on centrality indicators. These central stakeholders benefit from being at the centre of the information hub. However, does this structural position give them the power to advance the network according to their interests (Lazega, 2011)? Analysing the operationality of centrality makes it possible to understand whether the role of central stakeholder offers a true opportunity of power or if it is more of a structural constraint – a trap due to the coincidental aggregation of relationships and information on this stakeholder without allowing them sufficient autonomy to make significant changes or without a real commitment to maintain this position (Burth *et al.,* 1991).

Two study sites with similar stakeholders (same type of organization leading the AR project and an identical aim) and similar network properties during the initiation stage allows this question about the power of central stakeholders to be addressed. Agde and Gruissan are both ARs projects that are initially promoted by fishermen's organizations on the Mediterranean coast. As these projects are costly and technical, the township chose to lead their deployment in order to help assemble the marine resources. The network analysis shows that both sites during the initiation stage are characterized by high connectivity, but diverged in the deployment stage, in which the Agde social network evolved to high connexity, whereas the Gruissan network retains the same properties.

The central stakeholders evolve differently, although in both cases, the central stakeholder at the initiation stage do not maintain this position. The central stakeholders of the Gruissan AR at the initiation stage are professional and commercial organizations. During the implementation stage, new stakeholders intervene, and the central position is taken by a commercial organization. This stakeholder may have used this position to develop and maintain their own professional network (this network has not been studied in detail; however, this stakeholder continues to be listed as the AR's referral contact, suggesting that they have managed to maintain their network). The case of the Agde AR is quite different: an environmental organization became the central stakeholder during the deployment stage and used its position to advance the network according to its interests, promoting the AR as a tool to create a marine protected area. The centrality power offered by the of the project leader is used in a conscious way in this case to increase collaborative relationships within the network. In the current Agde network, this stakeholder is no longer in a position of centrality as it has since been integrated into the organization managing the project, concentrating governance and power within the same organization.

Thus, depending on the stakeholders and their interests, a central position can be a real opportunity to acquire functional autonomy within the network or, on the contrary, be a constraint, assigning a central role to a stakeholder with differing interests. This constraint may drive these stakeholders to leave the network or modify their role afterwards.

Cooperation and Collaboration are Key

Cooperation can be defined as an exchange of opinions between different stakeholders in order to build a common project (Bouamrane, 2006). The cooperation relies on shared reflection that helps decision-making. When cooperation takes place during the initiation stage, it can have an impact on the continuation of the project by promoting (McAfee *et al.,* 2021):

- the involvement of all stakeholders, from civil society to national authorities.
- innovative ideas.
- stakeholder buy-in on final decisions.

Several of the studied AR projects implements cooperation during the initiation stage. This is particularly the case with the Oléron and Marseille sites, which have both established a large cooperation committee. Specific connections for cooperation may also be set up, for example, to allow technical exchanges between specialists. In the Marseille AR, two specific channels for cooperation have been created, the scientific council and the development council (which focuses on education, awareness-raising and communication). Recommendations from these councils are then discussed in the full committee, which takes the final decision. These different channels for cooperation may be retained during the following stages of implementation and management.

However, cooperation does not ensure stakeholder involvement. Although promoted by national authorities to ensure public interest, it can affect project development by imposing the dominant stakeholder's interest, leading to a confrontation between stakeholders that may have long-term consequences or even stop the project (Sébastien, 2006). In one example in Senegal, cooperation did not involve all stakeholders, resulting in a misunderstanding of the fishing rules on the AR site (Brochier *et al.,* 2021). To avoid this, all stakeholders should be integrated into the cooperation committee: i.e., direct end users of marine resources, those affected directly by the project, regulatory stakeholders, indirect end users such as teachers or scientists, and individuals interested in the AR project (Sébastien, 2006; Ramos *et al.,* 2011b). When cooperation was properly carried out, there is a better chance of establishing connections of trust between stakeholders that could lead to collaboration.

Collaboration can be defined as a win-win relationship in which each stakeholder actively participates in a project and reaps benefits. It is key to AR projects as these are complex and require knowledge in many disciplines. Collaboration is mainly established between the project initiator (landowner and/or project manager) and public institutions. Like cooperation, this type of relationship actively engages stakeholders in AR projects (McAfee *et al.,* 2021). It contributes to the development of the network and can become a structuring element of the network.

Network Structure: Consequence or Vector of Management Actions

The network analysis highlights two main types of network structure during the management stage: high connectivity (G4) and high connexity (G2). Network properties are dependent on characteristics such as the number of stakeholders and the number of relationships. During the management stage, a variety of skills may be required to ensure scientific studies, create fishing regulations, preserve and protect AR sites, etc. The measures implemented can have an impact on the network properties. This raises the question of whether the structure of the network is a cause or a consequence of the management measures implemented. Or formulated another way: are the management measures a real choice or are they conditioned by opportunities related to the presence and involvement of stakeholders within the networks?

Some management measures have an influence on the structure of the network because they involve other stakeholders than the project manager. These include communications, cooperation and sometimes the development of the AR. While communication and development can be initiated by another stakeholder, cooperation is a measure emanating from and conducted by the project manager by choice or obligation. It is thus a vector that influences the structure of the network by integrating new stakeholders and creating links with them.

Other management measures may be reinforced or suggested by stakeholders other than the project manager within the network: two examples are monitoring and scientific study. Monitoring of AR sites requires significant technical and human resources. As this measure can be very expensive and restrictive for the project manager, it is little deployed (40% in the studied sites). However, the network has provided some sites with an opportunity to implement this. This is the case of the Vallauris site, which is monitored by professional fishermen who ensure the exclusionary zone of the site is respected when they pass near the AR. In this case, the presence of this stakeholder within the network made it possible to implement this management action. Other users, such as recreational divers, can also get involved in management measures. Connection to the project managers through the network allows them, for example, to suggest and implement citizen science actions and thus gather complementary data to scientific study.

This recent involvement of recreational divers can also result in the decreasing participation of commercial organizations. While the latter are involved in the initiation and implementation stages, they tend to gradually give way to recreational stakeholders in the management stage. This phenomenon can be explained by the increasing involvement of leisure organizations in citizen science. In the Etretat AR, for example, a diving organization was involved from the beginning to conduct a marine fauna inventory. If financial resources do not allow commercial activities to be involved in the management stage, the use of citizen science deployed by recreational organizations is an opportunity to maintain and monitor ARs.

Network Parameters of Successful AR Projects

Assessments of coastal solutions such as ARs are necessary to provide knowledge and improve their effectiveness and relevance. As ARs are deployed for human needs, their success can be measured in part by the involvement and participation of stakeholders and the sustainability of the network structure.

The network properties, integrating the involvement and participation of stakeholders are defined as the connexity and the connectivity. The analysis has shown that the high connexity have been expressed in network with high involvement and participation of stakeholders. Thus, the high connexity of the network can be used as a criterion of successful AR projects. In this study, ARs projects classified in G2 can be considered to be successful.

A network that prioritizes collaboration and information sharing promotes system resilience (Schaffer *et al.,* 2012), so even if a stakeholder leaves the network, while the structure of the network will be modified, it will stay functional. Sustainability can also be favoured through the decentralization of the stakeholders, which ensures higher adaptability of the network (Bodin *et al.,* 2016). This necessitates multiple high centrality stakeholders that are different than the project initiator. This decentralization is very marked in the Mediterranean, where stakeholders have strong links with stakeholders of other ARs sites, forming highly connected sub-networks.

Thus, network parameters of ARs projects ensuring the sustainability, resilience of the system and the success of the project can be defined as:

- The multiple high centrality stakeholders that are different than the project initiator.
- The high number of consultation and collaboration links.
- The high connexity of the network.

However, these parameters have to be put in perspective with social goals defined for each project and also ecological results.

FUTURE RESEARCH DIRECTIONS

Social network analysis is a valuable tool for providing quantified indicators that can be used to assess AR projects. But the difficulty of assessing the results of ARs arises partly from unclear and unquantified initial objectives and partly from the lack of social studies (Becker *et al.,* 2018). Hence, specifying and defining the objectives of these structures seems to be an essential first step in the assessment process (Claudet, 2006). Following that, a social network analysis can be used to provide indicators. As the purposes of ARs and their stakeholders are remarkably diverse, a social-ecological approach may also be a valuable tool to improve management of coastal areas under multiple pressures

CONCLUSION

This application of social network modelling to artificial reef projects in France – the first study to do so – allowed a visualization of (1) stakeholder networks and (2) structural metrics regarding the nodes and the links of the network. The use of indicators such as centrality or the number of nodes provided quantitative elements allowing a comparison of the evolution of AR social networks during the three stages of ARs projects: initiation, implementation and management. These kinds of indicators help to identify stakeholders' positions within a network and to compare successful and failed experiments to identify the social criteria that leads to success.

The end users of ARs are the direct and indirect beneficiaries of these structures. They can be recreational divers, leisure or professional fishermen or scientists and environmental organizations and indirect end users such as teachers or scientists, and individuals interested. Their positions within the ARs social network evolves through the project stages. They are highly representative during the initiation stage, but their participation is passive, with an informational role. During the management stage, their involvement increases and they interact through cooperation links. They have a position within the network that allows them to influence ongoing management.

This original approach offers new perspectives to assess the success of an AR project. As defined by Millon *et al.* (2000), an AR project is successful when the structure is used. Using the quantitative indicators defined in this study to assess stakeholder involvement, the greater the number of stakeholders interacting within an AR network, the more successful the project is. Thus, the hight number of collaboration and consultation links but also the high connexity of the network could be considered as network parameters of ARs projects ensuring the sustainability, resilience of the system and the success.

This approach is thus a promising tool for future AR project developers to qualify and quantify the effectiveness of their project.

ACKNOWLEDGMENT

This research is co-funded by National Association for Technical Research and Atlantique Landes Récifs. We thank all interviewed stakeholders for their time.

REFERENCES

Assemblée générale des nations-unies. (2019). United Nations Decade on Ecosystem Restoration.

Auzas, M., & Maurouard, J. (2015). *Immersion des récifs (Deployment of Artificial Reefs). Rapport technique (Tehnical report)*. EMCC.

Baine, M. (2001). Artificial reefs: A review of their design, application, management and performance. *Ocean and Coastal Management*, *44*(3-4), 241–259. doi:10.1016/S0964-5691(01)00048-5

Becker, A., Taylor, M. D., Folpp, H., & Lowry, M. B. (2018). Managing the development of artificial reef systems:The need for quantitative goals. *Fish and Fisheries*, *19*(4), 1–13. doi:10.1111/faf.12288

Bernard, G., Bonhomme, P., & Charbonnel, E. (1999). *Valorisation de la rade Sud de Marseille-Aménagement en récifs artificiels de la baie du Prado (Development of the South Bay of Marseille, Deployment of artificial reefs in the Prado Bay). Rapport technique (Tehnical report)*. Ville de Marseille – GIS Posidonie.

Bodin, Ö., Robins, G., McAllister, R. R. J., Guerrero, A., Crona, B., Tengö, M., & Lubell, M. (2016). Theorizing benefits and constraints in collaborative environmental governance: A transdisciplinary social-ecological network approach for empirical investigations. *Ecology and Society*, *21*(1), 40. doi:10.5751/ES-08368-210140

Borja, A., Elliott, M., Andersen, J. H., Berg, T., Carstensen, J., Halpern, B. S., Heiskanen, A.-S., Korpinen, S., Lowndes, J. S. S., & Martin Gand Rodriguez-Ezpeleta, N. (2016). Overview of Integrative Assessment of Marine Systems: The Ecosystem Approach in Practice. *Frontiers in Marine Science*, *3*, 20. doi:10.3389/fmars.2016.00020

Bortone, M., Samoilys, M., & Francour, P. (2000). Fish and macroinvertebrate evaluation methods. In *Artificial reef evaluation with application to natural marine habitats* (pp. 127–164). CRC.

Bouamrane, M. (2006). Biodiversité et acteurs: des itinéraires de concertation (Biodiversity and actors: consultation areas). Réserves de biosphère - Notes techniques 1 (Technical notes 1). UNESCO.

Brochier, T., Brehmer, P., Mbaye, A., Diop, M., Watanuki, N., Terashima, H., Kaplan, D., & Auger, P. (2021). Successful artificial reefs depend on getting the context right due to complex socio-bio-economic interactions. *Scientific Reports, 11*(1), 16698. doi:10.103841598-021-95454-0 PMID:34404822

Burt, R. S. (1991). Structure version 4. 2. Center for the Social Sciences, Columbia University.

Cárcamo, P. F., Garay-Flühmann, R., & Gaymer, C. F. (2014). Collaboration and knowledge networks in coastal resources management: How critical stakeholders interact for multiple-use marine protected area implementation. *Ocean and Coastal Management, 91*, 5–16. doi:10.1016/j.ocecoaman.2014.01.007

Castetz, N. (2008). A Etretat, un récif artificiel pour revigorer les fonds. In *Etretat, an artificial reef to improve marine life*. Libération.

Charbonnel, E., & Bachet, F. (2010). Artificial reefs in the Cote Bleue Marine Park: assessment after 25 years of experiments and scientific monitoring. In *Global Change: Mankind-Marine Environment Interactions* (pp. 73–79). Springer. doi:10.1007/978-90-481-8630-3_13

Chen, J.-L., Chuang, C.-T., Jan, R.-O., Liu, L.-C., & Jan, M.-C. (2013). Recreational Benefits of Ecosystem Services on and around Artificial Reefs: A Case Study in Penghu, Taiwan. *Ocean and Coastal Management, 85*, 58–64. doi:10.1016/j.ocecoaman.2013.09.005

Claudet, J. (2006). *Aires marines protégées et récifs artificiels: méthodes d'évaluation, protocoles expérimentaux et indicateurs* [Marine protected areas and artificial reefs: assessment methods, experimental designs and indicators] [PhD Thesis]. Perpignan University.

Cresson, P. (2013). *Fonctionnement trophique des récifs artificiels de la baie du Prado (Marseille, France) : Origine et devenir de la matière organique* [Trophic functionnig of the artificial reefs of the Prado Bay (Marseilles, France): origin and fate of organic matter] [PhD Thesis]. Aix-Marseille University.

Cresson, P., Le Direach, L., Rouanet, É., Goberville, E., Astruch, P., Ourgaud, M., & Harmelin-Vivien, M. (2019). Functional traits unravel temporal changes in fish biomass production on artificial reefs. *Marine Environmental Research, 145*, 137–146. doi:10.1016/j.marenvres.2019.02.018 PMID:30857649

De Juan, S., Gelcich, S., & Fernandez, M. (2017). Integrating stakeholder perceptions and preferences on ecosystem services in the management of coastal areas. *Ocean and Coastal Management, 136*, 38–48. doi:10.1016/j.ocecoaman.2016.11.019

Delmas, E., Besson, M., Brice, M., Burkle, L., Dalla Riva, G., Fortin, M., Gravel, D., Guimarães, P. R. Jr, Hembry, D. H., Newman, E. A., Olesen, J. M., Pires, M. M., Yeakel, J. D., & Poisot, T. (2018). Analysing ecological networks of species interactions. *Biological Reviews of the Cambridge Philosophical Society, 94*(1), 16–36. doi:10.1111/brv.12433 PMID:29923657

Dumoulin Kervran, D., & Pépin-Lehalleur, M. (2012). *Agir-en-réseau. Modèle d'action ou catégorie d'analyse [Act as a network. Action model or category of analysis]. Presses Universitaires de Rennes, coll.* Res Publica.

Fabi, G., Manoukian, S., & Spagnolo, A. (2006). Feeding behaviour of three common fishes at an artificial reef in the northern Adriatic Sea. *Bulletin of Marine Science, 78*(1), 39–56.

FAO. (2015). Practical guidelines for the use of artificial reefs in the Mediterranean and the Black Sea. Studies and Reviews, General Fisheries Commission for the Mediterranean, 96.

Fletcher, S., Bateman, P., & Emery, A. (2011). The governance of the Boscombe Artificial Surf Reef, UK. *Land Use Policy, 28*(2), 395–401. doi:10.1016/j.landusepol.2010.08.002

Folpp, H.R., Lowry, M., Gregson, M. & Suthers I.M. (2011). Colonization and community development of fish assemblages associated with estuarine artificial reefs. *Brazilian Journal of Oceanography, 59*, 55-67.

Francis, O. P., Kim, K., & Pant, P. (2019). Stakeholder assessment of coastal risks and mitigation strategies. *Ocean and Coastal Management, 179*, 104844. doi:10.1016/j.ocecoaman.2019.104844

Freeman, L. C. (1979). Centrality in Social Networks Conceptual Clarification. *Social Networks, 1*(3), 215–239. doi:10.1016/0378-8733(78)90021-7

Freeman, R. E. (1984). *Strategic management: A stakeholder approach.* Pitman Publishing.

Friedmana, R. S., Guerrero, A. M., McAllisterd, R. R. J., Rhodes, J. R., Santika, T., Budihart, T., Indrawan, S., Hutabarat, J. A., Kusworoh, A., Yogaswara, H., Meijaarrd, E., St. John, F., Struebig, M. J., & Wilson, K. A. (2020). Beyond the community in participatory forest management: A governance network perspective. *Land Use Policy, 97*, 104738. doi:10.1016/j.landusepol.2020.104738

Grimble, R., & Wellard, K. (1997). Stakeholder methodologies in natural resource management: A review of principles, contexts, experiences and opportunities. *Agricultural Systems, 55*(2), 173–193. doi:10.1016/S0308-521X(97)00006-1

Ingold, K. (2017). How to create and preserve social capital in climate adaptation policies: a network approach. *Ecological Economics, 131*, 414–424. doi:.2016.08.033 doi:10.1016/j.ecolecon

IPBES. (2019). Summary for policymakers of the global assessment report on biodiversity and ecosystem services of the Intergovernmental Science-Policy Platform on Biodiversity and Ecosystem Services. IPBES Secretariat.

Jensen, A. (2002). Artificial reefs of Europe: Perspective and future. *ICES Journal of Marine Science, 59*, 3–13. doi:10.1006/jmsc.2002.1298

Kaufmann, J. C. (2016). L'entretien compréhensif [Comprehensive interview]. Collection 128 tout le savoir, 4ème ed. Armand Colin.

Keller, K., Steffe, A. S., Lowry, M., Murphy, J. J., & Suthers, I. M. (2016). Monitoring boat-based recreational fishing effort at a nearshore artificial reef with a shore-based camera. *Fisheries Research, 181*, 84–92. doi:10.1016/j.fishres.2016.03.025

Kirkbride-Smith, A. E. (2014). *The economic, social and conservation benefits of recreation-orientated artificial reefs* [PhD thesis]. University of Hull.

Lacroix, D., Charbonnel, E., Dao, J.-C., Véron, G., Lagardère, J.-P., Mellon, C., Covès, D., & Buestel, D. (2002). Les récifs artificiels. Aménagement du littoral marin et repeuplement [Artificial reefs. Marine coastal development and imrovment of marine life]. IFREMER Publication.

Lacroix, D., Mora, O., Menthiere, N. D. E., & Bethinger, A. (2021). *La montée du niveau de la mer d'ici 2100: Scénarios et consequences [Sea level rise by 2100: Scenarios and consequences].* Quae.

Lagesse, C., Bonnin, P., Bordin, P., & Douady, S. (2016). Méthodologie de modélisation et de caractérisation des réseaux spatiaux. Application au reseau viaire de Paris [Modelling and Characterization of Spatial Networks: An Application to the Paris Street Network]. *Flux, 105*(105), 33–49. doi:10.3917/flux.105.0033

Lazega, E. (2014). *Réseaux sociaux et structures relationnelles* [Actors networks and relational structures]. Presses Universitaires de France. doi:10.3917/puf.lazeg.2014.01

Lee, M. O., Otake, S., & Kim, J. K. (2018). Transition of artificial reefs (ARs) research and its prospects. *Ocean and Coastal Management, 154*, 55–65. doi:10.1016/j.ocecoaman.2018.01.010

Lima, J. S., Atalah, J., Sanchez-Jerez, P., & Zalmon, I. R. (2020). Evaluating the performance and management of artificial reefs using artificial reef multimetric index (ARMI). *Ocean and Coastal Management, 198*, 105350. doi:10.1016/j.ocecoaman.2020.105350

Mavraki, N., Degraer, S., & Vanaverbeke, J. (2021). Offshore wind farms and the attraction–production hypothesis: Insights from a combination of stomach content and stable isotope analyses. *Hydrobiologia, 848*(7), 1639–1657. doi:10.100710750-021-04553-6

McAfee, D., Reinhold, S.-L., Alleway, H. K., & Connell, S. D. (2021). Environmental solutions fast-tracked: Reversing public scepticism to public engagement. *Biological Conservation, 253*, 108899. doi:10.1016/j.biocon.2020.108899

Meffe, G., Nielsen, L., Knight, R., & Schenborn, D. (2002). *Ecosystem Management: Adaptive, Community-Based Conservation.* Island Press.

Milon, J. W., Holland, S. M., & Whitmarsh, D. J. (2000). Social & economic valuation methods. In W. Seaman (Ed.), *Artificial reef evaluation with application to natural marine habitats* (pp. 165–194). CRC Press. doi:10.1201/9781420036633.ch6

Molino, G. D., Kenney, M. A., & Sutton-Grier, A. E. (2020). Stakeholder-defined scientific needs for coastal resilience decisions in the Northeast U.S. *Marine Policy, 118*, 103987. doi:10.1016/j.marpol.2020.103987

Mueller, E. (2012). How to manage networks? The role of network attributes and incentives in network governance. *International Journal of Entrepreneurship and Small Business, 15*(1), 57–75. doi:10.1504/IJESB.2012.044589

Nakamura, M. (1985). Evolution of artificial fishing reef concept in Japan. *Bulletin of Marine Science, 37*, 271–278.

Neves dos Santos, L., & Zalmon, I. R. (2015). Long-term changes of fish assemblages associated with artificial reefs off the northern coast of Rio de Janeiro, Brazil. *Journal of Applied Ichthyology, 31*, 15–23. doi:10.1111/jai.12947

Newman, M. E. J. (2003). The Structure and Function of Complex Networks. *SIAM Review, 45*(2), 167–256. doi:10.1137/S003614450342480

Ospina-Alvarez, A., de Juan, S., Alós, J., Basterretxea, G., Alonso-Fernández, A., Follana-Berná, G., Palmer, M., & Catalán, I. A. (2020). MPA network design based on graph theory and emergent properties of larval dispersal. *Marine Ecology Progress Series, 650*, 309–326. doi:10.3354/meps13399

Pereira, P. H. C., Dos Santos, M. V. B., Lippi, D. L., Silva, P. H. D. E. P., & Barros, B. (2016). Difference in the trophic structure of fish communities between artificial and natural habitats in a tropical estuary. *Marine and Freshwater Research, 68*(3), 473–483. doi:10.1071/MF15326

Pınarbaşı, K., Galparsoro, I., & Borja, I. (2019). End users' perspective on decision support tools in marine spatial planning. *Marine Policy, 108*, 103658. doi:10.1016/j.marpol.2019.103658

Pioch, S., De Monbrison, D., & Simard, F. (2019). *Artificial reefs in France: state of the art and last innovative projects. International Conference on Fisheries Engineering*, Nagasaki, Japan.

Ramos, J., Lino, P. G., Himes-Cornell, A., & Santos, M. N. (2019). Local fishermen's perceptions of the usefulness of artificial reef ecosystem services in Portugal. *PeerJ, 6*, e6206. doi:10.7717/peerj.6206 PMID:30647998

Ramos, J., Oliveira, M. T., & Santos, M. N. (2011). Stakeholder perceptions of decision-making process on marine biodiversity conservation on Sal island (Cape Verde). *Brazilian Journal of Oceanography, 59*(spe1), 95–105. doi:10.1590/S1679-87592011000500012

Ramos, J., Santos, M., Whitmarsh, D., & Monteiro, C. C. (2008). A socio-economic toolbox of artificial reef projects. *Proceeding of Conference: 14th IIFET.*

Reed, M. S., Graves, A., Dandy, N., Posthumus, H., Hubacek, K., Morris, J., Prell, C., Quinn, C. H., & Stringer, L. C. (2009). Who's in and why? A typology of stakeholder analysis methods for natural resource management. *Journal of Environmental Management, 90*(5), 1933–1949. doi:10.1016/j.jenvman.2009.01.001 PMID:19231064

Roa-ureta, R. H., Santos, M. N., & Leitão, F. (2019). Modelling long-term fisheries data to resolve the attraction versus production dilemma of artificial reefs. *Ecological Modelling, 407*(C), 1–1. doi:10.1016/j.ecolmodel.2019.108727

Rouse, S., Porter, J. S., & Wilding, T. A. (2020). Artificial reef design affects benthic secondary productivity and provision of functional habitat. *Ecology and Evolution, 10*(4), 2122–2130. doi:10.1002/ece3.6047 PMID:32128143

Salaün, J. (2022). *Vers une approche socio-écologique d'évaluation de la performance de projets d'aménagements côtiers avec des récifs artificiels, Etude de cas sur trois régions biogéographiques maritimes en France métropolitaine* [Towards a socio-ecological approach to evaluating the performance of coastal development projects with artificial reefs, Case study on three maritime biogeographic regions in metropolitan France] [PhD Thesis]. Paul Valéry University.

Salaün, J., Pioch, S. & Dauvin J-C. (2020). Description of a theoretical social-ecological approach to manage artificial reefs. *Vie et Milieu / Life & Environment, 70*(3-4), 129-135.

Santos, M. N., & Monteiro, C. C. (2007). A fourteen-year overview of the fish assemblages and yield of the two oldest Algarve artificial reefs (southern Portugal). *Hydrobiologia, 580*(1), 225–231. doi:10.100710750-006-0451-2

Scarcella, G., Grati, F., Bolognini, L., Domenichetti, F., Malaspina, S., Manoukian, S., Polidori, P., Spagnolo, A., & Fabi, G. (2015). Time-series analyses of fish abundance from an artificial reef and a reference area in the central-Adriatic. Sea. *Journal of Applied Ichthyology, 31*, 74–85. doi:10.1111/jai.12952

Schaffer, V., & Lawley, M. (2012). An analysis of the networks evolving from an artificial reef development. *Current Issues in Tourism, 15*(5), 497–503. doi:10.1080/13683500.2011.638704

Schumacher, J., Schernewski, G., Bielecka, M., Loizides, M. I., & Loizidou, X. I. (2018). Methodologies to support coastal management - a stakeholder preference and planning tool and its application. *Marine Policy, 94*, 150–157. doi:10.1016/j.marpol.2018.05.017

Sébastien, L. (2006). *Humains et non-humains en pourparlers: l'Acteur en 4 Dimensions Proposition théorique et méthodologique transdisciplinaire favorisant l'émancipation de nouvelles formes de gouvernances environnementales Application au domaine de l'eau sur trois territoires: la Plaine du Forez, les pentes du Kilimandjaro et les Barthes de l'Adour* [Humans and non-humans in talks: the Actor in 4 Dimensions Theoretical and transdisciplinary methodological proposal promoting the emancipation of new forms of environmental governance Application to the field of water in three territories: the Plaine du Forez, the slopes of the Kilimanjaro and the Barthes de l'Adour] [PhD Thesis]. Jean Monnet University.

Simard, F. (1997). Socio-economic aspects of artificial reefs in Japan. *Proceedings of the 1st EARNN Conference*, 233-240.

Simon, T., Joyeux, J.-C., & Pinheiro, H. T. (2013). Fish assemblages on shipwrecks and natural rocky reefs strongly differ in trophic structure. *Marine Environmental Research, 90*, 55–65. doi:10.1016/j.marenvres.2013.05.012 PMID:23796542

Simpson, S., Brown, G., Peterson, A., & Johnstone, R. (2016). Stakeholder perspectives for coastal ecosystem services and influences on value integration in policy. *Ocean and Coastal Management, 126*, 9–21. doi:10.1016/j.ocecoaman.2016.03.009

Smith, J. A., Lowry, M. B., Champion, C., & Suthers, I. M. (2016). A designed artificial reef is among the most productive marine fish habitats: New metrics to address 'production versus attraction'. *Marine Biology, 163*(9), 1–8. doi:10.100700227-016-2967-y

Streich, M. K., Ajemian, M. J., Wetz, J. J., & Stunz, G. W. (2018). Habitat-specific performance of vertical line gear in the western Gulf of Mexico: A comparison between artificial and natural habitats using a paired video approach. *Fisheries Research, 204*, 16–25. doi:10.1016/j.fishres.2018.01.018

Sutton, S., & Bushnell, S. (2007). Socio-economic aspects of artificial reefs: Considerations for the Great Barrier Reef Marine Park. *Ocean and Coastal Management, 50*(10), 829–846. doi:10.1016/j.ocecoaman.2007.01.003

Taormina, B., Claquin, P., Vivier, B., Navon, M., Pezy, J. P., Raoux, A., & Dauvin, J. C. (2022). A review of methods and indicators used to evaluate the ecological modifications generated by artificial structures on marine ecosystems. *Journal of Environmental Management, 310*, 114646. doi:10.1016/j.jenvman.2022.114646 PMID:35248990

Tessier, A. (2013). *Caractérisation des Récifs Artificiels du Golfe du Lion: de l'écologie aux usagers* [Characterization of the Artificial Reefs of the Gulf of Lion: from ecology to users] [PhD thesis]. Perpignan University.

Tessier, A., Bodilis, P., Charbonnel, E., Dalias, N., Francour, P., Lenfant, P., & Seaman, W. (2015). Assessment of French artificial reefs: due to limitations of research, trends may be misleading. Hydrobiologia, 753.

Trouillet, B. (2006). La mer-territoire ou la banalisation de l'espace marin (golfe de Gascogne) [The sea-territory or the trivialization of the marine space (Bay of Biscay)]. *M@ppemonde, 84*(4), 1-17.

Véron, G., Denis, J., Gérard, A., Thouard, E., & Thébaud, O. (2008). *Les récifs artificiels, état des connaissances et recommandations [Artificial reefs, state of the art and recommendations]*. Ifremer.

Vivier, B., Dauvin, J. C., Navon, M., Rusig, A. M., Mussio, I., Orvain, F., & Claquin, P. (2021). Marine artificial reefs in the world, a literature analysis of their designs, objectives and effectiveness. *Global Ecology & Distribution, 27*, e01538. doi:10.1016/j.gecco.2021.e01538

Wasserman, S., & Faust, K. (1994). *Social network analysis. Methods and applications* (Vol. 8). Cambridge university press. doi:10.1017/CBO9780511815478

Wu, Z., Tweedley, J. R., Loneragan, N. R., & Zhang, X. (2019). Artificial reefs can mimic natural habitats for fish and macroinvertebrates in temperate coastal waters of the Yellow Sea. *Ecological Engineering, 139*, 105579. doi:10.1016/j.ecoleng.2019.08.009

Chapter 3
A Stakeholder Analysis of the Artificial Reefs in Edremit Bay, Turkey:
Social, Economics, and Management Aspects

Sezgin Tunca
https://orcid.org/0000-0001-6840-1466
Independent Researcher, Turkey

Jose Luis Sanchez Lizaso
University of Alicante, Spain

Vahdet Ünal
Ege University, Turkey

ABSTRACT

Artificial reefs (ARs) have gained considerable attention in Turkey, due to their positive biological and socioeconomic implications. The most recent and comprehensive AR deployment was completed in Edremit Bay. The authors investigated the stakeholders' perception and satisfaction along with social, economic, and management concerns about the AR deployment for the period October-December 2014. They first defined user and non-user AR stakeholders and their interests in the project. They represented multiple interactions among stakeholders about the AR deployment by focusing on user types, attitudes, or behaviors towards the AR deployment. They queried existing and potential problems about the project and the deployment area to produce consistent management solutions. They suggest that public participation and multiple collaboration would enhance communication and deal among the stakeholders towards finding solutions for the existing problems. Future studies should reveal social, economic, and management aspects of the stakeholders by the medium and long-term monitoring studies.

DOI: 10.4018/978-1-6684-2344-8.ch003

INTRODUCTION

In the literature, biological investigations were intensely conducted to promote the sustainability of different ecosystems including seas, lakes, and streams (Charles, 2001). Based on these investigations, habitat creation and improvement by the deployment of artificial habitats were encouraged in the coastal marine areas where intensive human activities exist (Seaman and Sprague, 2001; Bortone et al., 2011). In general, the term artificial habitats include both artificial reefs (ARs) and fish aggregating devices (FADs). ARs are benthic materials, while FADs are materials positioned at or under the water surface. Therefore, AR can be defined as "*one or more natural and human-made objects deployed in a marine ecosystem to influence physical, biological and socio-economic processes related to living marine resources*" (Milon et al., 2000). This definition can be extended due to the existence of AR deployments in other freshwater ecosystems such as lakes and dams. Artificial shelters have been deployed on the seafloor to conserve sensitive aquatic ecosystems or to increase and contribute to the productivity of resources, which had positive outcomes for commercial and recreational fishing (Seaman and Sprague, 2001).

In addition to fisheries enhancement, ARs generate multiple ecosystem services for recreational users such as SCUBA divers, making ARs economically and socially influential (Roberts et al., 1985; Milon, 1988a,b; 1989a,b Ditton and Baker, 1999; Ditton et al., 2001; Pendleton, 2004; Oh et al., 2008; Morgana et al., 2009). It is noteworthy that ecosystem services and goods provided by ARs cannot be traded in the real market but, they have an economic value called non-market value (Hanemann, 1994; Bateman and Willis, 2002; Haab et al., 2002). Water supply for agricultural production, food supply, sustainability of natural gases in the air, cutting risks of outbreaks, sustainability of wildlife, recreational fishing, scuba diving, hiking, trekking, or hunting in a forest is other examples of non-market goods and services (Hanemann, 1994).

In the Mediterranean Riparian states, there is a limited number of investigations focusing on the socio-economic and management implications of artificial reefs (ARs). This caused uncertainty and knowledge gaps on the effects of AR deployments on stakeholders (AR users and non-users) including commercial fishers (CF), recreational fishers (RF), SCUBA divers, tourism entrepreneurs, fishing equipment shops, and local people. Uncertainty and knowledge gaps also prevent further discovery of potential AR areas and applicability of new projects in the future.

The AR practices in Turkey have increased in recent years, during which considerable attention has been paid to the limited biological and technical research (Lök, 1997; Düzbastılar and Tokaç, 2003; Lök and Gül, 2005; Ulaş, 2007). The first projected AR deployment dates to 20 years in Turkey (Lök et al., 2012). Local governmental and non-governmental organizations, where local public organizations actively involved in the decision-making process (Lök and Tokaç, 2000), steadily supported the AR projects in which concrete blocks and scrap materials, such as old navy vessels and aircraft were deployed on the seabed (Lök et al., 2002). The first economic study focusing on the ARs in the Turkish Seas aimed to analyze investments for the AR projects (Tiryakioğlu, 2008). In addition, limited but valuable scientific research guided future AR improvements.

For studies conducted outside of Turkey, the United States and Japan are leading developed countries, although not much consideration was paid to socioeconomic implications. AR deployments using different shaped, sized materials have also been successful in the European Seas (Bombace et al., 2000). In Turkey as well as in the rest of the Mediterranean states, the lack of socioeconomic studies focusing on the AR stakeholders caused a significant knowledge gap on the various implications of the AR deployments on user and non-user groups, such as commercial and recreational fishers, recreational divers, tourism operators, and local people while such knowledge gap prevented the proper improvement of the AR projects in potential areas. Therefore, socioeconomic studies on ARs should be conducted to enable baseline information for scientists and decision-makers.

Few investigations were conducted on the AR stakeholders and existing scientific literature concerning the AR deployments principally focused on biological perspectives although human interactions of the AR deployments have a major influence on local people and visitors. Therefore, our main goal in this study was to enable baseline and enhanced information on the stakeholders within the context of AR management.

EARLIER LITERATURE ON SOCIOECONOMIC EVALUATIONS OF AR

The existing scientific literature concerning AR deployments takes into consideration biological perspectives. Although social and economic aspects of AR structures had the highest direct influence and importance for the people and visitors at localities where the AR was deployed, there are quite a few scientific studies on the socioeconomic and management aspects of the ARs. Hereby, we summarized the socioeconomic effects of AR on the stakeholders all around the world while the studies conducted in the United States constitute most of the earlier studies. Use capacity of the AR in Florida was primarily investigated by Schug (1978), who calculated annual expenditures of the AR users as $181,000-253,000 with a benefit/cost ratio greater than one that explains positive economic returns of the ARs. Hanni and Mathews (1977) surveyed the costs concerning the ARs near to Florida coasts to calculate economic gains of ARs by recreational fishers and divers, while Milon (1988a) measured the AR benefits by using a travel cost (TC) approach. Milon (1988b) similarly quantified the economic effects of the AR program on users and non-users by using a mail survey on recreational boat owners and divers that finally valued the annual user benefits of the ARs by the contingent valuation (CV) approach. Samples (1989) proved positive economic gains by an economic analysis of recreational and commercial use of artificial habitats. Milon (1989b) explored the demand for offshore AR sites by finding out high demand and economic value via a mail survey. Oh, et al. (2008) calculated the total consumer surplus and non-market economic value given by artificial- and natural-reef divers. Brock (1994) discovered recreational diving demand in and around the ARs despite the existence of high fishing pressure in the area. Economic impact of ARs has also been estimated by Bell et al. (1998). Hushak et al. (1999) calculated consumer surplus by the recreational activities on the ARs by using a TC method. Milon et al. (2000) particularly described methods to evaluate the social and economic impacts of the AR deployments. McKean and Taylor (2000) performed surveys to estimate WTP amounts and costs of recreational users. Pendleton (2004) defined user groups, their economic benefits, and costs related to ships and oil platforms. Johns (2004) estimated economic values ($3.6 million), expenses, employment, and revenues of Martin County ARs in Florida. Leeworthy et al. (2006) investigated the use types and economic activity of a sunken ship. The authors highlighted the significant positive influence of the ARs on reducing pressure on the natural reefs while boosting economic benefits. Sutton and Bushnell (2007) conducted a socioeconomic assessment for the ARs by highlighting the complex and contrasting social face of AR deployments. Adams et al. (2004) described a general overview of the economic benefits of Florida AR and valuation methodologies to measure the impact and value of AR. Morgana et al. (2009) used a web-based TC survey to estimate recreational diving demand on an ex-navy vessel. Tiryakioğlu (2008) conducted a benefit-cost analysis of two projects ex-navy vessel and concrete blocks by concluding that sinking a scrap vessel was economically more beneficial than deploying concrete blocks.

Relevant to management aspects with more holistic evaluations, there exist few studies drawing interdisciplinary hotspots relevant to ARs. Johns et al. (2001) measured economic impacts and values (WTP amounts, total sales, incomes, employment, and tax revenues, rehabilitation, improvement, and socioeconomic characteristics) of the ARs by focusing on artificial and natural reef users on the Southwest of Florida. McGinnis et al. (2001) represented use types of oil structures as artificial habitats by accounting political, economic, and ecological aspects. Still, ARs enhance recreational fishing and diving opportunities. For instance, Hiett and Milon (2002) estimated demand, expenditures, and economic impact related to recreational fishing and diving near oil and gas structures in addition to the emphasis on the importance of fishers' and divers' participation in the decision-making process. Ramos et al. (2006a) revealed use types of ARs by surveying commercial and recreational fishers comparatively. Ramos et al. (2006b) represented AR diving decisions by assessing biological, geographical, atmospheric, economic, and incentive characteristics. Ramos et al. (2007) investigated the beliefs of AR stakeholders based on environmental, social, and economic factors. Survey groups consisted of commercial fishery organizations, diving clubs, angler associations, fishery and environment managers, natural and social scientists, local fishery council representatives, and environmental groups.

Ramos et al. (2011) also investigated stakeholders of the ARs in Algarve and their use and any other relevance with the ARs. They concluded that relationships with the AR structures can be public, private, or joint. They also discovered the influence of the AR program on stakeholders and their power to impact the AR outcomes. Whitmarsh et al. (2008) analyzed the potential and availability of AR in the context of coastal fishery management, and the economic benefits of the AR were evaluated. Tunca et al. (2012) revealed a multi-criteria decision-making case for multiple AR stakeholders including commercial fishers, recreational fishers, and residents through efficient use and share of the AR sites compared with no reef site and marine protected areas. Tunca et al. (2013) also assessed users' beliefs on the ARs by pointing out CF's agreement on the economic benefits of the ARs while RF and residents agreed on the biological benefits of the ARs.

MATERIAL AND METHOD

Study Area

The AR project was the pilot project of the Turkish National Artificial Reef Program that was financed by The Ministry of Food, Agriculture, and Livestock. The main goals of the project were supporting small-scale fisheries, conserving marine biodiversity, spawning and nursery areas (e.g., *Posidonia* meadows). The deployment of 7,000 concrete units with an approximate total volume of 25,000 m^3 was realized from 2011 to 2012 with a total budget of US $2 million. The AR units were deployed in a 0.5 km wide area along 15 km of coastline, covering a total marine area of 7.5 km^2.

The AR deployment zone has sandy-muddy and lives bottom characteristics. The AR blocks were deployed in seven rectangular zones (Figure 1). In each AR zone, there exist 115 AR clusters, and each AR cluster has 30 AR blocks. AR zones stretch along a 15 km shoreline, and ARs were planned to be deployed in depths between 17 and 32 m (Lök et al, 2013). In the design of AR blocks, the principal goals were to have sufficient strength and weight for being resistant against trawling and other dragging forces, to stop and give harm to illegal fishing gears, and to conduct its function in any type of fall on the seabed.

Figure 1. Location of the ARD area in Edremit Bay

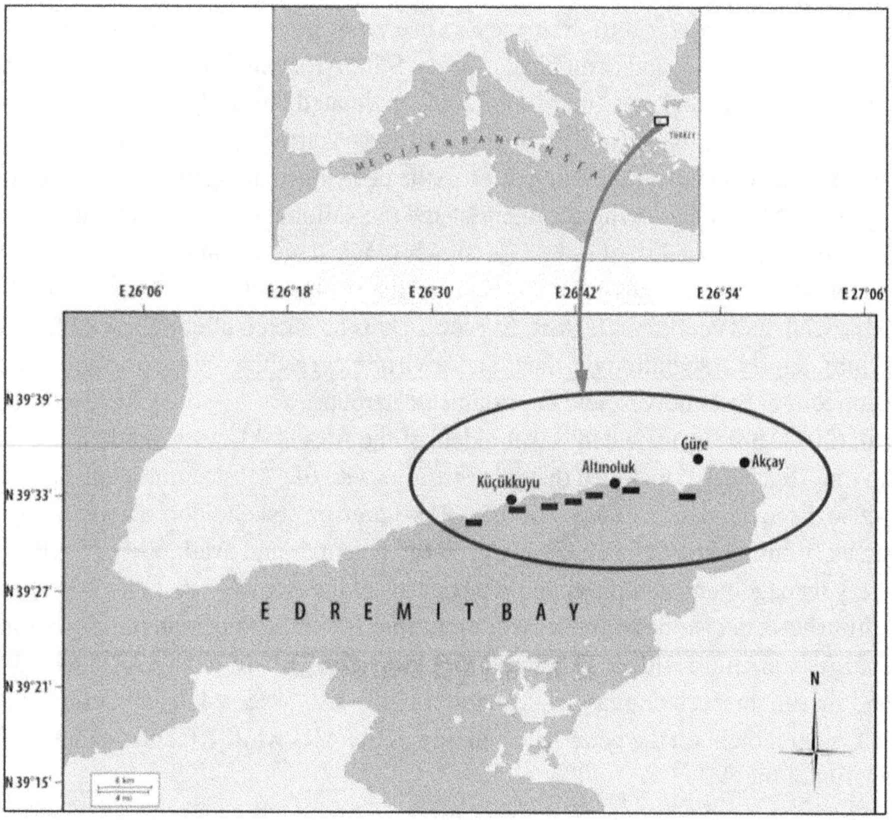

Two AR designs were used in the project. The first one is the production design and, it has a cubic shape with circular windows on all sides. This design meets habitat requirements for phyto- and zoobenthic organisms with its circular windows, flat areas, and by also creating dark areas in the blocks. Then, 30 AR clusters meet the biological and environmental demands of demersal and pelagic species (Lök et al, 2013). The second AR block design was for habitat protection. The protection AR blocks were used to protect sensible habitats like sea meadows (*Posidonia oceanica*) and to protect the AR production zones and AR blocks from the destruction of illegal fishing activities (i.e., bottom trawling).

Data Collection

In the study, 30 CF and 50 RF in each town, with total numbers of 90 CF and 150 RF, were investigated via on-site face-to-face interviews. In addition, 2 diving centers, 9 fishing equipment shops, and 3 heads of fishery cooperatives were surveyed. Furthermore, earlier studies at the national and international levels targeting the socioeconomics of ARs were used. The sampling was conducted from October to December 2014 in Altınoluk, Güre, and Küçükkuyu. The data was collected from CF and the RF (shore-based and boat-based) before, after, and during the fishing and recreational fishing activity via on-site face-to-face interviews using questionnaire forms on fishing sites or access points. A fixed number of 30 CF and 50 RF were randomly interviewed at each town to get a representative sample by site and commercial and recreational fishing modalities. Besides, the heads of fishery cooperatives, the owners of diving centers, and fishing equipment shops were queried via specifically designed questionnaire forms in each town. Basic statistics of the collected fisheries and socioeconomics data were discussed in the following sections.

Data Analysis

The AR deployments can influence numerous human activities, as well as diverse stakeholders. Stakeholder groups are recreational fishers, recreational divers, professional fishers, professional divers, resource managers, scientists, environmental groups (Milon et al., 2000). Stakeholder analysis is a proper tool to show the characteristics and relevance, conflicts of associated groups. In this study, the data analysis was constructed based on a detailed stakeholder analysis adopted from Ramos et al. (2011) by primarily focusing on satisfaction and perception levels relating to the AR deployment, alterations after the AR deployment, problems, suggestions, and management thoughts on the ARs. Furthermore, basic characteristics, interactions, problems, and conflicts among the stakeholders were given. Such information was supposed to be useful for managers to understand each group in especially constructing necessary management rules as well as throughout all phases of the AR project.

RESULTS

Stakeholder Analysis

The ARs in Edremit Bay influence multiple stakeholders with diverse and conflicting values and opinions. To effectively account for each stakeholder to promote a participative decision-making approach, managers primarily need to have a full understanding of the relevance, views, values of the stakeholders (Table 1). Therefore, in the following sections, we deeply explained the stakeholders and interaction attitudes, behaviors, problems, and conflicts relevant to the use and non-use roles of the AR deployment.

The ARs were deployed by the Ministry of Food, Agriculture, and Livestock of the Republic of Turkey with a direct attempt by the head of Altınoluk Fishery Cooperative (2011) and the scientists from Ege University, Faculty of Fisheries. Apart from Altınoluk Fishery Cooperative that directly supported the AR project, there were also two other fishing towns underexposure to the deployment area, one is Küçükkuyu of which the head of the cooperative did not agree on any deployment in their fishing zone, and the other is Güre that had not a fishery cooperative during the planning phase of the ARs. In 2013, they also founded a fishery cooperative, and the head of the cooperative did not agree with the AR deployment close to their fishing grounds. There was a wide range of AR stakeholders including shore-based and boat-based RF, and SCUBA divers, government authorities (management agencies including fisheries, coast guards, and environmental protection agencies), researchers (from universities and institutes), local community, and private businesses (e.g., dive operators, fish equipment shops).

In this project, the ARs were proposed and promoted primarily by a small group of small-scale local fishers and officials from the ministry with the principal aims of promoting fishing and diving interests as well as conserving sensitive habitats from harmful activities (i.e., bottom trawlers).

Table 1. Identification of the artificial reef stakeholders in Edremit Bay

Class	Stakeholder	Institutional Sector	Interests in the Project
I	Fishery Cooperatives	Cooperative	Create more fishing opportunities for their partners
I	Charter boat anglers	Private (firms)	Catch certainty or just recreation
I	Onshore anglers	Private (individuals)	Catch certainty or just recreation
I	Spearfishers	Private (individuals)	Catch certainty or just recreation
I	Diver club and associations	Private (firms)	Creation of extra diving sites
II	Fisheries research institutes	Public (local administration)	Demonstrate the AR potential, onsite observations, data collection
II	Directorate of fisheries and aquaculture	Public (local administration)	Protect stocks, avoid user conflicts
II	Directorate of environmental management	Public (local administration)	Stability of the materials, allocation of deployment site
II	Ports authorities	Public (local administration)	Involve in fisheries management design
II	Universities	Public (local administration)	Diversify and improve the knowledge base
II	City councils	Public (local government)	Creation of jobs and new income sources (both construction and use of the AR blocks)
II	Environmental agencies	Public (NGOs)	Coastal and stock protection, correct selection of materials

Commercial Fishers

Commercial fisheries by around 150 professional fishers in three fishing towns (Altınoluk, Küçükkuyu, and Güre) have great social and economic values. Also, the continuous rise of recreational catch amounts was assumed to have the potential to create serious conflicts between CF and RF. Furthermore, the inexistence of a full collective agreement among the CF on the AR deployment causes conflicts among the CF themselves. Then, uncertainty on the use rights of the ARs would be considered a significant issue of debate and should be carefully accounted for by the authorities.

Recreational Fishers and Divers

In general, RF and SCUBA divers are assumed as the direct users of the ARs, and they are, like the CF, the primary users of the ARs in the Edremit Bay. The RF and divers in the region are residents and domestic tourists accessing the ARs by either private or commercial vessels.

Before the AR deployment, the RF, and the divers in three towns had optimistic opinions towards the AR deployment while the diversity in attitudes, preferences, and opinions of the RF and divers about the ARs were illustrated in detail (Tunca, 2011). In three towns, there were less than five hundred shore-based and boat-based RF (~10% boat-based; ~50 vessels just exist in Altınoluk), excluding high seasons, who were active both in summer and winter seasons.

The RF complain about the high port prices that were taken by the fishery cooperatives that have rights to run the ports in their regions. Especially, the boat-based RF occasionally had conflicts with the CF due to overlapping times of the AR site use.

In the region, there was only one diving association which was founded in 2009, and it was the only active diving association serving its members during the summer periods. Since its foundation, it has been organizing recreational dive tours under the Altınoluk Diving Center.

During that time, hundreds of dives in the AR site were realized by this association. The club has been organizing group diving activities by five individuals in each dive starting on the shore. The association had around twenty-five members, and it had a founder diving instructor who had fifteen years of diving experience. The members commonly had various levels of the diving license of CMAS (World Confederation of Underwater Activities). Apart from SCUBA diving activities organized by the members, the club organized site introducing dives for native and foreign tourists. According to the personal communication with the head of the association, the divers were ignored during the AR project planning stage, and this situation resulted in the loss of great direct and indirect economic gains by enhanced diving tours to the ARs. Furthermore, the head of the club said that the association would organize specified diving tours to the AR site (Diving Center Founder, Pers. Comm., Tunca 2011). Today, the diving club was run as a commercial diving center with low economic gains.

Tourism and Related Businesses

Tourism is one of the largest commercial activities in Edremit Bay, generating a considerable amount of income for the residents. Interests such as diving and charter fishing, and diving equipment stores, and other indirect businesses such as hotels and restaurants potentially get benefits from the ARs in the region. There are few RF Charter operators in the Edremit Bay with a considerable number of non-licensed tour organizers who were mostly artisanal fishers and more than five fishing equipment shops, restaurants, and cafes which were demanded by non-local RF and divers. These stakeholders could also potentially suffer negative consequences if the AR deployment causes alterations in the site use patterns and pressures in Edremit Bay, particularly if changes result in moving tourism dollars outside the region rather than attracting and promoting alternative tourism types in the region.

Management Agencies

Public fisheries, maritime, and environment agencies responsible for managing the users and their activities in the AR site are one of the major stakeholders of the ARs project. The management power of the ARs in Edremit Bay is at the General Directorate of Fisheries and Aquaculture under the Ministry of Agriculture and Forestry. The directorate handles setting policy and making decisions on the management of the ARs. The directorate is also responsible for seeking compatibleness of the AR deployment and management with the existing national and international legislations.

Environmental Organizations

The ARs may attract a great deal of attention from environmental organizations as in other cases around the world. Environmentalists may place a high value on the ARs in Edremit Bay and they are concerned primarily with resource preservation and protection. In our case, there is no participation from an environmental organization neither in the planification nor in the future management of ARs in Edremit Bay. This may raise more questions, doubts, and uncertainties about the environmental implications of the materials used for the AR blocks, the deployment sites of the ARs.

Wider Community

People in the wider community locally, nationally, and internationally hold diverse values concerning the ARs in Edremit Bay. Although exploring and understanding the diversified values of the ARs given by the wider community can be difficult, it is known that the public holds strong values related to the protection of the unique environment of Edremit Bay. Edremit Bay, especially the coastal zone including Altınoluk, Güre, and Küçükkuyu, is a hotspot destination for national and international tourists due to their aesthetic and natural beauties. The AR deployment might be perceived as environmental degradation of the aesthetic values in the region.

As seen, there are a great variety of stakeholder types related to the ARs in Edremit Bay. Table 2 summarizes the identified stakeholders and their interaction with the ARs, whereas Table 3 shows the problems identified.

Table 2. Interaction of the artificial reef stakeholders with the artificial reef project

Class	Stakeholder	AR interaction	More likely attitudes and/or behaviors
I	Fishery Cooperatives	Conservation and future availability value	Preserving the ARs because they are beneficiaries of common fishing grounds while only catching necessary and sizeable fish species
I	Charter/boat anglers	Extractive value	Using the AR sites to potentially obtain larger specimens
I	Onshore anglers	Non-users	Assuming the ARs were not deployed for shore anglers
I	Spearfishers	Extractive value	Using the AR sites to potentially obtain larger specimens
I	Diver club and associations	Non-extractive value	ARs are additional diving spots but, their use will only be occasional due to similar design shapes
II	Fisheries research institute	All values	Data collection (e.g., biological, ecological, socioeconomic)
II	Directorate of fisheries and aquaculture	Non-users / Governing body	Do not bother much if there are no or few signs of conflicts among users
II	Directorate of environmental management	Non-users / Governing body	Do not bother much if there are no or few signs of conflicts among users
II	Ports authority	Non-users	They engage in the consultation process
II	University	Conservation value	Trying to get data from ARs
II	City councils	Non-users	Partial social and economic problems solved if the ARs generate new income sources
II	Environmental agencies	Conservation value	Opposing the ARs if there are any signs of pollution or overfishing

Table 3. Recognized problems of the artificial reef stakeholders

Class	Stakeholder	Problems
I	Fishery Cooperatives	- Uncertainty on use rights – future concerns - Conflicts with boat anglers on the use of artificial reef site
I	Boat/Charter anglers	- Uncertainty on use rights – future concerns - Conflicts with commercial fishers on the use of artificial reef site - Disregarded by the project
I	Onshore anglers	- Disregarded in the project (Mostly the residents)
I	Spearfishers	- Uncertainty on use rights – future concerns - Possible conflicts with commercial fishers on the use of artificial reef site - Disregarded by the project
I	Diving club/associations	- Uncertainty on use rights – future concerns - Possible conflicts with commercial fishers and anglers on the use of artificial reef site - Disregarded by the project
II	Fisheries research institutes	None
II	Directorate of Fisheries and Aquaculture	-Inexistence of prospects -Inexistence of a management plan and a specific responsible body
II	Directorate of Environmental Management	-Inexistence of prospects -Inexistence of a management plan and a specific responsible body (In cooperation with the Directorate of Fisheries and Aquaculture)
II	Ports authority	None
II	University	- Lack of contribution by researchers from multi-disciplines
II	City councils	- Disregarded by the project
II	Environmental agencies	- Disregarded by the project

DISCUSSION

Social

The literature represented various positive and negative social implications that often encompass the deployment and management of ARs. Likewise, both positive and negative social implications of the AR deployment were also found in our study. The ARs are used to make new dive opportunities where none existed beforehand or to improve effectively existing diving and fishing. Recreational angling is naturally demanded in the AR deployment sites; however, the results showed that the AR in the Edremit Bay likewise improve angling. ARs make or upgrade recreational anglers in many ways. First and foremost, they add to the mixture of angling and/or diving that exists inside a zone by giving distinctive sorts of structure and attracting several types of marine life as shown in the past investigations of Radonski et al. (1985) and Chou (1997). Furthermore, in the current study, there is a huge demand by boat and shore-based RF to fish around the AR site, and the anglers also have a considerable social impact by hired or private charter tours among friends. After the AR deployment, there is an extensive rise in the number of angling days and fishers at the local level. Moreover, the ARs in Edremit Bay enables more angling and diving opportunities so that they were deployed close to the shoreline, empowering individuals who are restricted by experience, boat size/horsepower, time, or money to appreciate recreational angling or diving (Radonski et al., 1985). The AR in Edremit Bay enhance the recreational experience and catch rate by attracting or enhancing biomass and expanding the likelihood of watching and/or catching fish as proved by McGlennon and Branden (1994). The introduction of new species with the ARs has enhanced fishers' RF experience in this study. Besides, the ARs in Edremit Bay could redistribute site use in the marine and territorial area, then reducing user overcrowding (Radonski et al., 1985); however, due to the absence of a management plan, overcrowding is an important reason for potential conflict among users have already been seen in certain sites of the ARs in Edremit Bay.

Overcrowding particularly occurred in Altınoluk where the ARs was closely deployed and, numerous anglers and commercial fishers from neighboring towns; Güre, Küçükkuyu, Akçay, and Edremit had few conflicts up to now by owning a risk for future conflicts.

While the positive effects of the AR deployment on recreation opportunities were reported in numerous places (Asafu-Adjaye and Tapsuwan, 2008; Ditton *et al.*, 1999; 2001; 2002; Hiett and Milon, 2002; Milon, 1988a; 1988b; 1989a; 1989b; Milon *et al.* 2000; Oh *et al.*, 2008; Pendleton, 2004, Sutton and Bushnell, 2007), it is hard to anticipate whether comparative and competitive results will be acknowledged following a couple of more years for the Edremit Bay. Having said that, Edremit Bay hosts a small zone of sensitive habitat such as coral reefs that means it enables a rich marine ecosystem for commercial and recreational marine activities. Enhancement of recreational activities as described above expanded opportunities that are prone to be the results most recognizable for divers and fishers in Edremit Bay.

Moreover, because of the limited but valuable previous data on the users' demands and tendencies towards use types and their benefits of the ARs in Edremit Bay, CF, RF, and recreational divers in Altınoluk, it is easy to anticipate that opportunities provided by the ARs have been mostly supported before and after the deployment, while the valuation whether the ARs will attract sufficient demand to create social and monetary advantages have proved its economic potential (Tunca, 2011; Ünal et al., 2013).

Perception

As in the studies by Santos and Monteiro (1997a; 1997b), high catch rates of fishing in the AR are perceived as enhanced fisheries, especially among Altınoluk CF. In addition, Altınoluk CF may be the most satisfied group because of enhanced fishing. RF in each town supported the AR and after the CF, they are the most satisfied group who also support the AR deployment. On the other side, dissatisfaction of the CF in Güre and Küçükkuyu seems to be the biggest failure of the current AR deployment. The respondents interviewed have agreed that the ARs are useful materials that positively contribute to marine life. Moreover, perception of increases in the catch numbers, only fishers in Altınoluk have positive ideas in contrast to the rest because of the lack of inclusion of all stakeholders in the management process.

Now, especially, Güre and Küçükkuyu fishers have low satisfaction levels with pessimistic beliefs about fishery enhancement. Based on the individual perception questions on the effects of the ARs on user groups, the ARs were perceived to increase the number of users and diversify their activities by introducing diving, boat-based recreational fishing, and other recreational opportunities. There is also an optimistic belief on the effects of the ARs by the residents. The presence of the ARs obviously creates a significant rise in the number of trips that will be contributing to economic activity at the micro-level.

Additionally, after the deployment of the ARs, an increase in diving demand, the number of diving charters, and economic activity due to recreational activities. Moreover, the ARs were suggested as a tool to enhance commercial fishing demand in these three towns. Therefore, increased willingness to conduct commercial fishing was perceived as an increased commercial harvest by the CF.

Harvest Impact

Fisheries catch rates on the AR sites were found as higher than catch rates in non-reef sites (Santos and Monteiro, 1997a; 1997b), prompting the impression of fisheries enhancement. Notwithstanding, there is an important discussion about whether enhancement in catch rates happens due to raised production in existing biomass (i.e., increased stock size) or through increased attraction and aggregation of existing fish stocks (Pears and Williams, 2005; Bohnsack, 1989; Pickering and Whitmarsh, 1997) while these findings are valid for certain species as each species has its own lifestyles and cycles. For example, in our case, significant rises in catch amounts of economically important species such as *P. saltatrix, B. Belone, S. officinalis, M. surmuletus, O. vulgaris, L. vulgaris, M. merluccius, S. aurata, S. scombrus* were seen in daily catches of the CF during the field studies. According to the statements of the CF, there were also increases in catch amounts of the species including *P. saltatrix, B. belone, S. aurata, P. erythrinus, S. sarda, M. merluccius* (Ünal et al. 2013). In Altınoluk, according to the observations of CF, the greatest increases in catch realized in *P. saltatrix, B. belone, L. vulgaris*, and *O. vulgaris, M. Surmuletus/M. Barbatus, S. Sarda, S. Salpa* as well as increases in disappeared or highly decreased species such as *D. puntazzo, Z. faber, S. aurata* and *D. dentex*. CPUE value of CF in Altınoluk were calculated as 4.1 kg h^{-1} that was higher than 1.7 kg h^{-1} in Güre and 1.1 kg h^{-1} in Küçükkuyu with the same daily fishing effort as 6 to 7 h. Even if CF in Altınoluk was found as the most satisfied group among all CF surveyed, they were also not happy about the failures of AR management as CF in Güre and in Küçükkuyu. All CF in three towns agreed on the constitution of an AR management committee with the involvement of all stakeholders, directed by the responsible ministry. *P. saltatrix, B. belone, O. vulgaris, L. vulgaris, S. aurata* as in the case of the CF were the mostly caught species in recreational catch. Reasonable increases in the recreational catch composition of the RF were found in the catch amounts of species such as *P. saltatrix, B. belone, O. vulgaris, L. vulgaris, S. aurata, D. dentex, D. annularis, D. puntazzo, Z. faber.*

No statistical difference was found among the RF in three towns although, the statistical differences were found between the CPUEs of the RF in Altınoluk and in Küçükkuyu as well as the CPUEs of the RF in Güre and Küçükkuyu. The RF in Altınoluk were highly satisfied with the RF in Güre that two-thirds of the RF was entirely satisfied with the ARs while the RF in Küçükkuyu had lower levels of satisfaction. As in the case of the CF, most of the RF in three towns agreed that the ARs are not managed well. Fishing experiments also proved that there were statistically significant differences between the CPUE amounts in the ARs and the CPUE amounts in the non-ARs. The results were given based on the catch amounts of two boat-based recreational fishers being one from the Altınoluk on the ARs and the other is from Küçükkuyu on the Non-ARs, were used to show the catch efficiency of the ARs. The most important species in catch composition were found as *P. saltatrix, D. vulgaris, D. puntazzo, P. erythrinus, S. umbra, S. aurata, B. boops* in AR site and *D. vulgaris, P. erythrinus, B. boops, S. aurata* in non-ARs. Mean CPUE values were calculated as 0.59 kg h^{-1} for the ARs and 0.43 kg h^{-1} for non-ARs with reasonable differences in fishing efforts as 6.9 h in non-ARs and 4.6 h in the ARs. The ARs were assumed to increase the catchability of certain species (GFCM, 2014). In Edremit Bay, the found increases in catch amounts of certain species will result in overfishing with intensive habitat destruction in a certain area by anchoring. Besides, ghost fishing has the potential to be a future threat to the marine ecosystem because the CF will lose their nets if they sustain their net fishing activities close to the ARs. According to the personal communications with the CF in Altınoluk, one fisherman has already left his net on the ARs. Especially, there can be an increase in the time of accommodation of migratory species (e.g., *P. saltatrix*) in the Edremit Bay due to the existence of the ARs. Moreover, CPUE amounts of both

CF and RF and of those collected by fishing experiments proved that the ARs have a direct impact on both commercial and recreational catch. Catch amounts per species registered by the fishery cooperative also proved significant increases over the last few years after the AR deployment. This resulted in excessive amounts of catch in this species for the year 2013; however, it is uncertain whether this impact is continuous or not. The ARs can induce overfishing pressure that was built on the collection/attraction of existing stocks without expanding general stock size (Bohnsack and Sutherland, 1985; Jebreen, 2005). This result is a counter to the targets of the ARs and, it would have direct social and financial impacts including decreased angling, lower quality recreational angling, negative economic effects on the groups and organizations that back the commercial and recreational fishing industries as well as decreased natural values of the Edremit Bay. The RF in Queensland put a high value on the protection of fisheries resources and highly support fisheries management intended to avoid overfishing by the RF (Sutton, 2005). Hence, the RF group would support the use of the ARs for angling in the Edremit Bay if the ARs were introduced to enhance overfishing.

Economic Impact and Value

According to the socio-economic study conducted in 2011 in Altınoluk just before the deployment of ARs, the RF stated a daily average cost as 6.4 € for transport, 3 € for bait, 2.1 € for fishing gear, and 2.6 for other expenses. A considerable amount of cost for repair and maintenance was calculated as 212.8 € while, annual transport costs were 223.1 € (Tunca, 2011). After three years of the first socioeconomic study, a considerable increase in RF expenditures was found in Altınoluk. Especially, bait costs showed a significant increase that was due to the increasing inflation rates during the years. It is noteworthy that the ups in the number of fishing days have increased the economic impact of the ARs.

The ARs have the potential to be highly valued by users and to inject significant amounts of new money into local economies; however, what is not clear is the extent to which these positive outcomes will be realized through AR development in Edremit Bay. The non-market economic value of ARs in Edremit Bay was also shown by the earlier socioeconomic survey conducted in Altınoluk. The contingent valuation results showed that the non-market value of the ARs was € 1.5 Million (1 € 2.34= 2.34 TL: 2011 Annual Mean Exchange Rate).

As noted previously, the AR deployment elsewhere generally occurs against a backdrop of degraded natural systems, or in the areas where natural reef habitat is limited. The Edremit Bay similarly hosts desired conditions for fishing and diving in a healthy and rich natural ecosystem. The studies conducted in the areas where natural and the AR habitats co-exist suggest that the users prefer natural reefs over ARs, place a higher value on natural reefs, and use natural reefs more than ARs (Johns et al., 2001; Pendleton, 2004). The studies also suggested that there is a declining marginal value with an increasing number of reefs (artificial or natural) in a marine area (Johns et al., 2001; Pendleton, 2004). Considering the small natural reef system existing next to the AR site, our results suggest that the value placed on the ARs in Edremit Bay may be lower than that proved in other areas. The potential for the ARs in Edremit Bay to produce substantial positive economic impacts on local communities is also uncertain. But, even if these uncertainties exist, the ARs in Edremit Bay proved their economic contribution to the local people by increases in the number of individual fishing days, higher catch value as well as its contribution to the fishery cooperative, diving activities, and fishing equipment markets in Altınoluk, Küçükkuyu, and Güre. As discussed previously, positive economic impacts will result only if the ARs attract new or added money flow to local or regional areas. Even if, the ARs are used extensively, the economic impact could be minimal if most use comes from locals or from visitors to the area who miss other opportunities during their trip to visit the ARs. Moreover, positive economic impacts in communities close to the ARs could be offset by negative impacts in other communities if the AR deployment results in a redistribution of tourism dollars among communities rather than attracting more tourism dollars into the region (Enemark, 1999). Lastly, there are significant economic impacts induced by the AR development that will depend on whether the recreation opportunities created by the ARs were sufficiently valued to attract more visitors to the region or increase the expenditures of visitors that were already there. The difficulty in the prediction of the exact economic impacts and benefits of the ARs in the Edremit Bay is worth noting that is due to lack of relevant reliable information from the AR stakeholders in Altınoluk, Küçükkuyu and Güre.

Management

The ARs in Edremit Bay were characterized by an open-access or non-property management under which use benefits are accessible to anybody besides a few zone closures. Although open-access management is no more as broadly implemented, the presumption of open-access still wins on a broad range (Cocklin et al., 1998). The AR deployment in Edremit Bay changes the management of property and asset use rights, which could be beneficial for a few users and hinder others. As the project's principal goal was to support small-scale fisheries, the priority should be given to the CF; however, the high numbers of the RF around the ARs may undervalue the CF because of principally high participation supported by high catch satisfaction, economic impact, and value. At this point, the AR deployments on the commercial fishing grounds can naturally prevent the CF's gains and privilege to benefit in the region that was formally exposed to a direct access regime.

In Europe, the ARs were regularly used to avoid trawling from sensitive habitats (Baine and Side, 2003); in our case, the AR cubic structures for production were circled by specifically designed trawl preventing the AR structures. This was thought to naturally limit the access of trawlers into the AR areas, but it did not keep them far away from the ARs. During the survey, it was found that a number of trawlers have been fishing so close to the ARs as well as attracting fish. Purse seiners, as in the case of the trawlers, have also been fishing around the ARs by using high fishing pressure to principally capture pelagic schooling species. Changes to property and asset use rights can be realized through the management of the ARs. The CF were often barred from using the ARs (Branden et al., 1994), and numerous ARs deployed and managed for diving excluded both commercial and recreational fishing (Dowling and Nichol, 2001; QEPA, 2005) Therefore, the same situation may occur for our case because there is high potential for future conflicts among small scale, large scale commercial fishers, and recreational fishers. Furthermore, restriction on the AR use in Edremit Bay is a state of discord on the grounds that such limitations may influence the institutional and human use attributes in the region (Cocklin et al., 1998). Rezoning of the Edremit Bay AR use rights showed a considerable level of stakeholder contention when resource-use rights are assumed to change. Existing restrictions especially prepare grounds for conflicts because of the lack of management participation of principal AR users in the region, the CF, and the RF. Particularly, the exclusion of these two stakeholders can be understood by a few fishers as further disintegration of fishing rights in the fishing zone. Like other types of aquatic environments, the ARs may require proper management either to assure that they enable the desired outcomes for both the biological resources and users. Effective management can reduce risks due to fishing gears damages in the ecosystem, injuries of recreational divers visiting the ARs, decomposed materials, or movement of the reef units off-site. Therefore, a comprehensive management plan should have been designed and improved after the AR deployment. This plan could have aimed to introduce the AR locations on nautical charts to avoid damages to fishing gears and to provide diver safety guidelines to prevent injuries to people diving at the ARs, as well to set up technical measures aimed to regulate access and exploitation rights in the AR site. Physical, biological, and socio-economic monitoring is always a key element of the management plan as it allows assessing the structural performance of the AR on time, assessing whether the AR provides the expected benefits from the ecological and environmental point of view by evaluating the efficiency of the applied control measures. The involvement of stakeholders in AR management is crucial. Professional and recreational fishers and divers can provide support in reef monitoring and evaluation. Applied research is another key element in the AR management programs helping in monitoring the activities conducted at the ARs, evaluating the efficiency of the adopted management measures, and, where necessary, finding actions to be undertaken and alternative management options. The ARs for recreational use are normally conveyed to meet a demand which originates from the absence of such opportunities in a specific area. The ARs in Edremit Bay turn into the main point of conflict in and across the stakeholder groups (Samples, 1989). Conflict is characterized as an objective interference credited to another's behavior (Jacob and Schreyer, 1980), and has been reported on various AR sites around the world (Samples, 1989; Murray and Betz, 1994). Conflict around ARs in the Edremit Bay, as similar to the cases around the world, occurred and have the potential to occur from various issues including (1) Crowding and congestion; (2) stock impacts; and (3) use type.

CONCLUSION

Deployment of the ARs fishery in Edremit Bay affected stakeholders in the region particularly small-scale CF and RF. Proposals are needed that will support and enhance the AR fishery while considering the principal functions of the AR for the conservation of aquatic biodiversity. These include, but are not restricted to: (1) maintaining the AR fishery and ecosystem in the face of other aquatic resource developments; (2) investing in the small scale fishery and recreational fishery sectors, e.g. human capital, to promote AR fishery and conservation interests by an integrated approach; (3) identifying a mechanism for AR management by which the fishing operations do not conflict among themselves and with environmental issues; (4) taking into consideration the use of some management tools such as shared AR zone for different types of fishing techniques and other recreational fishing uses (e.g. private boat-based fishers, recreational charters), besides closed areas in the AR zone as in the present case, to promote management.

After years of AR deployment, our analysis revealed that still, many problems exist concerning AR management in the region. However, these problems can be eradicated by a good, local management scheme. Further, two important stakeholders, fishery cooperatives and diving charter organizations in each town were determined to have positive and negative socio-economic implications and willingness to contribute to the future management of the ARs even if they also agreed on the lack of management regulations. These problems were also more difficult to overcome if management scales are large or a management plan does not exist. In addition, the CF and the RF were ready to contribute to the AR management because they were aware of its proven biological and harvest contribution in the area and, most of them are dependent on the local marine ecosystem as a source of livelihood. Then, reconciliation of small-scale commercial and recreational fisheries and the AR deployment concerns could be possible, because small-scale commercial and recreational fisheries sectors have future potential for greater socioeconomic activity as well as lobbying and engagement.

Finally, an AR management plan should be planned with the active participation of all users and interested parties. There would also be a need for updated AR guidelines that are easily understandable by all stakeholders. These guidelines should be in line with national and international agreements and regulations to ease the establishment of comprehensive management schemes for the ARs. The AR deployment guidelines should be incorporated into national and local level policy.

REFERENCES

Adams, C., Lindberg, B., & Stevely, J. (2004). *The Economic Benefits Associated with Florida's Artificial Reef.* Available from: https://edis.ifas.ufl.edu/fe649

Asafu-Adjaye, J., & Tapsuwan, S. (2008). A contingent valuation study of scuba diving benefits: Case study in Mu Ko Similan Marine National Park. *Tourism Management, 29*(6), 1122–1130. doi:10.1016/j.tourman.2008.02.005

Baine, M., & Side, J. (2003). Habitat modification and manipulation as a management tool. *Reviews in Fish Biology and Fisheries, 13*(2), 187–199. doi:10.1023/B:RFBF.0000019480.95010.67

Bateman, I., & Willis, K. G. (2002). *Valuing Environmental Preferences: Theory and Practice of Contingent Valuation Method in the US, EU, and Developing Countries.* Oxford University Press.

Bell, F. W., Bonn, M. A., & Leeworthy, V. R. (1998). *Economic Impact and Importance of Artificial Reefs in Northwest Florida.* NOAA Paper Contract Number MR235 Office of Fisheries Management and Assistance Service, Florida Department of Environmental Administration.

Bohnsack, J. A. (1989). Are high densities of fishes at artificial reefs the result of habitat limitation or behavioural preference? *Bulletin of Marine Science, 44*(2), 631–645.

Bohnsack, J. A., & Sutherland, D. L. (1985). Artificial reef research: A review with recommendations for future priorities. *Bulletin of Marine Science, 37*, 11–39.

Bombace, G., Fabi, G., & Fiorentini, L. (2000). Artificial reefs in the Adriatic Sea. In A. C. Jensen, K. J. Collins, & A. P. M. Lockwood (Eds.), *Artificial Reefs in European Seas* (pp. 31–63). Kluwer Academic Publications. doi:10.1007/978-94-011-4215-1_3

Bortone, S. A., Brandini, P. P., Fabi, S., & Otake, S. (2011). *Artificial reefs in fishery management.* CRC Press. doi:10.1201/b10910

Branden, K. L., Pollard, D. A., & Reimers, H. A. (1994). A review of recent artificial reef developments in Australia. *Bulletin of Marine Science, 55*(2–3), 982–994.

Brock, R. E. (1994). Beyond fisheries enhancement: Artificial reefs and ecotourism. *Bulletin of Marine Science, 55*(2–3), 1181–1188.

Charles, A. (2001). *Sustainable Fishery Systems.* Wiley Press.

Chou, L. M. (1997). Artificial reefs of southeast Asia–do they enhance or degrade the marine environment? *Environmental Monitoring and Assessment*, *44*(1/3), 45–52. doi:10.1023/A:1005759818050

Cocklin, C., Craw, M., & McAuley, I. (1998). Marine reserves in New Zealand: Use rights, public attitudes, and social impacts. *Coastal Management*, *26*(3), 213–231. doi:10.1080/08920759809362353

Ditton, R. B., & Baker, T. L. (1999). Demographics, Attitudes, Management Preferences, and Economic Impacts of Sport Divers using Artificial Reefs in Offshore Texas Waters. Report prepared for the Texas Parks and Wildlife Department through a research contract with Texas A&M University.

Ditton, R. B., Osburn, H. R., Baker, T. L., & Thailing, C. E. (2002). Demographics, attitudes, and reef management preferences of sport divers in offshore Texas waters. *ICES Journal of Marine Science*, *59*, 186–191. doi:10.1006/jmsc.2002.1188

Ditton, R. B., Thailing, C. E., Riechers, R., & Osburn, H. (2001). The Economic impacts of sport divers using Artificial Reefs in Texas Offshore Waters. *Proceedings of the Annual Gulf and Caribbean Fisheries Institute*, *54*, 349-360.

Dowling, R. K., & Nichol, J. (2001). The HMAS Swan artificial dive reef. *Annals of Tourism Research*, *28*(1), 229–232. doi:10.1016/S0160-7383(00)00019-0

Düzbastılar, F. O., & Tokaç, A. (2003). Determination of effects of artificial reef size on local scouring phenomena resulting from wave action *Ege. Journal of Fisheries and Aquatic Science*, *20*(3-4), 373–381.

Haab, T. C. (2002). *Valuing Environmental and Natural Resources, Economics of Non-Market Valuation*. Edward Elgar Publishing Limited. doi:10.4337/9781843765431

Hanemann, W. M. (1994). Valuing the Environment Through Contingent Valuation. *The Journal of Economic Perspectives*, *8*(4), 19–43. doi:10.1257/jep.8.4.19

Hanni, E., & Mathews, H. H. (1977). *Benefit-Cost Study of Pinellas County Artificial Reefs. FLSGP-T-77-005, Florida Sea Grant College Program*. University of Florida.

Hiett, R., & Milon, J. W. 2002. Economic Impact of Recreational fishing and Diving Associated with Offshore Oil and Gas Structures in the Gulf of Mexico. Minerals Management Service Document MMS Study.

Hushak, L. J., Kelch, D. O., & Glenn, S. J. (1999). The Economic Value of the Lorain County, Ohio, Artificial Reef. Ohio Sea Grant College Program. *American Fisheries Society Symposium*, *22*, 348–362.

Jacob, G., & Schreyer, R. (1980). Conflict in outdoor recreation: A theoretical perspective. *Journal of Leisure Research*, *12*(4), 368–380. doi:10.1080/00222216.1980.11969462

Jebreen, E. (2005). *An investigation into the effects of artificial reefs on fish stocks*. Queensland Department of Primary Industries. http://www.dpi.qld.gov.au/far/9279.htmS

Johns, G. (2004). *Socioeconomic Study of Reefs in Martin County, Florida. Report prepared for Martin County, Florida* .

Johns, G., Leeworthy, V. R., Bell, F. W., & Bonn, M. A. (2001). *Socioeconomic Study of Reefs in Southeast Florida. Report prepared for Miami-Dade County, Florida* .

Leeworthy, V. R., Maher, T., & Stone, E. A. (2006). Can Artificial Reefs Reduce or Alter User Pressure on Adjacent Natural Reefs? *Bulletin of Marine Science, 78*, 29–37.

Lök, A. (1997). Last developments on production versus attraction hypothesis in artificial reefs. *Ege Journal of Fisheries and Aquatic Sciences, 14*(1-2), 239-242.

Lök, A. (2012). Artificial reef applications in Turkey. In *GFCM Scientific Advisory Committee (SAC) 12th Session of the Sub-Committee on Marine Environment and Ecosystems (SCMEE)*, 19.

Lök, A., & Gül, B. (2005). Evaluation of fish fauna associated with experimental artificial reefs in Hekim Island in Izmir Bay (Aegean Sea, Turkey). *Ege Journal of Fishery and Aquatic Sciences, 22*(1-2), 109–114.

Lök, A., Metin, C., Duzbastılar, F. O., Ulaş, A., & Özgül, A. (2013). *Fish Community Structure and Monitoring of Physico-Chemical Parameters in Edremit Bay Artificial Reef Area. Ege University Underwater Research and Application Center Research Report for Turkish Government*. Ministry of Food, Agriculture and Livestock.

Lök, A., Metin, C., Ulaş, A., Düzbastılar, F. O., & Tokaç, A. (2002). Artificial reefs in Turkey. *ICES Journal of Marine Science, 59*, 192–195. doi:10.1006/jmsc.2002.1221

Lök, A., & Tokaç, A. (2000). Turkey: A new region for artificial habitats. In Artificial Reefs in European Seas. Kluwer Academic Publications.

McGinnis, M., Fernandez, L., & Pomeroy, C. (2001). *The Politics, Economics, and Ecology of Decommissioning Offshore Oil and Gas Structures.* DOI Minerals Management Service Document, MMS Publication 2001-006.

McGlennon, D., & Branden, K. L. (1994). Comparison of Catch and Recreational Anglers Fishing on Artificial Reefs and Natural Seabed in Gulf St. Vincent, South Australia. *Bulletin of Marine Science, 55*(2–3), 510–523.

McKean, J. R., & Taylor, R. (2000). *Outdoor Recreation Use and Value: Snake River Basin of Central Idaho. Idaho*. Experiment Station Bulletin.

Milon, J. W. (1988a). *The Economic Benefits of Artificial Reefs: An Analysis of the Dade County, Florida Reef System. Sea Grant Extension Program, University of Florida. SGR-90, Florida Sea Grant College Program*. University of Florida.

Milon, J. W. (1988b). A Nested Demand Shares Model of Artificial Marine Habitat Choice By Sport Anglers. *Marine Resource Economics, 5*(3), 191–213. doi:10.1086/mre.5.3.42628883

Milon, J. W. (1989a). Artificial Marine Habitat Characteristics and Participation Behavior by Sport Anglers and Divers. *Bulletin of Marine Science, 44*, 853–862.

Milon, J. W. (1989b). Contingent Valuation Experiments for Strategic Behavior. *Journal of Environmental Economics and Management, 17*(3), 293–308. doi:10.1016/0095-0696(89)90022-3

Milon, J. W., Holland, S. M., & Whitmarsh, D. J. (2000). *Artificial Reef Evaluation with Application to Natural Marine Habitats*. Social and Economic Evaluation Methods.

Morgana, O. A., Massey, D. M., Huth, W. L., & Hall, R. (2009). Demand for Diving on Large Ship Artificial Reefs. *Marine Resource Economics*, *24*(1), 43–59. doi:10.1086/mre.24.1.42629644

Murray, J. D., & Betz, C. J. (1994). User views of artificial reef management in the southeastern US. *Bulletin of Marine Science*, *55*(2–3), 970–981.

Oh, C., Ditton, R. B., & Stoll, J. R. (2008). The Economic Value of Scuba-Diving Use of Natural and Artificial Reef Habitats. *Society & Natural Resources*, *21*(6), 455–468. doi:10.1080/08941920701681953

Pears, R., & Williams, D. McB. (2005). *Potential effects of artificial reefs on the GREAT BARRIER reef. Cooperative Research Centre for the great barrier Reef World Heritage Area technical report 60*. James Cook University.

Pendleton, L. H. (2004). Creating Underwater Value: The Economic Value of Artificial Reefs for Recreational Diving, California Artificial Reef Enhancement Program. The San Diego Oceans Foundation.

Pickering, H., & Whitmarsh, D. (1997). Artificial reefs and fisheries exploitation: A review of the 'attraction versus production' debate, the influence of design and its significance for policy. *Fisheries Research*, *31*(1-2), 39–59. doi:10.1016/S0165-7836(97)00019-2

Queensland Environment Protection Agency (QEPA). (2005). *Sink the Brisbane. Environmental Protection Agency/Queensland Parks and Wildlife Service*. http://www.epa.qld.gov.au/about_the_epa/coming_events/sink_the_brisbaneS

Radonski, G. C., Martin, R. G., & DuBose, W. P. (1985). Artificial reefs: the sportfishing perspective. In F. M. D'Itri (Ed.), *Artificial reefs: marine and freshwater applications* (pp. 529–536). Lewis Publishers, Inc.

Ramos, J., Santos, M. N., Whitmarsh, D., & Monteiro, C. C. (2006a). Patterns of Use in an AR System: A Case Study in Portugal. *Bulletin of Marine Science*, *78*, 203–211.

Ramos, J., Santos, M. N., Whitmarsh, D., & Monteiro, C. C. (2006b). The Usefulness of the Analytic Hierarchy Process for Understanding Reef Diving Choices: A Case Study. *Bulletin of Marine Science*, *78*, 213–219.

Ramos, J., Santos, M. N., Whitmarsh, D., & Monteiro, C. C. (2007). Stakeholder perceptions regarding the environmental and socio-economic impacts of the Algarve artificial reefs. *Hydrobiologia*, *580*(1), 181–191. doi:10.100710750-006-0454-z

Ramos, J., Santos, M.N., Whitmarsh, D., & Monteiro, C.C. (2011). Stakeholder analysis in the Portuguese Artificial Reef Context: Winners and Losers. *Brazilian Journal of Oceanography*, *59*, 133-143.

Roberts, J. K., Thompson, M. E., & Pawlyk, P. W. (1985). Contingent Valuation of Recreational Diving at Petroleum Rigs, Gulf of Mexico. *Transactions of the American Fisheries Society*, *114*(2), 214–219. doi:10.1577/1548-8659(1985)114<214:CVORDA>2.0.CO;2

Samples, K. C. (1989). Assessing recreational and commercial conflicts over artificial fishery habitat use: Theory and practice. *Bulletin of Marine Science*, *44*(2), 844–852.

Santos, M. N., & Monteiro, C. C. (1997a). The OLHAO artificial reef system (south Portugal): Fish assemblages and fishing yield. *Fisheries Research*, *30*(1-2), 33–41. doi:10.1016/S0165-7836(96)00567-X

Santos, M. N., & Monteiro, C. C. (1997b). Comparison of the catch and fishing yield from an artificial reef system and neighboring areas off Faro (Algarve, south Portugal). *Fisheries Research, 39*(1), 55–65. doi:10.1016/S0165-7836(98)00169-6

Schug, D. M. (1978). *Political and Economic Aspects of Artificial Reefs in Pinellas County, Florida* (Thesis). Department of Marine Science, University of South Florida, Tampa, FL.

Seaman, W., Jr., & Jensen, A. C. (2000). Purposes and Practices of Artificial Reef Evaluation. In Artificial habitats for marine and freshwater fisheries. Academic Press.

Seaman, W., & Sprague, L. M. (Eds.). (2000). *Artificial habitats for marine and freshwater fisheries.* Academic Press.

Sutton, S. G. (2005). *An assessment of the social characteristics of Queensland's recreational fishers. Cooperative Research Centre for the great barrier Reef World Heritage Area technical report 65*. James Cook University.

Sutton, S. G., & Bushnell, S. L. (2007). Socio-economic aspects of artificial reefs Considerations for the Great Barrier. *Ocean and Coastal Management, 50*(10), 829–846. doi:10.1016/j.ocecoaman.2007.01.003

Tiryakioğlu, F. Ö. (2008). *Socio-Economic Evaluation of Artificial Reefs In Aegean Sea, Turkey* (Master's dissertation). University of Portsmouth Business School.

Tokaç, A., Ünal, V., Tosunoğlu, Z., Akyol, O., Özbilgin, H., & Gökçe, G. (2010). *Aegean Sea Fisheries.* IMEAK Deniz Ticaret Odası İzmir Şubesi Yayınları.

Tunca, S. (2011). *Assessing the Socio-Economic Effects of Artificial Reef Deployments in the Northern Aegean Sea: Altınoluk Case* (MSc dissertation). Ege University Graduate School of Natural and Applied Sciences.

Tunca, S. (2012). Decisions of Stakeholders for the Proposed Artificial Reef Deployment: Analytic Hierarchy Process Approach. *Ege J Fish Aqua Sci, 29*(1), 21–29.

Tunca, S. (2013). Benefits of artificial reefs in Altınoluk (Northern Aegean), Turkey: Assessment of potential users' Opinions via Fuzzy Pair Wise Comparison Approach. *Ege J Fish Aqua Sci, 30*(3), 99–104.

Tunca, S. (2014). Perception and demand for artificial reefs by relevant local groups in Altınoluk (Turkey). *Ege J Fish Aqua Sci, 31*(1), 5–10.

Tunca, S., Miran, B., & Ünal, V. (2016). Non-Market Economic Value of Artificial Reefs in Edremit Bay (Eastern Mediterranean, Turkey). *Ege Journal of Fisheries and Aquatic Sciences, 33*(2), 97–107. doi:10.12714/egejfas.2016.33.2.02

Ulaş, A. (2007). A pre-study for determining of efficiency of fish sampling methods on artificial reefs. *Ege Journal of Fisheries and Aquatic Sciences, 24*(3-4), 287–293.

Ünal, V., Tunca, S., & Miran, B. (2013). Socioeconomic Impact of Edremit Bay Artificial Reefs Research Report. Ministry of Food, Agriculture and Livestock, General Directorate of Fisheries and Aquaculture, National Artificial Reef Monitoring Project.

Whitmarsh, D., Santos, M. N., Ramos, J., & Monteiro, C. C. (2008). Marine habitat modification through artificial reefs off the Algarve (southern Portugal): An economic analysis of the fisheries and the prospects for management. *Ocean and Coastal Management*, *51*(6), 463–468. doi:10.1016/j.ocecoaman.2008.04.004

Section 2
Management of Fisheries in Local Communities

Chapter 4
Artificial Reef Contributions to Sustainability:
A Case Study on Small-Scale Fisheries in Michoacan, Mexico

Jose Antonio Santamaria Reyes
Universidad Michoacana de San Nicolás de Hidalgo, Mexico

José Odón García
Universidad Michoacana de San Nicolás de Hidalgo, Mexico

ABSTRACT

For 59 million people in the world, fishing represents their main source of income. In addition, for 3.2 billion people, fishery resources provide 20% of the protein of animal origin in their diet. One tool that could contribute to recovering stocks in a sustainable way is artificial reefs. In Mexico, 750 structures were deployed in Barra de Pichi and 500 in Caleta de Campos municipality of Lázaro Cárdenas, Michoacán, between 2016 and 2018. The objective of this work is to evaluate the level of sustainability of the artisanal fishing system between Barra de Pichi and Caleta de Campos, in relation to the artificial reefs (ARs) and with the purpose of assessing its usefulness as a public policy tool. The framework for the evaluation of natural resource management systems incorporating sustainability indicators (MESMIS) was used as a methodological guide, comparing two groups: AAusers (users ARs) and non-AAusers (non-users of ARs). The results showed that the artisanal fishing activity carried out by ARs users is more sustainable than non-ARs users.

INTRODUCTION

Fishing activity along with others such as agriculture and hunting have sustained human groups throughout history. At present, the fishing activity provides 59 million jobs, food, contributing with 20% of protein of animal origin to feed 3,200 million people. However, fish stocks have gone from 10 to 33.1%, since 1974 to 2015 at levels of overexploitation (FAO, 2018). Consequently, alternatives have been sought to

DOI: 10.4018/978-1-6684-2344-8.ch004

protect and increase marine biodiversity of fisheries importance. In this sense, Artificial Reefs (ARs) have been used for this purpose. AR structures are submerged deliberately built or placed on the seabed to imitate some of the functions of natural reefs, such as protecting, regenerating, concentrating and / or increasing the populations of living marine resources (London Convention and Protocol / UNEP, 2009).

The emergence of sustainable development, as a social and political movement of humanity, has led to the orientation of scientific, economic, social, and political efforts to sustainable trajectories (Gallopín et al., 2001; Martens, 2006; Salas-Zapata et al., 2011; Lang et al., 2012). Hence the need for such studies to be based on the notion of sustainability (UN, 1987). To measure it, the three main pillars on which it is based must be covered, the economic, the social and the environmental (Brundtland, 2016).

The evaluation of artisanal fishing using ARs requires ethnographic studies, which are generally not used by public administrators, even though it is the fishermen who have the knowledge due to their direct and regular contact with these resources (Johannes, 2002; Drew, 2005; Silvano et al., 2009). However, several studies include ethnographic interviews and in situ sampling to support the information provided by the fisherman's experience and relate ecological systems to fishing activity. In addition, other studies have examined the role of fishermen in the interrelation of people with natural resources and changes in the environment (Folke et al., 1998; Davidson-Hunt and Berkes, 2003; Moreno et al., 2007; Begossi et al., 2011).

Most of the tools to measure sustainability worldwide have emerged from the Earth Summit of Rio de Janeiro in 1992. At least 17 have been counted (Singh et al., 2009). At the international level, there is a methodological guide for the design of composite indicators of sustainable development (Schuschny and Soto, 2009), the ecological footprint (Badii, 2008), the sustainability barometer (Prescott-Allen, 1997), the index of human development (Mancero, 2001), the sustainable economic well-being index (Daly and Cobb, 1993), the genuine progress indicator (Anielski, 2001) and the environmental sustainability index (Environmental Performance Index, 2016). Additionally, there are methods designed for multiple purposes supported by participatory processes, one of which has made the greatest progress is the one obtained by the interdisciplinary group of Appropriate Rural Technology (GIRA), when developing the Framework for the Evaluation of Resource Management Systems incorporating Sustainability Indicators (MESMIS), which is an evaluation proposal in search of a sustainable development of rural areas (Masera and López-Ridaura, 2000; Astier et al., 2002).

The objective of this research is to identify the economic, environmental, and social impact of Artificial Reefs (ARs) on the level of sustainability of artisanal fishing, according to the perception of fishermen between Barra de Pichi and Caleta de Campos in the municipality of Lázaro Cárdenas Michoacán, Mexico. Probably the social dimension is the one with the greatest influence on the level of sustainability, followed by the environmental and economic dimension. If the hypothesis is confirmed, the ARs placed in the study area can contribute positively if the artisanal fishermen of the region have the capacity to organize themselves, establish rules, norms and carry out processes of awareness of sustainable management in the fishing grounds with and without ARs.

BACKGROUND

According to their use, ARs are divided into two groups as protection and production. ARs have become a strategy for the conservation and increase of marine biodiversity, with the purpose of reestablishing fish stocks (Yamane, 1989; Revenga et al., 1997; Jensen, 2002; Texas Parks and Wildlife Department Artificial Reef Program, 2016).

In the case of Mexico, the Program for the Promotion of Fisheries and Aquaculture Productivity (PFPPA) aims to ensure that the Aquaculture Fisheries Economic Units (UEPA) increase their productivity within a framework of sustainability. The Fisheries and Aquaculture Regulation and Surveillance component contemplates in its subcomponent IV, the placement of ARs. These structures have been placed in various States of Mexico such as Quintana Roo, Colima, Campeche, Jalisco, Baja California Sur, Veracruz, Yucatán, and Michoacán (SAGARPA, 2010).

Some 4,796 artificial reefs have been installed in the State of Michoacán (*op. cit.*). It is estimated that, in Aquila and Coahuayana, municipalities on the Michoacan coast, there are 3,543 ARs (data as obtained through personal interviews) and a sunken ship in Faro de Bucerías. In the municipality of Lázaro Cárdenas (Barra de Pichi) there are 750 ARs and in the Habillal there is a stranded ship. On the other hand, in Caleta de Campos there is a sunken ship, and in August 2018 around 500 ARs were placed.

In 2010 CONAPESCA and the Government of Campeche analyzed the existing regulation to improve coordination between the productive sectors and the institutions in charge of implementing the program, specifically the placement of ARs. It was proposed to create an official standard for ARs placement and the need for each project to be monitored and evaluated according to its objective and to measure its social and biological impact. However, no evaluation that measured the positive or negative effects of ARs on Mexican fisheries could be identified. Therefore, the evaluation of ARs and its social impact are a challenge in the management of public policies.

MATERIAL AND METHODS

Study Area

The State of Michoacán has three municipalities that make up the coastal zone with an approximate length of 262 km facing the Pacific Ocean, between the sea and the Sierra Madre del Sur, Coahuayana, Aquila and Lázaro Cárdenas (Figure 1). They cover a territorial extension of 4,150 km^2 with a population of 222,441 inhabitants (INEGI, 2016). The fishing villages are scattered along the coast and their population has quite different socio-cultural characteristics depending on their ethnic origin.

The bathymetric characteristics of the Michoacan coast allow us to consider that only the portion of coastal waters in the first two kilometers perpendicular to the coast, are suitable for the generation of natural reefs and / or the installation of artificial reefs, that is, there is a shortage of hard sea bottoms (INECC, 2007).

The study area is 47 km between Barra de Pichi (17 ° 58'24.47 "N) and Caleta de Campos (18 ° 4'26.93" N). In the Barra de Pichi locality, 750 ARs units distributed over 34,389 m^2 were deployed, at 1.10 km from the coastline and at a depth of 20 m. The beached ship Betula (17 ° 59'12.08 "N) is located about 50 m from the coastline and reaches a depth between 3 and 7 m, in front of the town of El Habillal. In Caleta de Campos is the sunken ship Quetzalcóatl E10 (18 ° 4'7.94 "N), located at 900 m from the cove at a depth of approximately 30 m and in August 2018, 500 ARs were deployed. The data described above was corroborated by exploratory work in the field during 2017 and 2018. The artificial reef unit deployed has a weight of 480 kg with holes of 30 cm in diameter, and with an average height of 80 cm.

Figure 1. Municipality of Lázaro Cárdenas, Michoacán and location of ARs on the coast (Source: Direct observation and map based on INEGI, 2016)

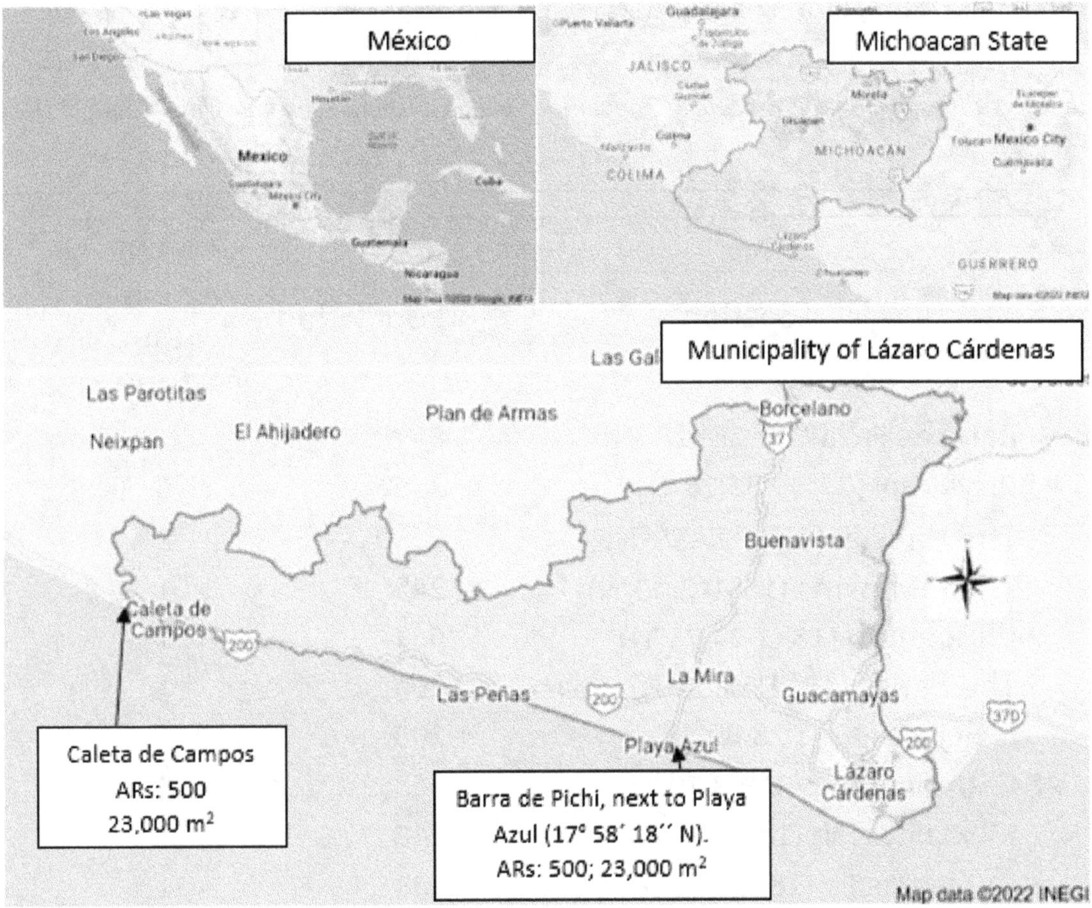

Small-Scale Fishing between Barra de Pichi and Caleta de Campos

Small-scale fishing is carried out in the coastal zone, smaller boats with a maximum of 15 m in length are used, outboard motor from 40 to 60 hp that are designed for catching, but not for conservation, which limits the capacity for autonomy (Breton and López Estrada, 1989; Lobato González, 1996; Alcalá, 1986). There are 5 target species that can be found: (1) Red snapper; (2) Horse mackerel; (3) Sierra; (4) Grunt; (5) Snapper. These species are the ones in greatest demand by the consumer (SAGARPA, 2010). There are other species of interest such as lobsters that are at a critical level of exploitation.

The fishermen are integrated into cooperatives, unions, or federations of fishermen. In 2007 there were 1,522 registered fishermen distributed along the Michoacan coastal strip (Marín, 2004). By 2009 there were 1,314 registered fishermen (CONAPESCA, 2009). At least 50% of them alternate their fishing activity with extensive livestock farming, agriculture, and commerce (INEGI, 2016).

The fishing villages are scattered along the coast and their population has quite different social and cultural characteristics from the point of view of their ethnic origin. The municipality of Lázaro Cárdenas with the largest population (183,185 inhabitants) has the largest number of riverine fishermen (948), of which 240 are located between Barra de Pichi and Caleta de Campos. These fishermen are distributed in localities and economic fishing units, see Table 1. According to INEGI (2016) there are 20 Fisheries Economic Units (FEU), each one has an average of 12 members.

Table 1. Localities of the study area, their population, and the Fishing Economic Units (Source: INEGI, 2016)

Location	Population	Fishing Economic Unit (FEU)
Barra de Pichi (17° 58′ 18′′N)	8	
Playa Azul (17°58'55.28"N)	3054	5
The Habillal (18° 0'27.88"N)	1530	2
The Pumpkins (18° 0'2.00"N)	245	1
Chucutitan (18° 1'24.71"N)	681	1
The Peñas (18° 1'15.00"N)	206	1
El Bejuco (18° 1'28.89"N)	200	1
Chuquiapan (18° 3'8.79"N)	363	2
Mexcalhuacán (18° 3'17.29"N)	248	1
La Manzanilla (18° 3'49.00"N)	189	1
Caleta de Campos (18° 3'17.29"N)	2041	5
Total	8756	20

Data Collection and Measurement

The method was designed based on the construction of indicators, considering the perception of fishermen, from an adaptation of the MESMIS method (Masera and López-Ridaura, 2000; Sepúlveda, 2002; FAO, 2004; Venegas, 2004). Research variables were identified and justified with an ethnographic exercise. Likewise, Likert-type scales were operationalized and used for its measurement (Weiers, 1986; Kerlinger, 1997; Rojas, 1997; Hernández et al., 2014).

To determine the level of sustainability of artisanal fishing in relation to artificial reefs, it was necessary to construct sustainability indicators according to the MESMIS method as a guide that includes social, economic, and environmental aspects. The method allows it to be done transversally, from production units to regions (Masera and López-Ridaura, 2000; Astier and Hollands 2007). It is suggested to define the critical points of the system based on the weaknesses identified. This work was carried out with ethnographic work, participant observation, informal interviews, and semi-structured dialogues (Malinowsky, 1984; Restrepo, 2015).

The researcher and a collaborator accompanied the fishermen to carry out the fishing activity, the following were identified as critical points: the fishing catch, the time in fishing hours, the time in fishing days, the costs of the process, the fishing gear, the abundance of target species, location of fishing grounds, surveillance of fishing and industrial fishing vessels. Subsequently, the attributes were associated with their respective dimensions and based on them the "diagnostic criteria" were identified and the indicators were defined (Masera and López-Ridaura, 2000; Astier et al., 2002). There are six attributes identified that represent sustainability for this study: productivity, stability, reliability, adaptability, self-sufficiency, and equity (Masera and López-Ridaura, 2000).

The process was strengthened with the identification and justification of indicators through a method of weighting and frequencies. The criteria of mental map, survey of experts, frequency of variables and applied models were used. The results were contrasted and adjusted to those obtained through ethnographic work using the triangulation method (Yeasmin and Rahman, 2012).

The variables were subsequently operationalized and once the questionnaire was designed, a pilot test was applied. As a result of the process, for the three dimensions (economic, environmental, and social), 15 diagnostic criteria and 28 indicators were obtained.

The fishermen interviewed were chosen by snowball sampling (Salomón, 2016; Espinosa, 2018), considering the following criteria: (1) be artisanal coastal marine fishermen; (2) place of fishing, within the study area, and (3) be active or working fishermen.

According to the population of fishermen (N = 240), the sample resulted in n = 148, using the formula for finite populations (Herrera, 2011). Due to the insecurity situation that is registered in the study area, 57 interviews could be carried out. Despite being a third of the estimated sample, the number of interviewees is within the range proposed by Mason (2010) as representative for qualitative research (between 30 and 60). In this type of research, above the mentioned range, there are no contributions of new information, since up to the proposed limit the saturation principle is fulfilled (Crouch and Mckenzie, 2006).

The questionnaire was divided into two parts. The first one captures data regarding the name, age, gender, level of education, years of fishing practice, town of residence, which cooperative it belongs to or whether it is independent, the sale price of its product and whether it uses ARs. The second one collects the data based on the indicators derived from the attributes and each one has a question. For this research, the equity indicator is determined by the use or non-use of ARs and serves as a classification variable. This allows the sample to be separated into two groups.

Likert-type scales were used (Ocaña Moral et al., 2013). To a statement that goes from a negative to a positive sense and is assigned a value, the important thing is to maintain the same level for all items, which can be observed in Figure 2 (Kerlinger, 1997; Hernández et al., 2014).

Figure 2. Sustainability scale

1	2	3	4	5
Not sustainable	Very little Sustainable	Moderately sustainable	Almost Sustainable	Sustainable

Once the survey was defined, the reliability test or Cronbach's alpha coefficient was performed. The minimum acceptable value is 0.70; below this value the internal consistency of the scale used is low. For its part, the maximum expected value is 0.90 (Hernández et al., 2014) . In this case the result was 0.871.

The Kolgomorov-Smirnov (K-S) test was used, which is valid for samples of between 50 and 1000 cases. In this case, a significance level $\alpha = 0.05$ was established, since the p-value <0.05 in all the indicators, it was accepted that the distribution is not normal. Considering the above, to find the level of association between the variables, the non-parametric Spearman test (Rho) was used, which does not fit a normality distribution (Barrera, 2014; Lizama & Boccardo, 2014).

The Spearman coefficient equation is:

$$r_s = 1 - (6 \sum d_1^2 / n(n^2 - 1))$$

According to the analysis carried out, the correlations with an $\alpha < 0.05$ are significant and would indicate "correlation", for the interpretation of the results it is suggested to consider the "degree of relationship" as suggested by Hernández et al., (2014). If the alpha error or significance level is 0.000 (zero) or $\alpha > 0.05$, it would indicate that there is a strong correlation, and that it is most likely true (Barrera, 2014; Lizama and Boccardo, 2014).

RESULTS

The results are presented by dimension in Table 2, 3 and 4. The comparison and results by groups: Artificial Reef Users (USAR) and NO Users of AR (NOUSAR), are presented in Table 5.

Table 2. Results in the economic dimension

Attribute	Diagnostic criteria	USAA	NOUSAA
Stability	Biodiversity of fishing interest	3.04	2.17
	Compliance with fishing regulations	2.77	2.52
Reliability	Protection of fishery resources and coasts	2.76	2.34
	AA exploitation	4.42	1.79
	Average scalar value per group	3.25	2.21

Table 3. Results in the environmental dimension

Attribute	Indicator	USAA Scalar value	NOUSAA Scalar value
Productivity	Kilos of catch	3.06	2.58
	Catch cost	4.18	2.75
	Fishing days	2.09	2.00
	Fishing hours	4.39	3.58
	Percentage of sale	4.70	4.58
	Percentage of sale	2.12	2.00
	Market	4.30	3.79
	Total / average scalar value per group	3.33	2.77

Table 4. Scalar evaluation of diagnostic criteria, social dimension

Dimension	USAA	NOUSSA
Economical	3.33	2.77
Environmental	3.25	2.21
Social	3.76	2.95
general	3.45	2.64

Table 5. Comparative table by dimension and group

Attribute	Diagnostic criteria	USAA	NOUSAA
Adaptability	Strengthening the management of ARs	2.05	2.08
Self-sufficiency	Subsidies	4.55	4.52
Equity	Equity	4.70	2.25
	Average scalar value per group	3.77	2.95

It was possible to identify some attributes and variables that negatively affect the sustainability of the system, see Table 6.

Table 6. Attributes and indicators that affect artisanal fishing activity in the use of ARs and their possible associated causes

Attributes	Scalar value of the diagnostic criterion	Indicator	Causes
Fishing days	2.09	Fishing days	Fishing pressure
Sale size	2.12	Sale size	Demanding market of a specific size
Compliance with fishing regulations	3.12	Type of fishing gear used	Use of fishing gear not allowed
	2.42	Fishing catch size	Catch of fish below the minimum size
Protection of fishery resources and coasts	1.39	Level of surveillance of fishery resources by fishermen	Lack of interest of fishermen to monitor their fishing resources
Strengthening the management of AA	1.09	Level of surveillance of fishery resources by the authorities	Lack of interest by the authorities to monitor fishery resources
	1.21	Number of courses for the management of AA	Lack of training in the proper management and exploitation of AA
	2.88	Level of participation of fishermen in the management of AA	Little interest of fishermen in managing AA to increase fish availability

Regarding the dimensions of sustainability, the economic one has a positive correlation with the artisanal fishing activity (rho = .666; α = .000). The environmental dimension had a strong positive correlation (rho = .811; α = .000). Regarding the social dimension, it is the one with the greatest influence on the sustainability of artisanal fishing in relation to ARs (rho = .833; α = .000), for the criteria Hernández et al. (2014). It is highlighted that social processes such as organization and democracy in decision-making were the ones with the greatest influence to achieve an optimal level of sustainability in artisanal fishing in relation to ARs (indicator 25: rho = .900; α =. 000 and indicator 28: rho = .510; α = .000), see Table 7.

Table 7. Correlations (rho) between level of sustainability, dimensions, and indicators [SPSS, Rho correlation analysis (Spearman)]

	Sustainability	Social dimension	Environmental dimension	Economic dimension
Sustainability		rho=.833 α=.000	rho=.811 α=.000	rho=.666 α=.000
Indicators:				
1. Kilos of catch per slaughter				rho=.649 α=.000
2.- Catch cost per slaughter	rho=.753 α=.000			rho=.803 α=.000
6.- Hours of catch per slaughter	rho=.474 α=.000			rho=.628 α=.000
12.- Type of fishing gear used			rho=.527 α=.000	
17.- Level of protection against industrial fishing	rho=.404 α=.000		rho=.479 α=.000	
22.- Level of availability of fishing resources in AA	rho=.778 α=.000		rho=.850 α=.000	
23.- Equity (Profit level)	rho=.815 α=.000	rho=.651 α=.000		
25.- Level of participation of fishermen in the design, implementation and monitoring of AA		rho=.541 α=.000		
28.- Level of democratic participation within fishing organizations	rho=.471 α=.000	rho=.511 α=.000		

DISCUSSION

The use of artificial reefs aims to increase fishing productivity in a framework of sustainability in the case of Mexico (DOF, 2017). However, the evaluations of programs such as the placement of ARs do not allow conclusions to be drawn due to the lack of homogeneity between the protocols and the data processing between the studies, in addition, the perception of the Fishermen has not been considered (Tessier et al., 2015). For example, in France, up to 2014 only 3 studies had been carried out that included perception (Collart and Charbonnel, 1998; CREOCEAN, 2003; Leleu et al., 2012). Thus, the evaluation of the economic and social problems related to fishermen is important because the main objective of most ARs is to improve artisanal fishing (Barnabé, et al., 2000).

In the case of the Michoacan coast, the scarcity of natural rocky reefs is a limitation in the fishing production. Therefore, artificial reefs can potentially increase such production. However, if exploitation in the area has already reduced stocks to unsustainable levels, then the additional artificial reefs will have no effect on fish production, they will simply cause a redistribution of existing biomass (Bohnsack, 1989; Polovina, 1991). In this case, prior studies are needed to determine the level of exploitation of the different fisheries and to be considered for the deployment of the structures.

In the short term, the positive impact of an artificial reef is due to the aggregation of biomass and not to in situ production. It is the presence of prey fish that attracts other pelagic and demersal species of fishing interest, supporting the usefulness of the structures as a fishing management tool (Delgadillo-Gárzon et al., 2009). In this case, ARs could contribute in a positive way to increase the availability of fish, but very little to the real increase in biomass.

Some studies indicate that the carrying capacity at a sustainable level can be achieved between one and five years depending on various factors such as environmental, physical, biological and exploitation (Bohnsack et al., 1991). In another study, Relini et al. (1995) suggests that in ARs the processes of increase and decrease in biomass are intermittent or cyclical, showing a greater abundance of some species for certain periods and a decrease in others.

The ARs of Barra de Pichi were placed during 2016 and early 2017, the Quetzalcóatl and Betúla (sunken ships) are more than 10 years old. This suggests that the ARs in Barra de Pichi have not reached their sustainable carrying capacity. On the other hand, the boats have allowed the development of a great biodiversity and it is well known that the fishing activity around them is intense. The variability that they could have, according to Relini et al. (1995), implies that there are probably changes in the dominant species and therefore the species of fishing interest. Therefore, proper management of AR placement becomes vitally important, establishing rules, regulations and monitoring that allows decisions to be made to maintain them over time.

CONCLUSIONS

The artificial reefs placed between Barra de Pichi and Caleta de Campos have a positive impact on the sustainability of artisanal fishing in the studied area. The USAR (scalar score between 3 and 4) carries out more sustainable practices than the NOUSAR (score between 1 and 3).

For these structures in the short term (1 to 3 years), probably due to a redistribution effect, there was the perception of an increase in biomass produced in the ARs, in this sense if they comply with increasing productivity. However, it does not necessarily guarantee the availability of fishery resources in the medium or long term, its contribution such as recruitment depends on the pressure exerted by the fishermen and the status of the fisheries involved.

The way in which the program's resources are allocated generates conflicts among the fishermen. When a group of organized fishermen have the initiative to place artificial reefs, they intend to benefit from that effort. However, other fishermen consider that the structures also belong to them, considering that the origin of the money was public. This generates conflicts, greater pressure on the ARs and lack of interest in its conservation. In this sense, the placement of ARs is recommended involving fishermen in a general way, without benefiting in a particular way only some economic fishing units, but with agreements, rules and norms established by all users.

The placement of ARs is not accompanied by processes of awareness from the point of view of sustainable management, nor by a process of vigilance. Therefore, the user exerts pressure on uncontrolled fishing resources, which could lead to overexploitation and therefore the sustainability of the artificial system can be questionable.

On the other hand, the lack of financial support for gasoline and diesel contributes positively to the sustainability of coastal fishing by reducing the pressure on fishing resources. In this case it is advisable to invest in more AR instead of subsidizing fuels. By increasing the areas covered with AR near the coast, a subsidy on gasoline and marine diesel would not be justified. In addition, by reducing the pressure on natural reefs, their recovery is promoted, thereby making an important contribution to the sustainability of artisanal fishing.

According to the results, users and non-users of artificial reefs catch practically the same number of fish. The difference is in the costs generated, since the first group only must move between 1 and 2 kilometers from the coast. On the other hand, the second group must travel several kilometers to reach the fishing zone, for this reason the hours dedicated to the capture oscillate between 12 and 18 hours. On the other hand, users of artificial reefs work between 7 and 8 hours. So, the fisherman gets to live with his family and carry out other activities. It is the benefit that fishermen obtain by making use of artificial reefs on the Michoacán coast, increasing their well-being.

REFERENCES

Alcala, G. (1986). Lázaro Cárdenas, Michoacán: Living conditions and perspective of the fishermen. Michoacán Studies. The College of Michoacán, Mexico.

Anielski, M. (2001). Measuring the sustainability of nations: The genuine progress indicator system of sustainable wellbeing accounts. In *The Fourth Biennial Conference of the Canadian Society for Ecological Economics: Ecological Sustainability of the Global Market Place*. Pembina Institute for Appropriate Development.

Astier, M.; Maass, M. & Etchevers, J. (2002). Derivation of soil quality indicators in the context of sustainable agriculture. *Agriscience, 36*(5).

Badii, M. H. (2008). La huella ecológica y sustentabilidad (Ecological footprint and sustainability). *Daena: International Journal of Good Conscience, 3*(1), 672-678.

Barnabe, Charbonnel, Marinaro, Ody, & Francour. (2000). Artificial reefs in France: analysis, assessments and prospects. In Artificial Reefs in European Seas. Kluwer Academics Publishers.

Begossi, A., May, P.H., Lopes, P.F., Oliveira, L.E.C., Vinha, V., & Silvano, R.A.M. (2011). Compensation for environmental services from artisanal fisheries in SE Brazil: policy and technical strategies. *Ecol. Econ., 71*, 25–32. .2011.09.008 doi:10.1016/j.ecolecon

Bohnsack, J. A. (1989). Are high densities of fishes at artificial reefs the result of habitat limitation or behavioral preference? *Bulletin of Marine Science*, (44), 631–645.

Bohnsack, J. A., Johnson, D.L., & Ambrose, R.E. (1991). *Ecology of artificial reef habitats and fishes.* Academic Press.

Breton, Y., & Lopez Estrada, E. (1989). Social sciences and fisheries development. Models and methods applied to the case of Mexico. National Institute of Anthropology and History-Secretariat of Fisheries, Mexico.

Brundltand, R. (2016). *Sustainability and Development.* Retrieved on June 16, 2016, from https://sustentabilidadydesarrollo.com/2011/06/25/82/

Collart, D. & E. Charbonnel, (1998*). Impact des re ´cifs artificiels de Marseillan et d'Agde sur le milieu marin et la pe ^che professionnelle-Bilan du suivi 1996/1997.* Contrat Conseil Re ´gional Languedoc-Roussillon et Conseil Ge ´ne ´ral de l'He ´rault. CEGEL et GIS Posidonie, France.

CREOCEAN. (2003). *Impact des re ´cifs artificiels sur le milieu marin et la pe ^che professionnelle dans le golfe d'AiguesMortes.* CREOCEAN.

Crouch, M., & McKenzie, H. (2006). The logic of small samples in interview based qualitative research. *Social Sciences Information. Information Sur les Sciences Sociales, 45*(4), 483–499. doi:10.1177/0539018406069584

Daly, H. E., & Cobb, J. B. (1993). For the common good: Reorienting the economy towards community, the environment, and a sustainable future. Economic Culture Fund.

Davidson-Hunt, I. J., & Berkes, F. (2003). *Nature and Society through the Lens of Resilience: toward a Human-in-Ecosystem Perspective. Navigating Social-ecological Systems: Building Resilience for Complexity and Change.* Cambridge University Press.

Delgadillo-Garzón & Camilo. (2009). Impact of two artificial reefs on the diurnal artisanal fishing of the Gulf of Morrosquillo, Caribbean of Colombia. *Revista de Biología Tropical, 57*(4), 993–1007. PMID:20073330

DOF. (2017). Rules of operation for the Program for the Promotion of Fisheries and Aquaculture Productivity. Official Gazette of the Federation.

Drew, J. A. (2005). Use of traditional ecological knowledge in marine conservation. *Conservation Biology, 19*(4), 1286–1293. doi:10.1111/j.1523-1739.2005.00158.x

Espinosa, T. P. (2018). *Snowball Sampling.* Department of probability and statistics. UNAM. http://www.dpye.iimas.unam.mx/patricia/muestreo/datos/trabajos%20alumnos/Proyectofinal_Bola%20de%20Nieve.pdf

FAO. (2004). *Application of development strategies focused on people within the scope of FAO.* FAO.

FAO. (2016). The State of World Fisheries and Aquaculture 2016. Contributing to food security and nutrition for all. Rome, Italy: FAO.

FAO (2018). *The State of World Fisheries and Aquaculture 2018. Meeting the Sustainable Development Goals.* License: CC BY-NC-SA 3.0 IGO.

Folke, C., Berkes, F., & Colding, J. (1998). Ecological practices and social mechanisms for building resilience and sustainability. In F. Berkes & C. Folke (Eds.), *Linking Social and Ecological Systems: Management Practices and Social Mechanisms for Building Resilience*. Cambridge University Press.

Gallopín, G. C. (2001). Science for the 21st Century: From Social Contract to the Scientific Core. *Int. Journal Social Science, 168*, 219–229.

Gonzalez, L., & Maria, P. (1996). Reflections on coastal fishing. Effort and Capture. Technology and overexploitation of living marine resources, El Colegio de México, México.

Hernandez Sampieri, R., Fernández Collado, C., & Baptista Lucio, P. (2014). Investigation methodology. McGraw Hill.

Herrera, C. M. (2011). *Formula for calculating a sample of finite populations*. Retrieved from: https://investigacionpediahr.files.wordpress.com/2011/01/formula-para-cc3a1lculo-de-la-programa-poblaciones-finitas-var-categorica.pdf

INECC. (2007): Retrieved from: http://www2.inecc.gob.mx/publicaciones2/libros/402/cuencabalsas.html

INEGI. (2016). Retrieved on January 1, 2016, from https://www.inegi.org.mx/

Jensen, A. (2002). Artificial reefs of Europe: perspective and future. *ICES Journal of Marine Science, 59*, 3-13. Obtenido de: https://www.researchgate.net/publication/228903123Artificial_reefs_of_Europe_Prospective_and_future

Johannes, R.E. (2002). The renaissance of community based marine resource management in oceania. *Annu. Rev. Ecol. Systemat., 33*, 317–340. .ecolsys.33.010802.150524 doi:10.1146/annurev

Kerlinger, F. N. (1997). Behavior Research. McGraw Hill.

Lang, D. J. (2012). Transdisciplinary research in sustainability science: practice, principles, and challenges. *Sustainability Science, 1*, 25-43.

Leleu, K., Alban, F., Pelletier, D., Charbonnel, E., Letourneur, Y., & Boudouresque, C. F. (2012). Fishers' perceptions as indicators of the performance of Marine Protected Area (MPAs). *Marine Policy, 36*(2), 36. doi:10.1016/j.marpol.2011.06.002

Lizama, P., & Boccardo, G. (2014). *Variable Association Guide* (Pearson and Spearman in SPSS). University of Chile Faculty of Social Sciences (FACSO) Department of Sociology.

London Convention and Protocol/UNEP. (2009). *London Convention and Protocol/UNEP Guidelines for the Placement of Artificial Reefs*. Author.

Mancero, X. (2001). *The measurement of human development: Elements of a debate*. ECLAC.

Marin, G. (2004). The end of all the earth. College of Mexico.

Martens, P. (2006). Sustainability: Science or fiction? *Sustainability Science Practice Policy, 2*(2), 36–41. doi:10.1080/15487733.2006.11907976

Masera, O., & López-Ridaura, S. (2000). *Sustainability and peasant systems. Five evaluation experiences in rural Mexico.* MundiPrensa, GIRA, UNAM.

Mason, M. (2010). Sample size and saturation in PhD studies using qualitative interviews. Forum Qualitative Sozial for Schung/Forum Qualitative Social Research, 11.

Mondragon, B. (2014). Use of the Spearman correlation in an intervention study in physiotherapy. *Mov. cient., 8*(1), 98-104.

Moral, O., Teresa, M., Ferra, P., & Miguel, Q. L. (2013). Elaboration and validation of a scale of beliefs of students in compulsory secondary education regarding the environment. *Faculty Journal of Curriculum and Teacher Training, 17.*

Moreno, G., Dagorn, L., Sancho, G., García, D., & Itano, D. (2007). Using local ecological knowledge (LEK) to provide insight on the tuna purse seine fleets of the Indian Ocean useful for management. *Aquatic Living Resources, 20*(4), 367–376. doi:10.1051/alr:2008014

Polovina, J. J. (1991). *Fisheries applications and biological impacts of artificial habitats*. Academic Press.

Prescott-Allen, R. (1997). *Sustainability barometer. Measurement and communication of well-being and sustainable development* (No. 333.716 P929E). IUCN.

Relini, G., Relini, M., & Torchia, G. (1995). La Barrier artificiale di Loano. *Biologia Marina Mediterranea*, (2), 21–64.

Revenga, S., Fernández, F., González, J. L., & Santaella, E. (1997). Artificial reefs in Spain: the regulatory framework. In A. C. Jensen (Ed.), *European Artificial Reef Research* (pp. 161–174). Southampton Oceanography Centre.

Rojas, S. (1997). Guide to realize social investigations. Plaza and Valdez.

SAGARPA. (2010). *Enrichment of Seagrass Meadows by means of Artificial Reefs, as a Coastal Fishing Management and Conservation Strategy in the Municipalities of Aquila and Coahuayana, Michoacán*. SAGARPA.

Salas-Zapata, W. A. (2011, September). The emerging science of sustainability: From scientific practice towards the constitution of a science. *Interciencia, 36*(9), 699–706.

Schuschny, A., & Soto, H. (2009). *Methodological Guide. Design of composite indicators of sustainable development. Economic Commission for Latin America and the Caribbean*. ECLAC.

Sepulveda, S. (2002). Micro-regional Sustainable Development. In Sustainable development. Agriculture, Natural Resources and Rural Development. IICA.

Silvano, R. A. M., Gasalla, M. A., & Souza, S. P. (2009). Applications of Fishers' local ecological knowledge to better understand and manage tropical fitropica. In P. Lopes & A. Begossi (Eds.), *Current Trends Human Ecology* (pp. 76–100). Cambridge Scholars Publishing. doi:10.5848/CSP.0441.00004

Solomon, A.L. (2016). *Sampling techniques*. EAFIT University.

Tessier, Francour, Charbonnel, Dalias, Bodilis, Seaman, & Lenfant. (2015). Assessment of French artificial reefs: due to limitations of research, trends may be misleading. *Hydrobiologia, 753*, 1–29. doi:10.1007/s10750-015-2213-5

Texas Parks and Wildlife Department Artificial Reef Program. (2016). https://tpwd.texas.gov/landwater/water/habitats/artificial_reef/ 25/10/2017

UN. (1987). *Report of the world commission on environment and development*. Retrieved from: http://www.ecominga.uqam.ca/PDF/BIBLIOGRAPHIE/GUIDE_LECTURE_1/CMMAD-Informe-Comision-Brundtland-sobre-Medio-Ambiente-Development.pdf

Venegas, R. (2004). Property sustainability indicators. *Journal of Agroecology and Development*, (11/12).

Weiers, R. M. (1986). *Investigación de Mercados*. Prentice Hall Interamericana.

Yale Center for Environmental Law and Policy Yale University, Center for International Earth Science Information Network Columbia University, World Economic Forum, & Joint Research Centre European Commission. (2005). *2005 Environmental Sustainability Index (ESI)*. NASA Socioeconomic Data and Applications Center (SEDAC).10.7927/H40V89R6

Yamane, T. (1989). Status and future of artificial reef projects in Japan. *Bulletin of Marine Science*, *44*(2), 1038–1040.

Yeasmin, S., & Rahman, K. F. (2012). 'Triangulation' research method as the tool of social science research. *Bup J*, *1*(1), 154–163.

Chapter 5
Impacts of Artificial Reefs on Artisanal Fishers:
Southeastern Brazil

Juliano Silva Lima
Federal Institute of Sergipe, Brazil

Ilana R. Zalmon
University of North Rio de Janeiro, Brazil

ABSTRACT

The aim of this chapter is to report the perception of artisanal fishers regarding long-term implementation impacts of artificial reefs in the northern coast of Rio de Janeiro, Brazil. Data was collected through 60 interviews with local and regional artisanal fishers. The fishers stated an increase in fish abundance and species richness around the reef complex and consider it very important for local fisheries due to its ability to attract commercial species. These findings were clearly supported by their positive perception regarding the installation of reef balls and how these structures have benefited the local and regional fishing community over time (>25 years). In addition, the data confirms that the presence of different species in the reef complex is due to an increase of consolidated and stable substrate, which provides several types of prey and habitat for recruitment and shelter.

INTRODUCTION

The deployment of artificial reefs (AR) is a worldwide practice that aims to enrich fish stocks and mitigate the decline of fishery resources (Seaman & Lindberg, 2009; Becker et al., 2018). Since the 1960s until the 1990's, artificial reef had been deployed for fishery increases. After, artificial reef research has been centered on ecological topics. In contrast, studies focusing on socioeconomic aspects and the impacts on fishing communities are less frequent (Lima & Zalmon, 2020). However, they have contributed to better understand the role of different experimental structures and highlight the success of implementing artificial habitats to stakeholders (e.g., fishermen, divers, managers) (Ramos et al., 2011, 2019).

DOI: 10.4018/978-1-6684-2344-8.ch005

Although great advances have been made in AR science, many questions regarding performance, management, socioeconomic aspects, and impacts of these structures remain unanswered (Firth et al., 2016; Lee et al., 2018). AR are important elements used in coastal management planning in several countries (Bucaram et al., 2018; Reguero et al., 2018), although most studies on this topic have focused on marine environments in subtropical regions (Lima, Zalmon, et al., 2019).

The impacts of artificial structures on fishery resources have been mainly investigated through the lens of ecological aspects (Bull & Love, 2019; Liu et al., 2013). Although these studies demonstrate the application of these experimental structures for the recovery of aquatic habitats and to increment fish assemblages, studies that assess impacts on artisanal fishing since the implementation of AR are still scarce (Lima & Zalmon, 2020). Recent studies have highlighted the importance of fishing communities to understand the dynamics of these artificial habitats (Evans et al., 2017; Macusi, Katikiro, et al., 2017; Lima et al., 2018; Ramos et al., 2019). Historically, local ecological knowledge (LEK) of fishermen has been neglected by researchers and managers, despite the importance of this knowledge for fisheries management (Moreno et al., 2007; Macusi, Abreo, et al., 2017).

The compartmentalized approach seen in AR research, though essential to identify species' responses to attraction and colonization, does not allow a broader view, which is necessary for management actions (Hylkema et al., 2021). For instance, Brazilian policies for the implementation and management of AR (IBAMA/Brasil, 2020) have been scarcely applied, and the lack of dialogue between institutions and affected populations has yielded projects that are unrelated to local demands (Lima, Atalah, et al., 2020).

The lack of information about socioeconomic aspects and the effects of artificial structures on artisanal fisheries has made it difficult to manage these habitats considering a broader context. Thus, this chapter seeks to fill some gaps about artisanal fishermen and their relationship with AR implanted along the Brazilian coast in 1996. The aim of this chapter is to analyze the perception of artisanal fishermen regarding the influence of the AR on local and regional fishing and to evaluate its impact on fishery resources.

MAIN FOCUS OF THE CHAPTER

This chapter provides an overview of the perception of artisanal fishermen regarding the impacts of implementing an AR in the northern coast of the state of Rio de Janeiro, Brazil. This study was approved by the Human Research Ethics Committee of the Federal Institute of Sergipe, Brazil (CAAE 64939116.1.0000.8042).

Data was collected from 60 interviews carried out with local artisanal fishermen. The first interviewee was selected through collaboration with a local guide (president of the Z-1 Fishers' Colony). The snowball sampling technique was used to identify candidates for the second interviewee on, with each interviewee identifying the next interview candidate (Bailey, 2007). At times the snowball sampling was interrupted, a fisherman was randomly chosen in order to minimize the possible personal tendencies of indicating someone.

The interviews were conducted with a semi-structured questionnaire (Kendall, 2008), divided into three topics: (1) fishermen's profile, with questions regarding age, gender and fishing time in the region; (2) Artisanal fishing characteristics, with questions related to the vessels, the use of fishing gears and local direction to capture target species and (3) Perception of AR, with questions on their knowledge, the role of AR in artisanal fisheries and species found in areas with and without the AR influence.

For the taxonomic identification of fishery resources mentioned in the interviews, fish were either caught with gill nets or obtained from local markets after verifying where they came from (Froese & Pauly, 2021). Lobsters, oysters, and shrimps were sampled at local markets, upon confirmation with the fishermen about the site they were captured. Specimens were preserved in a solution of 10% formaldehyde. After identification, the species' common names were confirmed with the contribution of seven "local knowledge experts". These experts were defined as major fishing knowledge holders (Lima, Zappes, et al., 2019).

BACKGROUND

Artificial Reefs with Attractive Features for Artisanal Fisheries

Artisanal fisheries is an important economic activity in many fishermen communities, providing food security, employment and income, in addition to being part of the cultural identity of various social groups (Frid & Belmaker, 2019; Malcolm et al., 2021). In these practices, fishermen explore the coastline with small vessels (length: 5-13 meters) that have low autonomy and capture species for either subsistence or small-scale commercialization (Ramires et al., 2012; Oliveira et al., 2016). On the other hand, industrial fishery activities use medium (length: 13-20 meters) and large-sized vessels (length: >20 meters), operating on continental and oceanic waters, and their production is exclusively for trade (Boonstra et al., 2020).

Both the artisanal and industrial fisheries categories can be intensive when there is an uncontrolled exploitation of resources and they can affect environmental sustainability (Barbosa-Filho et al., 2020; Boonstra et al., 2020). Thus, coastal AR emerge as an alternative to maintain biodiversity by providing refuge and feeding areas for local species, and preventing intensive fishing and the use of bottom gears in shallow waters (Keller et al., 2017; Sun et al., 2017). In addition, these AR are also used to support artisanal fishermen because they can help in the maintenance and conservation of fish stocks (Baine & Side, 2003; Onyango et al., 2021).

Studies in Brazil and several other countries demonstrate the potential of AR to mitigate the decline of fish species (e.g., China: Wang et al., 2015; Italy: Scarcella et al., 2015; Mexico: Ajemian et al., 2015; Brazil: Lima, Sanchez-Jerez, et al., 2020). From an ecological standpoint, significant advances have been made regarding the role of AR in attraction-aggregation (Bohnsack, 1989; Smith et al., 2015) and/or production of fishery resources (Brickhill et al., 2005; Lindberg, 1997). From a socioeconomic perspective, the creation of AR has emerged as part of an alternative management strategy to increase productivity and income from aquatic systems (Hylkema et al., 2021; Ramos et al., 2021). Activities that apply this strategy include the creation of industrial fishing exclusion areas (Murillas-Maza et al., 2013) that benefit fishing communities (Lima, Zappes, et al., 2019; Ramos et al., 2019).

Initiatives such as the European Union's Water Framework Directive, the National Ocean Policy of the USA, and Australia's Oceans Policy have emphasized the use of comprehensive management strategies for aquatic environments, including guidelines on the strategic use of AR (Fabi et al., 2011; Pioch et al., 2020). In countries without regulations and guidelines on the deployment and monitoring of AR, it has been difficult to maintain the integrity of structures and there has been low efficiency in fishery production areas (Islam et al., 2014; Ramos et al., 2021).

Despite increased global interest to understand how AR can function in resource management, most studies have been conducted in temperate zones (Lee et al., 2018). Regional inequality regarding number of studies is another obstacle for a broader assessment of reef complexes under different geographic, political, and socioeconomic conditions (Ramos, Santos, et al., 2011; Sun et al., 2017). Considering that these reefs can help protect coastal areas, increase biodiversity, and enhance fishing, monitoring them in different countries becomes important in any attempt to fill existing gaps in data (Lima & Zalmon, 2020; Vivier et al., 2021).

The Artificial Reef on the Northern Coast of Rio de Janeiro

In southeastern Brazil, many artisanal fishing ports support the economy of traditional communities (Begossi, 2014; Musiello-Fernandes et al., 2017). The state of Rio de Janeiro (RJ) is ranked as the seventh largest fish producer in Brazil, and the northern coast of the state has six important artisanal fishing ports, of which the Guaxindiba port, located in the municipality of São Francisco de Itabapoana, is of particular relevance (FIPERJ, 2015). Most artisanal fishermen operate in coastal marine areas of the Campos Basin, covering areas near the shoreline to areas up to 120 km offshore (Lima et al., 2018).

The Guaxindiba port (21°29'S, 41°00'W) is characterized by shrimp, weakfish, croaker, catfish, and shark fisheries (Lima, Costa, et al., 2020). However, factors such as degradation of coastal environments, pollution of rivers and mangroves, oil activities, and the use of low-selectivity fishing gear have reduced fish stocks along the coast and are potential threats to artisanal fisheries in this area (Oliveira et al., 2016).

The northern coast of Rio de Janeiro is located in a transition zone between hot, oligotrophic waters and the cold, resurgent, nutrient-rich current from the South Atlantic (Valentin, 2001). Primary productivity (chlorophyll-a) is low, Secchi depth does not exceed 4 m, and strong bottom currents are common (Krohling & Zalmon, 2008). The seabed of the continental shelf in this region is composed of sand, mud, and biogenic sediment from the Paraíba do Sul and Itabapoana rivers (Murillo et al., 2009). During the rainy season (January to March), the estuarine plume of the Paraíba do Sul River stretches out 20 km offshore, which affects ecological processes in marine and local fish communities (Ovalle et al., 2013).

In 1996, AR were installed 5 km away from the community of Guaxindiba (21°29'S, 41°00'W) to improve artisanal fishing (Figure 1), with financial support by the Brazilian Agency for Research Development – CNPq and Research Foundation of Rio de Janeiro (FAPERJ). The reef modules were arranged in 12 groups of 3 units, 100 m from each other, and covering about 100,000 m^2. Each module of the reef is made by concrete, with a volume (1 m^3), weight 500 kg, designed with twelve holes, each measuring 20 cm in diameter (Santos et al. 2010). These reef balls were installed on a homogeneous flat bottom surrounded by rhodolith beds, approximately 9 m deep, creating effective spaces for attracting and providing shelter for many marine species (Zalmon et al., 2002; Santos & Zalmon, 2015).

The local sediment composition is mainly mud (60%), sand and gravel (~20% each) (Zalmon et al., 2012). The macrobenthos of the reef area comprised a total of 89 taxa, including 46 Annelida, 27 Mollusca, 13 Arthropoda (Crustacea) and 1 taxon each of Echinodermata, Echyura and Sipuncula. The annelids *Goniadides carolinae*, *Syllis* sp., *Lumbrineris latreilli*, *Exogone* sp. and *Paraonis gracilis gracilis* and the bivalve mollusks *Corbula caribaea* and *Crassinella martinicensis* represented 52% of the total abundance of infauna at the AR (Zalmon et al., 2014).

The installation of these modules was intended to increase the abundance and biomass of the local marine biota due to a previous lack of rocky and stable substrate for the settlement of benthic organisms (Santos et al., 2013). Studies in zoology and ecology related to fish and invertebrates associated with this AR have been carried out over the past 25 years in northern Rio de Janeiro (Zalmon et al., 2002; Brotto et al., 2006; Santos et al., 2011; Rocha et al., 2014; Santos & Zalmon, 2015; Lima, Sanchez-Jerez, et al., 2020). Some studies have shown that regional fishermen have used the local where the AR was deployed as an auxiliary fishing area to capture several target species (Lima et al., 2018; Lima, Zappes, et al., 2019).

Figure 1. Location of the Guaxindiba community - municipality of São Francisco de Itabapoana, and the artificial reef installed in northern Rio de Janeiro, southeastern Brazil (IGI, 2022)
Source: IGI, 2002

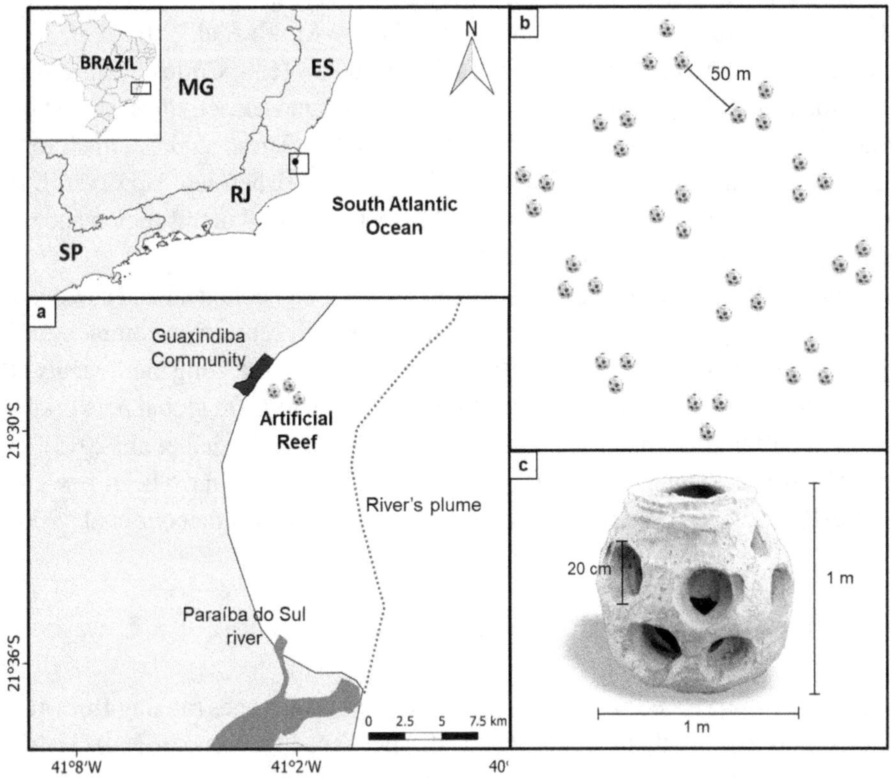

Knowledge of Artisanal Fishermen in the Management of Artificial Reefs

AR projects that disregard the perception of local communities tend to face greater difficulties when combining the demands of fishery resource exploitation with the maintenance of AR (Andriesse, 2018; Rouse et al., 2018). Ethnoecological studies have provided relevant information about the environment and exploited species because they consider the knowledge of subjects that are in direct and regular contact with natural resources (Ramos et al., 2019; Cisneros-Montemayor et al., 2020).

The LEK of artisanal fishermen provides knowledge about ecological, taxonomic, and ethological aspects of aquatic communities that are important for inventories, especially in places where scientific studies are scarce (Silvano & Begossi, 2012; Pinto et al., 2015; Macusi, Abreo, et al., 2017). Analyzing the interactions between fishermen and fishery resources is important to understand distinct forms of use, capture, categorization, and commercialization of target species (Gerhardinger et al., 2009). By using fishermen' knowledge it is possible to better understand, for example, aspects related to behavior and occurrence of species, as well as environmental impacts that affect them, which often go unnoticed by researchers (Lima & Zalmon, 2020).

Several studies have contributed to this subject, through daily observations of fishing activities, regular monitoring, and interviews with fishermen (Pinto et al., 2015; Ramos et al., 2019; Barbosa-Filho et al., 2020). These studies included ethnographic interviews and either *in situ* or fish market sampling to support the information provided by memory from the fishermen, and to characterize the socioecological systems related to the fishing activity (Moreno et al., 2007; Macusi, Abreo, et al., 2017). In addition, certain studies have examined the role of fishermen in understanding interrelations among people, natural resources, and temporal changes in the environment (Kirkbride-Smith et al., 2013).

Studies have been conducted along the coast of Rio de Janeiro (e.g., Côrtes et al., 2014; Fernandes et al., 2014; Oliveira et al., 2016; Lima, Zappes, et al., 2019) to analyze activities carried out by fishermen, socioeconomic aspects, and fishery resources management. These studies aimed to describe the main boats, fishing gear, and target species captured by artisanal fishermen. However, little is known about the relationship between artisanal fishermen and the AR deployed in this region and what are their impacts on local/regional fishing (Lima et al., 2018).

Although the guidelines for AR management depend heavily on ecological and socioeconomic concepts, the standards currently available for reef management are still subjective, complex, and difficult to access by managers. Another obstacle is related to the use of AR for protecting biodiversity (Pascaline et al., 2011). Some authors consider that areas with AR can be used to form global networks of protected areas, functioning as buffer zones against human impacts (Harmelin-Vivien et al., 2015), especially in overfished areas (Walker & Schlacher, 2014). Globally, however, there have been few circumstances where areas with AR have been used as biodiversity protection zones (Francour et al., 2001; Hylkema et al., 2021).

Description of the Artisanal Fishing Activities

The artisanal fishermen interviewed were mostly men (Table 1), with ages ranging from 40 to 78 years. The most common boats used by them are trawlers and canoes (paddle or motorized). Trawlers are used for short (up to 8 h) and long (up to 10 days) fishing trips. Motorized canoes are used for short fishing trips (up to 6 hours), and paddle canoes are used to transport cargo to trawlers, since there are no docks in the fishing community. The most used fishing gear is the bottom trawl net, followed by bottom line and hook, longline, bottom gillnet, purse seine net, trammel net, and surface gillnet (Table 1). The same fisher sometimes uses more than one fishing gear, and consequently the number of responses is greater than the number of interviewees.

Table 1. Description of the artisanal fishermen's profile (n = 60) and fishing gear in Guaxindiba community, northern Rio de Janeiro, Brazil (IGI, 2022)

Gender	n	%		Fishing gear	n	%
Male	59	98.3		Bottom trawl net	41	32.5
Female	1	1.7		Bottom line and hook	21	16.7
Age	**n**	%		Longline	19	15.1
40 - 49	20	33.3		Bottom gillnet	17	13.5
50 - 59	29	48.3		Purse seine net	12	9.5
60 - 69	8	13.3		Trammel net	9	7.1
70 - 79	3	5.0		Surface gillnet	7	5.6

Source: IGI, 2022. Several fishermen use more than one fishing gear.

The fishermen mentioned three distinct zones used for regional artisanal fishing: seashore (0 to 30 km from the shoreline), "malacacheta" (30 to 90 km), and open sea (over 90 km). In the seashore zone, fishermen indicated two important fisheries, the artificial reef (area with reef balls) and the estuarine area (mouth of the Paraíba do Sul River). The fishermen also highlighted some environmental conditions (wind, tides, and moon phases) as determining factors when planning fishing trips. Wind direction is the main predictor to conduct these activities near the AR (Figure 2).

Winds from the north, northeast, and southeast are considered the best for fishing (Table 2). Winds from the north lead several target species to the coast, while those from the northeast increase water turbidity, making it possible to catch shrimp and sharks. Winds from the southeast are less intense, which is good for fishing in areas farther from the coast. However, southern, southeastern, and continental winds cause the worst fishing conditions (Table 2). According to the fishers, wind from the south is very intense so they do not fish during this time (August and September). Though wind from the southeast is related to an increase in the availability of target species, strong bottom currents hamper the handling of fishing gear. Continental winds are considered calm, but they direct the target species away from the coast, which increases the time and cost of fishing.

Figure 2. Schematic view of the fishing area, types of boats, and wind classifications with their respective fishing characteristics according to artisanal fishermen in Guaxindiba community, northern Rio de Janeiro, Brazil (IGI, 2022)
Source: IGI, 2022

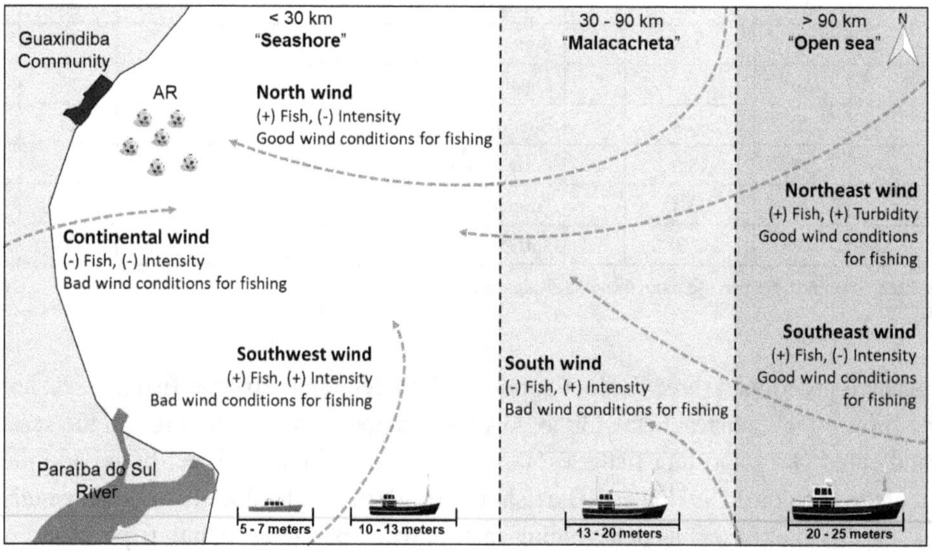

Table 2. Description of fishing conditions by wind type and tide according to the artisanal fishermen (n = 60) in Guaxindiba community, northern Rio de Janeiro, Brazil (IGI, 2022)

Wind	n	%	Fishing conditions	Wind speed	Fish catch rate
North	56	93.3	Excellent	12 - 19 km/h	High
Northeast	52	86.7	Excellent	20 - 28 km/h	High
Southeast	42	70.0	Good	1 - 5 km/h	Medium
Southern	41	68.3	Poor	38 - 49 km/h	Low
Southeastern	38	63.3	Poor	29 - 38 km/h	Medium
Continental	31	51.7	Fair	6 - 11 km/h	Low
Tide	**n**	**%**	**Fishing conditions**	**Tidal height**	**Fish catch rate**
Spring tide	52	86.7	Excellent	0.0 - 1.7 m	High
Neap tide	39	65.0	Good	0.5 - 1.2 m	Medium

Source: IGI, 2022

The artisanal fishermen considered spring tide or syzygy as the best condition to fish on the artificial reef compared to neap tide or quadrature (Table 2). The artisanal fishermen considered spring tide or syzygy as the best condition to fish on the artificial reef compared to neap tide or quadrature (Table 2). Full moon with a spring tide was reported as a favorable condition to fish for shrimp (*Artemesia longinaris* and *Xiphopenaeus kroyeri*) and when fish species move to regions nearest to the coast. Neap tide is suitable for catching species such as snooks (*Centropomus undecimalis* and *Centropomus parallelus*) and sharks (*Rhizoprionodon lalandii* and *Rhizoprionodon porosus*) because the moonlight is less intense on the water's surface. A waxing moon is characterized by fishing in regions further than 30 km from the coast due to weaker sea currents. Fishermen commonly associated a waning moon with the period of fish spawning in the estuary of the Paraíba do Sul River.

Characterization of Fishery Resources

The fishermen reported 27 species associated with the AR (AR: 24 fishes, two crustaceans and one mollusk) and 19 species were reported in the areas without the modules (AWM: 17 fishes and two crustaceans). According to the artisanal fishermen, since the AR was installed in 1996, 15 species of fish appeared in the region, mainly those with morphological characteristics adapted for the predation of larvae and juvenile fish, such as Pomatomidae and Centropomidae, and predation of organisms associated with substrates, such as Haemulidae and Sparidae.

The artisanal fishermen differentiate fish habits by their vertical distribution in the water column. Pelagic fish (AR: 7 spp.; AWM: 6 spp.) are represented by species that feed on the water surface and include species that move a lot in the water column, and bottom fish (demersal and benthic) (AR: 20 spp.; AWM: 13 spp.) are those which inhabit and feed within close proximity to soft and hard substrates (Figure 3a-b).

Figure 3a. Marine environment according to artisanal fishermen. Marine species related to the (A) artificial reef (AR) and (B) areas without modules (AWM). Water column distribution according to artisanal fishermen: pelagic, demersal and benthic (IGI, 2022). Source: IGI, 2022

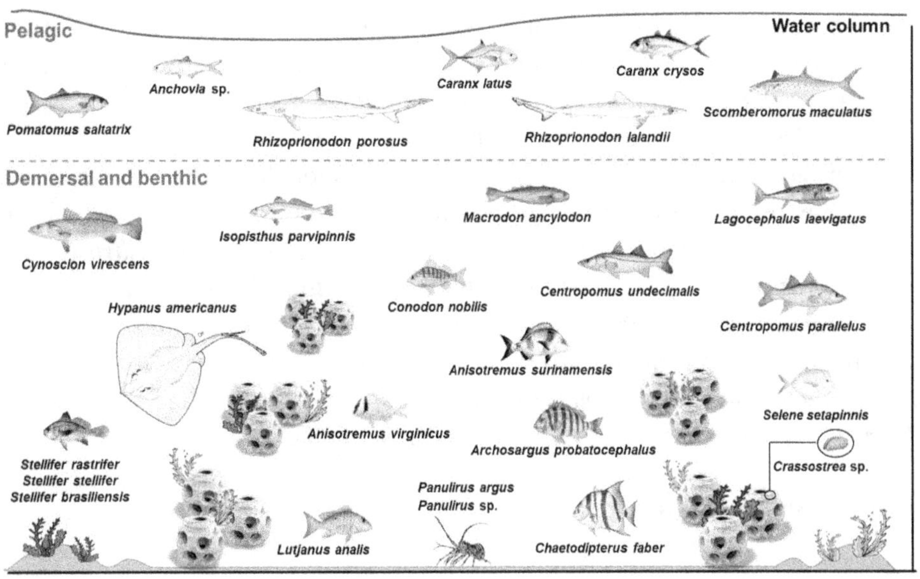

Figure 3b.

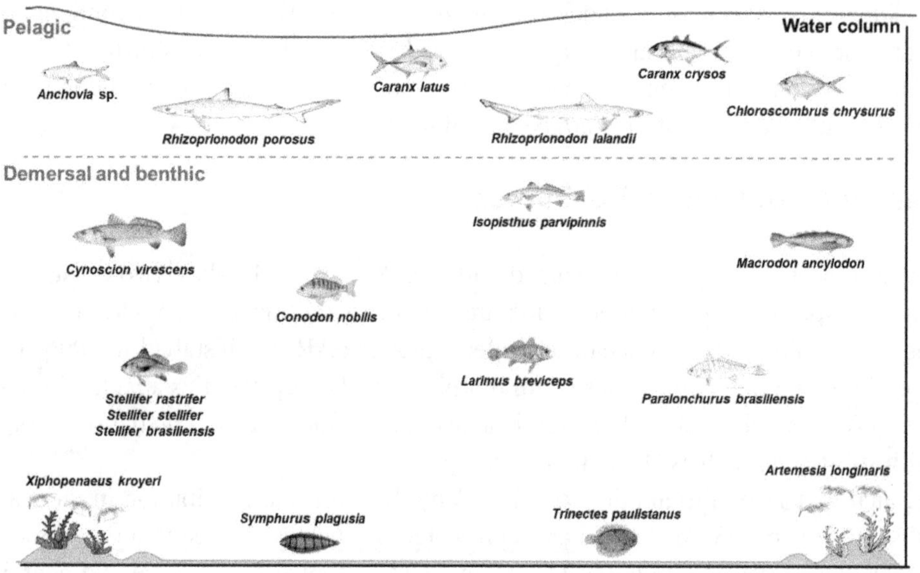

Based on the degree of fish reefiness (Nakamura, 1985), the species *Anisotremus surinamensis, Anisotremus virginicus, Archosargus probatocephalus, Lutjanus analis* and *Chaetodipterus faber* are classified as "Type A fishes", because they use the reef balls as habitats, particularly as feeding and reproduction areas, according the local fishermen. On the other hand, the species *Hypanus americanus, Lagocephalus laevigatus, Centropomus undecimalis, Centropomus parallelus, Cynoscion virescens, Isopisthus parvipinnis, Macrodon ancylodon, Conodon nobilis* and *Selene setapinnis* are classified as "Type B fishes", because they like to swim around the reef while remaining near the bottom. The other species *Rhizoprionodon porosus, Rhizoprionodon lalandii, Scomberomorus maculatus, Caranx latus, Caranx crysus* and *Pomatomus saltatrix* are classified as "Type C fishes", since they hover over the reef and access the modules by vertical migration in the water column.

Most species are associated with sandy sediment (18 spp.), though nine species were associated with hard substrate. These species are found only on the AR and have different ecological and economic characteristics, some of which are adapted to reef environments (e.g., the lobsters *Panulirus* sp. and *Panulirus argus*, the oyster *Crassostrea* sp. and the fishes *Lutjanus analis, Chaetodipterus faber, Anisotremus virginicus, Anisotremus surinamensis,* and *Archosargus probatocephalus*), some estuarine fish species of economic importance (e.g., *Pomatomus saltatrix, Lagocephalus laevigatus, Centropomus undecimalis, Centropomus parallelus,* and *Scomberomorus maculatus*), and others that are highly vulnerable (e.g., the sharks *Rhizoprionodon porosus* and *Rhizoprionodon lalandii*).

Importance of the Artificial Reefs

The artisanal fishermen stated an increase in fish abundance and species richness around the offshore AR (northern Rio de Janeiro, Brazil). They also consider the reef complex important for local fishing because it attracts commercial fish and due to the ecological role (feeding, breeding, and shelter) of these structures. For fishers, AR have several functions as an area for fish feeding, fish reproduction, promoting shelter, increasing fish weight, attracting fish to the coastline, research, and reducing industrial fishing (Table 3). The number of functions identified was greater than the number of respondents as some fishers described more than one function associated with the reef modules.

Table 3. Artificial reef functional role according to the artisanal fishermen (N=60) in Guaxindiba community, northern Rio de Janeiro, Brazil (IGI, 2022)

Artificial reef function	n	%
Fish feeding	26	22.4
Fish reproduction	21	18.1
Fish shelter	20	17.2
Increasing fish weight	15	12.9
Attracting fish to the coastline	13	11.2
Research	11	9.5
Reducing industrial fishing	10	8.6

Source: IGI, 2022. Several fishermen quoted more than one reef function.

AR were reported to have positive effects on local fishing, with 68.3% (n = 41) of fishermen stating that they increase the diversity of target species caught. These fishers suggested the installation of new reef modules when they identified the importance of these structures to increase species richness and, consequently, improve artisanal fishing conditions in the region. Several fishers (43.3%) state that they use AR as fishing grounds and attributed the presence of 13 species that had not occurred previously in the region to AR (Table 4). Among the fishery resources found in the AR, 19 are marketed at several fishing ports along the coast of Rio de Janeiro (e.g., *Panulirus* sp., *Panulirus argus*, *Centropomus undecimalis*, *Centropomus parallelus*, *Cynoscion virescens*, *Rhizoprionodon lalandii*, and *Rhizoprionodon porosus*) and nine species are considered non-target species (bycatch species). Nineteen species are found in areas without AR (Table 4), including nine target species, mainly the shrimps *Xiphopenaeus kroyeri* and *Artemesia longinaris*, and 10 fish species (bycatch) that are either discarded or marketed at low prices (value below one dollar/pound).

Due to the reef balls, only a few types of fishing gear are used there (e.g., surface gillnet, longline, and bottom line). Despite not being able to use bottom trawl nets, which are the most common equipment, the fishers do not feel hindered by AR, since artisanal fishers divide fishing regions according to fishery resources and fishing gear that can be used (Table 4). Shrimps traded in the northern coast of Rio de Janeiro (*Xiphopenaeus kroyeri* and *Artemesia longinaris*) are caught in areas without the reef modules, especially near the mouth of the Paraíba do Sul River, where it is possible to use bottom nets. These species are captured using bottom trawl nets and purse seine nets, which generate bycatches of a large number of juvenile fish or species with little or no commercial value.

Table 4. Local marine species captured on the artificial reef (AR) and areas without modules (AWM) according to artisanal fishermen (n = 60); Fishing gear: L - longline, BL - bottom line and hook, BT - bottom trawl net, SG - surface gillnet, BG - bottom gillnet, and PS - purse seine net(IGI, 2022)

Family/Species	Common name	AR	AWM	US$/Kg	Fishing Gear
Achiridae					
Trinectes paulistanus	Slipper sole		X	Discard	Bycatch
Carangidae					
Caranx crysos	Blue runner	X	X	1.7	L, SG
Caranx latus	Horse-eye jack	X	X	1.7	L, SG
Chloroscombrus chrysurus	Atlantic bumper		X	Discard	Bycatch
Selene setapinnis	Atlantic moonfish	X		0.2	Bycatch
Carcharhinidae					
Rhizoprionodon porosus	Caribbean sharpnose shark	X	X	4.2	L, BL, SG
Rhizoprionodon lalandii	Brazilian sharpnose shark	X	X	4.2	L, BL, SG
Centropomidae					
Centropomus undecimalis	Common snook	X		5.7	BL, BG
Centropomus parallelus	Fat snook	X		5.7	BL, BG
Cynoglossidae					
Symphurus plagusia	Duskycheek tonguefish		X	Discard	Bycatch
Dasyatidae					
Hypanus americanus	Southern stingray	X		0.6	L, BT, BG
Engraulidae					
Anchovia sp.	Anchovy	X	X	0.2	Bycatch
Ephippidae					
Chaetodipterus faber	Atlantic spadefish	X		1.7	BL, BG, RF
Haemulidae					
Anisotremus surinamensis	Black margate	X		2.9	BL, BG
Anisotremus virginicus	Porkfish	X		0.2	BL, BG
Conodon nobilis	Barred grunt	X	X	0.2	Bycatch
Lutjanidae					
Lutjanus analis	Mutton snapper	X		1.4	BL, L
Pomatomidae					
Pomatomus saltatrix	Bluefish	X		2.9	L, SG
Sciaenidae					
Stellifer rastrifer	Rake stardrum	X	X	0.2	Bycatch
Stellifer stellifer	Little croaker	X	X	0.2	Bycatch
Stellifer brasiliensis	Brazilian stardrum	X	X	0.2	Bycatch
Larimus breviceps	Shorthead drum		X	0.2	Bycatch
Paralonchurus brasiliensis	Banded croaker		X	0.2	Bycatch
Cynoscion virescens	Green weakfish	X	X	5.2	BL, BG, BT

Continued on following page

Table 4. Continued

Family/Species	Common name	AR	AWM	US$/Kg	Fishing Gear
Isopisthus parvipinnis	Bigtooth corvina	X	X	1.2	BT, BG, PS
Macrodon ancylodon	King weakfish	X	X	1.2	BT, BG, PS
Scombridae					
Scomberomorus maculatus	Atlantic Spanish mackerel	X		2.9	L, SG
Sparidae					
Archosargus probatocephalus	Sheepshead	X		2.9	BL, BG
Tetraodontidae					
Lagocephalus laevigatus	Smooth puffer	X		2.3	BL, L, BG
Other families					
Artemesia longinaris	Argentine stiletto shrimp		X	3.0	BT, PS
Xiphopenaeus kroyeri	Atlantic seabob		X	3.4	BT, PS
Panulirus sp.	Spiny lobster	X		8.6	BT, BG
Panulirus argus	Caribbean spiny lobster	X		8.6	BT, BG
Crassostrea sp.	Oyster	X		Discard	Bycatch

Source: IGI, 2022

SOLUTIONS AND RECOMMENDATIONS

The lack of basic information about the effects of AR on artisanal fisheries has made it difficult to manage these habitats at a broader scale (Lima, Atalah, et al., 2020; Ramos et al., 2021). In this context, the present chapter advances this discussion by analyzing the role of an AR from the perspective of artisanal fishers, elucidating the effects of AR in the increase of fishery resources. The experimental AR was deployed in 1996 and is located on the northern coast of the state of Rio de Janeiro, Brazil. Artisanal fishing is socially relevant in this community and directly influences the economy, targeting mainly shrimp and some fish species (e.g., *Isopisthus parvipinnis, Macrodon ancylodon, Cynoscion virescens,* and *Rhizoprionodon porosus*) (Lima, Zappes, et al., 2019).

Fishing efforts in northern Rio de Janeiro are concentrated along the coastal zone. Only the largest boats can reach the farthest fishing grounds ("malacacheta" and open sea) due to their greater autonomy (high speed, displacement capacity, and space to store fish). Living in this marine environment has enabled fishers to recognize and identify the best fishing grounds and environmental conditions to catch various target species. Their empirical knowledge about the dynamics of environmental parameters (winds, tides, currents, and moon phases) and the ethology of the target species allow fishers to understand the marine environment and, consequently, increase their chances of catching fishery resources (Begossi et al., 2016).

The LEK of artisanal fishers helped us to understand which species were attracted after the AR was installed and to understand species dynamics (Moreno et al., 2007; Ramires et al., 2015). Fishers' knowledge of the species associated with the reef modules is consistent and primarily learned through fishing, including the recognition of target species and relationship between the artificial reef and surrounding area. Twenty-seven species were recorded in the reef complex, and the spatial distribution of these species throughout the water column allowed some habitat categories to be assigned by the artisanal fishers. The fishers describe several benthic species, such as lobsters and oysters, maintaining direct contact with the reef balls, besides demersal species living around these structures and occasionally visiting their interior, while pelagic ones visit the AR to forage. The LEK of species align with data available in Froese and Pauly (2021) and suggests that the occurrence of different species in the AR is due to the increase of consolidate and stable substrates, which provided new prey and habitat for recruitment.

The implementation of the AR near the Guaxindiba fishing community resulted in benefits to the artisanal fishers and these structures emerge as an auxiliary fishing area to capture several commercial species (Lima et al., 2018). After the deployment of the reef balls, local fishers captured many target species (e.g., *Centropomus undecimalis*, *Centropomus parallelus*, *Hypanus americanus*, *Chaetodipterus faber*, *Anisotremus surinamensis*, *Lutjanus analis*, *Pomatomus saltatrix*, *Scomberomorus maculatus*, *Archosargus probatocephalus*, *Lagocephalus laevigatus*, *Panulirus argus* and *Panulirus* sp.) closer to the shoreline. The role of artificial reefs in the enrichment of fish stocks has also been demonstrated in others regions of the world (e.g., France: Tessier et al., 2015; Philippines: Macusi, Katikiro, et al., 2017; Scotland: Rouse et al., 2018; Portugal: Ramos et al., 2019).

Despite the frequent use of bottom trawling, this equipment is not used in the vicinity of the AR. Thus, this reef has provided refuge and feeding areas for local species, since the AR prevent the use of bottom gear and their negative impact on seabed health (Seaman & Lindberg, 2009). Practices that control the use of active gears should be encouraged for the Brazilian coast and the deployment of AR is suggested to contribute to the sustainable management of fishery resources (Lima, Atalah, et al., 2020).

In general, the artisanal fishers presented a positive perception about the AR, although there have been no social marketing initiatives to promote the benefits of the structures. This perception can be attributed to the proper installation area and the positive experiences of the fishers with reef balls. These fishers show interest in new AR, particularly due to high fishing demand. We highlight that fishing community participation is essential in the choice of additional areas for AR because artisanal fishers depend on the marine environment for sustenance and income generation. Choosing inappropriate locations and the non-involvement of fishers in the decision making process leads to conflicts between fishing communities and public authorities (Ramos et al., 2007; Islam et al., 2014). On the other hand, participatory management of the process of AR development and installation leads to successful actions (Evans et al., 2017; Whitmarsh et al., 2008).

Short-term measures for the management of experimental AR include activities with other fishing communities to share their knowledge about the role of AR, and the importance of maintaining these AR for local artisanal fisheries. Medium-term measures include mapping, with the local community, other areas for the installation of new AR and testing new materials and models capable of attracting other marine resources. Long-term measures involve the installation of new reefs and co-participatory management programs (managers and fishers) in the process of inspection and maintenance of AR. For the maintenance of fishing resources, we also suggest reducing the use of purse seine nets in the region, greater inspection of oil and port activities, and revitalization of local mangroves and both rivers. Such actions aim to contribute by providing information about the local population's relationship with artificial habitats, but also to fill gaps in the development of future operations for the deployment and monitoring of artificial reefs along the Brazilian coast.

FUTURE RESEARCH DIRECTIONS

Despite important advances in performance monitoring of AR, the lack of a standardized and global methodological protocol to assess the efficiency of these habitats in different regions has been an obstacle that needs to be overcome in future management studies (Becker et al., 2018; Onyango et al., 2021; Vivier et al., 2021). In response to this gap, substantial changes in the global management of AR with participation of local communities have been suggested in terms of implementation, monitoring and comprehensive management of these areas (Brochier et al., 2015; Lima & Zalmon, 2020).

Our analysis also reveals some important trends in AR research, such as a better understanding of fishers' interactions with AR and impacts to artisanal fishing. The implementation of this type of reef depends on more frequent and continuous studies combining the demands of fishery resource exploitation with the maintenance of local communities' interests (Andriesse, 2018; Rouse et al., 2018). Also, it is important to develop an evaluation protocol that includes socioeconomics aspects, conservation policies, and investment in science communication (Lima & Zalmon, 2020).

Future challenges imposed on the use and monitoring of AR offshore include improvements in interdisciplinary studies of fishing potential evaluation, classification of models and materials of reef types for different purposes (e.g., fishing, tourism, protected area), use of multimetric indices with ecological and socioeconomic data (e.g., ARMIr: Artificial reef global ranking index and AMRIe: Artificial reef local effect index), conservation issues, and reduction of conflicts with local communities (Lima, Atalah, et al., 2020; Ramos et al., 2021). The benefits of such approaches are evident in the improvement of management projects and the functional analysis of AR in different ecological, economic, and social contexts (Fletcher et al., 2011; Ng et al., 2013). In addition, it is helpful to support the co-participation of local people and public agencies in determining whether AR are an appropriate solution for local demands (Murillas-Maza et al., 2013).

Furthermore, strategies for the conservation of AR should include full protection of the structures and preventing illegal catches by artisanal and industrial fisheries (Ramos, Santos, et al., 2011; Tynyakov et al., 2017). Studies that project the occurrence and suppression of species, monitor fishing resources, and analyze the efficiency of reef complexes should be encouraged (Andriesse, 2018; Bortone et al., 2000). A conservation program in these areas is just one of the initiatives that provides an option for increasing fishery resources (Bucaram et al., 2018; Hylkema et al., 2021). Conservation programs combined with large-scale monitoring and a reduction in fishing fleets also have the potential to halt the current decline of fisheries and provide a sustainable increase in fishing production (Becker et al., 2018).

At a regional scale, despite the important advances of several studies carried out regarding the AR on the northern coast of Rio de Janeiro (Zalmon et al., 2002; Brotto et al., 2006; Santos et al., 2011; Rocha et al., 2014; Santos & Zalmon, 2015; Lima, Sanchez-Jerez, et al., 2020), some questions still remain unanswered. Future studies should prioritize the identification of aspects associated with the productivity of AR, revealing the capacity of AR to restore fish assemblages. It is also important to verify aspects related to the effect of fishing in areas with artificial structures over time and whether AR can be used as a protection area for vulnerable species. And finally, to analyze potential resilience of AR in the face of climate change processes. Such information can highlight the importance of implementing AR as areas that promote environmental goods and services, aiming for the production and/or conservation of fishery resources.

CONCLUSION

This chapter provides original information about the perception of artisanal fishermen regarding AR installed in the southern coast of Brazil. Fishermen have a positive perception of the AR, and this can be attributed to the location chosen for the AR, which allowed an increase in local fishing without conflicts with artisanal fishermen. In addition, our results indicate that the creation of a reef complex contribute to an increase in local and regional fish over time, and might provide new sites that potentially foster fishing in the northern coast of Rio de Janeiro. The data described in this study complement past data on ichthyofauna associated with AR in Rio de Janeiro (1996-2021), advance on issues related to artisanal fishing in artificial habitats, and identify the potential of adapted AR in attracting commercially important species. Finally, our results provide an important contribution to current and future plans for the management of AR by addressing how fishing communities interact with AR on the Brazilian coast.

ACKNOWLEDGMENT

The authors thank the Research Foundation of Rio de Janeiro (FAPERJ) and the Coordination for the Improvement of Higher Education Personnel (CAPES) for the PhD scholarship and the internship abroad scholarship, respectively. This research was supported by FAPERJ (E-26/200.941/2021)) and by the Brazilian Agency for Research Development-CNPq (301203/2019-9). This study was also partially financed by the CAPES (finance code 001).

REFERENCES

Ajemian, M. J., Wetz, J. J., Shipley-Lozano, B., Shively, J. D., & Stunz, G. W. (2015). An analysis of artificial reef fish community structure along the northwestern Gulf of Mexico shelf: Potential impacts of "rigs-to-reefs" programs. *PLoS One*, *10*(5), 1–22. doi:10.1371/journal.pone.0126354 PMID:25954943

Andriesse, E. (2018). Persistent fishing amidst depletion, environmental and socio-economic vulnerability in Iloilo Province, the Philippines. *Ocean and Coastal Management*, *157*(January), 130–137. doi:10.1016/j.ocecoaman.2018.02.004

Bailey, K. D. (2007). *Methods of Social Research* (4th ed.). Free Press.

Baine, M., & Side, J. (2003). Habitat modification and manipulation as a management tool. *Reviews in Fish Biology and Fisheries*, *13*(2), 187–199. doi:10.1023/B:RFBF.0000019480.95010.67

Barbosa-Filho, M. L. V., Seminara, C. I., Tavares, D. C., Siciliano, S., Hauser-Davis, R. A., & da Silva Mourão, J. (2020). Artisanal fisher perceptions on ghost nets in a tropical South Atlantic marine biodiversity hotspot: Challenges to traditional fishing culture and implications for conservation strategies: Ghost net impacts: Artisanal fisher perceptions and challenges to. *Ocean and Coastal Management*, *192*(March), 105189. Advance online publication. doi:10.1016/j.ocecoaman.2020.105189

Becker, A., Taylor, M. D., Folpp, H., & Lowry, M. B. (2018). Managing the development of artificial reef systems: The need for quantitative goals. *Fish and Fisheries*, *19*(4), 740–752. doi:10.1111/faf.12288

Begossi, A. (2014). Ecological, cultural, and economic approaches to managing artisanal fisheries. *Environment, Development and Sustainability, 16*(1), 5–34. doi:10.100710668-013-9471-z

Begossi, A., Salivonchyk, S., Lopes, P. F. M., & Silvano, R. A. M. (2016). Fishers' knowledge on the coast of Brazil. *Journal of Ethnobiology and Ethnomedicine, 12*(1), 20. doi:10.118613002-016-0091-1 PMID:27245047

Bohnsack, J. A. (1989). Are High Densities of Fishes at Artificial Reefs the Result of Habitat Limitation or Behavioral Preference? *Bulletin of Marine Science, 44*(2), 631–645.

Boonstra, W. J., Björkvik, E., Joosse, S., & Hanh, T. T. H. (2020). From Anthrome to Refugium? A Short History of Small-Scale Fisheries in the Anthropocene. Encyclopedia of the World's Biomes, 5, 180–187. doi:10.1016/B978-0-12-409548-9.11931-1

Bortone, S. A., Samoilys, M. A., & Francour, P. (2000). Fish and Macroinvertebrate Evaluation Methods. In W. Seaman Jr., (Ed.), *Artificial Reef Evaluation: With Application to Natural Marine Habitats* (pp. 127–164). CRC Press. doi:10.1201/9781420036633.ch5

Brickhill, M. J., Lee, S. Y., & Connolly, R. M. (2005). Fishes associated with artificial reefs: Attributing changes to attraction or production using novel approaches. *Journal of Fish Biology, 67*(SUPPL.), 53–71. doi:10.1111/j.0022-1112.2005.00915.x

Brochier, T., Auger, P., Thiam, N., Sow, M., Diouf, S., Sloterdijk, H., & Brehmer, P. (2015). Implementation of artificial habitats: Inside or outside the marine protected areas? Insights from a mathematical approach. *Ecological Modelling, 297*, 98–106. doi:10.1016/j.ecolmodel.2014.10.034

Brotto, D. S., Krohling, W., & Zalmon, I. R. (2006). Fish community modeling agents on an artificial reef on the northern coast of Rio de Janeiro - Brazil. *Brazilian Journal of Oceanography, 54*(39), 205–212. doi:10.1590/S1679-87592006000300004

Bucaram, S. J., Hearn, A., Trujillo, A. M., Rentería, W., Bustamante, R. H., Morán, G., Reck, G., & García, J. L. (2018). Assessing fishing effects inside and outside an MPA: The impact of the Galapagos Marine Reserve on the Industrial pelagic tuna fisheries during the first decade of operation. *Marine Policy, 87*(June), 212–225. doi:10.1016/j.marpol.2017.10.002

Bull, A. S., & Love, M. S. (2019). Worldwide oil and gas platform decommissioning: A review of practices and reefing options. *Ocean and Coastal Management, 168*(December), 274–306. doi:10.1016/j.ocecoaman.2018.10.024

Cisneros-Montemayor, A. M., Zetina-Rejón, M. J., Espinosa-Romero, M. J., Cisneros-Mata, M. A., Singh, G. G., & Fernández-Rivera Melo, F. J. (2020). Evaluating ecosystem impacts of data-limited artisanal fisheries through ecosystem modelling and traditional fisher knowledge. *Ocean and Coastal Management, 195*(February), 105291. Advance online publication. doi:10.1016/j.ocecoaman.2020.105291

Côrtes, L. H. de O., Zappes, C. A., & Di Beneditto, A. P. M. (2014). Ethnoecology, gathering techniques and traditional management of the crab Ucides cordatus Linnaeus, 1763 in a mangrove forest in southeastern Brazil. *Ocean and Coastal Management, 93*, 129–138. doi:10.1016/j.ocecoaman.2014.03.021

Evans, A. J., Garrod, B., Firth, L. B., Hawkins, S. J., Morris-Webb, E. S., Goudge, H., & Moore, P. J. (2017). Stakeholder priorities for multi-functional coastal defence developments and steps to effective implementation. *Marine Policy, 75*(October), 143–155. doi:10.1016/j.marpol.2016.10.006

Fabi, G., Spagnolo, A., Bellan-Santini, D., Charbonnel, E., Çiçek, B. A., García, J. J. G., Jensen, A. C., Kallianiotis, A., & dos Santos, M. N. (2011). Overview on artificial reefs in Europe. *Brazilian Journal of Oceanography, 59*(SPEC. ISSUE 1), 155–166. doi:10.1590/S1679-87592011000500017

Fernandes, L. P., Keunecke, K. A., & Di Beneditto, A. P. M. (2014). Produção e socioeconomia da pesca do camaração sete-barbas no norte do estado do Rio De Janeiro [Production and socioeconomic of the seven-barb shrimp fishery in the north state of Rio de Janeiro]. *Boletim do Instituto de Pesca, 40*(4), 541–555.

FIPERJ. (2015). Relatório da Fundação Instituto de Pesca do Estado do Rio De Janeiro. Secretaria de Desenvolvimento Regional, Abastecimento e Pesca.

Firth, L. B., Knights, A. M., Bridger, D., Evans, A. J., Mieszkowska, N., Moore, P. J., O'Connor, N. E., Sheehan, E. V, Thompson, R. C., & Hawkins, S. J. (2016). Ocean Sprawl: Challenges and Opportunities for Biodiversity Management in a Changing World. *Oceanography and Marine Biology: An Annual Review, 54*(September), 193–269.

Fletcher, S., Bateman, P., & Emery, A. (2011). The governance of the Boscombe Artificial Surf Reef, UK. *Land Use Policy, 28*(2), 395–401. doi:10.1016/j.landusepol.2010.08.002

Francour, P., Harmelin, J.-G., Pollard, D., & Sartoretto, S. (2001). Mediterranean region: Siting, usage, zonation and management. *Aquatic Conservation : Marine and Freshwater Ecocsystems, 11*(3), 155–188. doi:10.1002/aqc.442

Frid, O., & Belmaker, J. (2019). Catch dynamics of set net fisheries in Israel. *Fisheries Research, 213*(March), 1–11. doi:10.1016/j.fishres.2018.12.021

Froese, R., & Pauly, D. (2021). *FishBase.* www.fishbase.org

Gerhardinger, L. C., Godoy, E. A. S., & Jones, P. J. S. (2009). Local ecological knowledge and the management of marine protected areas in Brazil. *Ocean and Coastal Management, 52*(3–4), 154–165. doi:10.1016/j.ocecoaman.2008.12.007

Harmelin-Vivien, M., Cottalorda, J. M., Dominici, J. M., Harmelin, J. G., Le Diréach, L., & Ruitton, S. (2015). Effects of reserve protection level on the vulnerable fish species Sciaena umbra and implications for fishing management and policy. *Global Ecology and Conservation, 3*, 279–287. doi:10.1016/j.gecco.2014.12.005

Hylkema, A., Hakkaart, Q. C. A., Reid, C. B., Osinga, R., Murk, A. J., & Debrot, A. O. (2021). Artificial reefs in the Caribbean: A need for comprehensive monitoring and integration into marine management plans. Ocean and Coastal Management, 209(April), 105672. doi:10.1016/j.ocecoaman.2021.105672

Islam, G. M. N., Noh, K. M., Sidique, S. F., & Noh, A. F. M. (2014). Economic impact of artificial reefs: A case study of small scale fishers in Terengganu, Peninsular Malaysia. *Fisheries Research, 151*, 122–129. doi:10.1016/j.fishres.2013.10.018

Islam, G. M. N., Noh, K. M., Sidique, S. F., Noh, A. F. M., & Ali, A. (2014). Economic Impacts of Artificial Reefs on Small-Scale Fishers in Peninsular Malaysia. *Human Ecology, 42*(6), 989–998. doi:10.100710745-014-9692-2

Keller, K., Steffe, A. S., Lowry, M. B., Murphy, J. J., Smith, J. A., & Suthers, I. M. (2017). Estimating the recreational harvest of fish from a nearshore designed artificial reef using a pragmatic approach. *Fisheries Research, 187*, 158–167. doi:10.1016/j.fishres.2016.11.022

Kendall, L. (2008). The conduct of qualitative interview: research questions, methodological issues and researching online. In J. Coiro, M. Knobel, C. Lankshea, & D. Leu (Eds.), *Handbook of Research on New Literacies* (pp. 133–149). Lawrence Erlbaum Associares. doi:10.4324/9781410618894.ch5

Kirkbride-Smith, A. E., Wheeler, P. M., & Johnson, M. L. (2013). The Relationship between Diver Experience Levels and Perceptions of Attractiveness of Artificial Reefs - Examination of a Potential Management Tool. *PLoS One, 8*(7), e68899. Advance online publication. doi:10.1371/journal.pone.0068899 PMID:23894372

Krohling, W., & Zalmon, I. R. (2008). Epibenthic colonization on an artificial reef in a stressed environment off the north coast of the Rio de Janeiro State, Brazil. *Brazilian Archives of Biology and Technology, 51*(1), 213–221. doi:10.1590/S1516-89132008000100025

Lee, M. O., Otake, S., & Kim, J. K. (2018). Transition of artificial reefs (ARs) research and its prospects. *Ocean and Coastal Management, 154*(January), 55–65. doi:10.1016/j.ocecoaman.2018.01.010

Lima, J. S., Atalah, J., Sanchez-Jerez, P., & Zalmon, I. R. (2020). Evaluating the performance and management of artificial reefs using artificial reef multimetric index (ARMI). *Ocean and Coastal Management, 198*, 105350. Advance online publication. doi:10.1016/j.ocecoaman.2020.105350

Lima, J. S., Costa, I. D., & Zalmon, I. R. (2020). Length-weight relationship of fish species captured around an artificial offshore reef (northern Rio de Janeiro, Brazil). *Journal of Applied Ichthyology, 37*(2), 337–341. doi:10.1111/jai.14109

Lima, J. S., Sanchez-Jerez, P., Santos, L. N., & Zalmon, I. R. (2020). Could artificial reefs increase access to estuarine fishery resources? Insights from a long-term assessment. *Estuarine, Coastal and Shelf Science, 242*(May), 106858. Advance online publication. doi:10.1016/j.ecss.2020.106858

Lima, J. S., & Zalmon, I. R. (2020). Trends and Obstacles in Artificial Reef Research. In S. A. Bortone & S. Otake (Eds.), *Modern Fisheries Engineering - Realizing a Healthy and Sustainable Marine Ecosystem* (pp. 11–30). CRC Press. doi:10.1201/9780429328039-3

Lima, J. S., Zalmon, I. R., & Love, M. (2019). Overview and trends of ecological and socioeconomic research on artificial reefs. *Marine Environmental Research, 145*(January), 81–96. doi:10.1016/j.marenvres.2019.01.010 PMID:30837123

Lima, J. S., Zappes, C. A., Di Beneditto, A. P. M., & Zalmon, I. R. (2018). Artisanal fisheries and artificial reefs on the southeast coast of Brazil: Contributions to research and management. *Ocean & Coastal Management, 163*(July), 372–382. doi:10.1016/j.ocecoaman.2018.07.018

Lima, J. S., Zappes, C. A., Di Beneditto, A. P. M., & Zalmon, I. R. (2019). Ethnoecology and socioeconomic around an artificial reef: The case of artisanal fisheries from southeastern Brazil. *Biota Neotropica, 19*(2), e20180620. Advance online publication. doi:10.1590/1676-0611-bn-2018-0620

Lindberg, W. J. (1997). Can Science Resolve the Attraction- Production Issue? *Fisheries (Bethesda, Md.), 22*(4), 10–16. doi:10.1577/1548-8446-22-4

Liu, Y., Zhao, Y., Dong, G., Guan, C., Cui, Y., & Xu, T.-J. (2013). A study of the flow field characteristics around star-shaped artificial reefs. *Journal of Fluids and Structures, 39*, 27–40. doi:10.1016/j.jfluidstructs.2013.02.018

Macusi, E. D., Abreo, N. A. S., & Babaran, R. P. (2017). Local Ecological Knowledge (LEK) on Fish Behavior Around Anchored FADs: The Case of Tuna Purse Seine and Ringnet Fishers from Southern Philippines. *Frontiers in Marine Science, 4*(June), 1–13. doi:10.3389/fmars.2017.00188

Macusi, E. D., Katikiro, R. E., & Babaran, R. P. (2017). The influence of economic factors in the change of fishing strategies of anchored FAD fishers in the face of declining catch, General Santos City, Philippines. *Marine Policy, 78*(October), 98–106. doi:10.1016/j.marpol.2017.01.016

Malcolm, C. D., Olivas, M. L. B., & Dagostino, R. M. C. (2021). Reported capture, fishery perceptions, and attitudes toward fisheries management of urban and rural artisanal, small-scale fishers along the Bahía de Banderas coast, Mexico. *Environmental Challenges, 4*(April), 100110. doi:10.1016/j.envc.2021.100110

Moreno, G., Dagorn, L., Sancho, G., García, D., & Itano, D. (2007). Using local ecological knowledge (LEK) to provide insight on the tuna purse seine fleets of the Indian Ocean useful for management. *Aquatic Living Resources, 20*(4), 367–376. doi:10.1051/alr:2008014

Murillas-Maza, A., Moreno, G., & Murua, J. (2013). A socio-economic sustainability indicator for the basque tropical tuna purse-seine fleet with a FAD fishing strategy. *Economía Agraria y Recursos Naturales, 13*(2), 5–31. doi:10.7201/earn.2013.02.01

Murillo, V. C., Silva, C. G., & Fernandez, G. B. (2009). Nearshore Sediments and Coastal Evolution of Paraíba do Sul River Delta, Rio de Janeiro, Brazil. *Journal of Coastal Research, 56*, 650–654.

Musiello-Fernandes, J., Zappes, C. A., & Hostim-Silva, M. (2017). Small-scale shrimp fisheries on the Brazilian coast: Stakeholders perceptions of the closed season and integrated management. *Ocean and Coastal Management, 148*, 89–96. doi:10.1016/j.ocecoaman.2017.07.018

Nakamura, M. (1985). Evolution of artificial fishing reef concepts in Japan. *Bulletin of Marine Science, 37*(1), 271–278.

Ng, K., Phillips, M. R., Calado, H., Borges, P., & Veloso-Gomes, F. (2013). Seeking harmony in coastal development for small islands: Exploring multifunctional artificial reefs for São Miguel Island, the Azores. *Applied Geography (Sevenoaks, England), 44*, 99–111. doi:10.1016/j.apgeog.2013.07.013

Oliveira, P. C., Di Beneditto, A. P. M., Bulhões, E. M. R., & Zappes, C. A. (2016). Artisanal fishery versus port activity in southern Brazil. *Ocean and Coastal Management, 129*, 49–57. doi:10.1016/j.ocecoaman.2016.05.005

Onyango, H. O., Ochiewo, J. O., & Karani, N. J. (2021). Socio-economic prospects and problems in under-exploited offshore marine fisheries: The case of Fish Aggregating Devices (FADs) in Kenya coastal fisheries. *Regional Studies in Marine Science*, *44*, 101706. doi:10.1016/j.rsma.2021.101706

Ovalle, A. R. C., Silva, C. F., Rezende, C. E., Gatts, C. E. N., Suzuki, M. S., & Figueiredo, R. O. (2013). Long-term trends in hydrochemistry in the Paraíba do Sul River, southeastern Brazil. *Journal of Hydrology (Amsterdam)*, *481*(1), 191–203. doi:10.1016/j.jhydrol.2012.12.036

Pascaline, B., Catherine, S., Charbonnel, E., & Patrice, F. (2011). Monitoring of the artificial reef fish assemblages of golfe juan marine protected area (France, North-Western Mediterranean). *Brazilian Journal of Oceanography, 59*(SPEC. ISSUE 1), 167–176. doi:10.1590/S1679-87592011000500018

Pinto, M. F., Mourão, J. S., & Alves, R. R. N. (2015). Use of ichthyofauna by artisanal fishermen at two protected areas along the coast of Northeast Brazil. *Journal of Ethnobiology and Ethnomedicine*, *11*(1), 20. doi:10.118613002-015-0007-5 PMID:25888941

Pioch, S., Monbrison, D., & Simard, F. (2020). Artificial Reefs in France: Current State-of-the-Art and Recent Innovative Projects. In S. A. Bortone & S. Otake (Eds.), *Modern Fisheries Engineering* (1st ed., pp. 31–40). CRC Press. doi:10.1201/9780429328039-4

Ramires, M., Clauzet, M., Barrella, W., Rotundo, M. M., Silvano, R. A., & Begossi, A. (2015). Fishers' knowledge about fish trophic interactions in the southeastern Brazilian coast. *Journal of Ethnobiology and Ethnomedicine*, *11*(1), 19. doi:10.118613002-015-0012-8 PMID:25889226

Ramires, M., Clauzet, M., Rotundo, M. M., & Begossi, A. (2012). Artisanal Fishing and Fishermen of Ilhabela - Sao Paulo State - Brazil. *Boletim do Instituto de Pesca*, *38*(3), 231–246.

Ramos, J., Lino, P. G., Himes-Cornell, A., & Santos, M. N. (2019). Local fishermen's perceptions of the usefulness of artificial reef ecosystem services in Portugal. *PeerJ*, *2019*(1), 1–23. doi:10.7717/peerj.6206 PMID:30647998

Ramos, J., Oliveira, M. T., & Santos, M. N. (2011). Stakeholder perceptions of decision-making process on marine biodiversity conservation on Sal Island (Cape Verde). *Brazilian Journal of Oceanography, 59*(1), 95–105.

Ramos, J., Santos, M. N., Whitmarsh, D., & Monteiro, C. C. (2007). Stakeholder perceptions regarding the environmental and socio-economic impacts of the Algarve artificial reefs. *Hydrobiologia*, *580*(1), 181–191. doi:10.100710750-006-0454-z

Ramos, J., Santos, M. N., Whitmarsh, D., & Monteiro, C. C. (2011). Stakeholder analysis in the Portuguese artificial reef context: Winners and losers. *Brazilian Journal of Oceanography, 59*(SPEC. ISSUE 1), 133–143. doi:10.1590/S1679-87592011000500015

Ramos, J., Tuaty-Guerra, M., Almeida, M., Raposo, A. C., Gaudêncio, M. J., Silva, A. D., Rodrigues, N., Leandro, S. M., & Caetano, M. (2021). An artificial reef at the edge of the deep: An interdisciplinary case study. *Ocean and Coastal Management*, *210*(May), 105729. Advance online publication. doi:10.1016/j.ocecoaman.2021.105729

Reguero, B. G., Beck, M. W., Agostini, V. N., Kramer, P., & Hancock, B. (2018). Coral reefs for coastal protection: A new methodological approach and engineering case study in Grenada. *Journal of Environmental Management, 210*, 146–161. doi:10.1016/j.jenvman.2018.01.024 PMID:29339333

Rocha, D. F., Franco, M. A. L., Gatts, P. V., & Zalmon, I. R. (2014). The effect of an artificial reef system on the transient fish assemblages – south-eastern coast of Brazil. *Journal of the Marine Biological Association of the United Kingdom, 95*(03), 635–646. doi:10.1017/S0025315414001519

Rouse, S., Kafas, A., Catarino, R., & Peter, H. (2018). Commercial fisheries interactions with oil and gas pipelines in the North Sea: Considerations for decommissioning. *ICES Journal of Marine Science, 75*(1), 279–286. doi:10.1093/icesjms/fsx121

Santos, L. N., Brotto, D. S., & Zalmon, I. R. (2011). Assessing Artificial Reefs for Fisheries Management A 10-Year Assessment off Northern Coast of Rio de Janeiro. In S. A. Bortone, F. P. Brandini, G. Fabi, & S. Otake (Eds.), *Artificial reefs in fisheries management* (1st ed., pp. 125–140). CRC Press.

Santos, L. N., & Zalmon, I. R. (2015). Long-term changes of fish assemblages associated with artificial reefs off the northern coast of Rio de Janeiro, Brazil. *Journal of Applied Ichthyology, 31*, 15–23. doi:10.1111/jai.12947

Scarcella, G., Grati, F., Bolognini, L., Domenichetti, F., Malaspina, S., Manoukian, S., Polidori, P., Spagnolo, A., & Fabi, G. (2015). Time-series analyses of fish abundance from an artificial reef and a reference area in the central-Adriatic Sea. *Journal of Applied Ichthyology, 31*, 74–85. doi:10.1111/jai.12952

Seaman, W., & Lindberg, W. J. (2009). Artificial reefs. In Encyclopedia of Ocean Sciences (2nd ed., pp. 226–233). Elsevier Ltd. doi:10.1016/B978-012374473-9.00668-8

Silvano, R. A. M., & Begossi, A. (2012). Fishermen's local ecological knowledge on southeastern Brazilian coastal fishes: Contributions to research, conservation, and management. *Neotropical Ichthyology, 10*(1), 133–147. doi:10.1590/S1679-62252012000100013

Smith, J. A., Lowry, M. B., & Suthers, I. M. (2015). Fish attraction to artificial reefs not always harmful: A simulation study. *Ecology and Evolution. Bohnsack, 1989*(20), 4590–4602. doi:10.1002/ece3.1730 PMID:26668725

Sun, P., Liu, X., Tang, Y., Cheng, W., Sun, R., Wang, X., Wan, R., & Heino, M. (2017). The bio-economic effects of artificial reefs: Mixed evidence from shandong, China. *ICES Journal of Marine Science, 74*(8), 2239–2248. doi:10.1093/icesjms/fsx058

Tessier, A., Dalias, N., & Lenfant, P. (2015). Expectations of professional and recreational users of artificial reefs in the Gulf of Lion, France. *Journal of Applied Ichthyology, 31*, 60–73. doi:10.1111/jai.12951

Tynyakov, J., Rousseau, M., Chen, M., Figus, O., Belhassen, Y., & Shashar, N. (2017). Artificial reefs as a means of spreading diving pressure in a coral reef environment. *Ocean and Coastal Management, 149*, 159–164. doi:10.1016/j.ocecoaman.2017.10.008

Valentin, J. L. (2001). The Cabo Frio Upwelling System, Brazil. In U. Seelinger & B. Kjerfve (Eds.), *Coastal Marine Ecosystems of Latin America* (Vol. 144). Springer. doi:10.1007/978-3-662-04482-7_8

Vivier, B., Dauvin, J. C., Navon, M., Rusig, A. M., Mussio, I., Orvain, F., Boutouil, M., & Claquin, P. (2021). Marine artificial reefs, a meta-analysis of their design, objectives and effectiveness. *Global Ecology and Conservation*, *27*, e01538. doi:10.1016/j.gecco.2021.e01538

Walker, S. J., & Schlacher, T. A. (2014). Limited habitat and conservation value of a young artificial reef. *Biodiversity and Conservation*, *23*(2), 433–447. doi:10.100710531-013-0611-4

Wang, Z., Chen, Y., Zhang, S., Wang, K., Zhao, J., & Xu, Q. (2015). A comparative study of fish assemblages near aquaculture, artificial and natural habitats. *Journal of Ocean University of China*, *14*(1), 149–160. doi:10.100711802-015-2455-x

Whitmarsh, D., Santos, M. N., Ramos, J., & Monteiro, C. C. (2008). Marine habitat modification through artificial reefs off the Algarve (southern Portugal): An economic analysis of the fisheries and the prospects for management. *Ocean and Coastal Management*, *51*(6), 463–468. doi:10.1016/j.ocecoaman.2008.04.004

Zalmon, I. R., Boina, C. D., & Almeida, T. C. M. (2012). Artificial reef influence on the surrounding infauna-north coast of Rio de Janeiro State, Brazil. *Journal of the Marine Biological Association of the United Kingdom*, *92*(6), 1289–1299. doi:10.1017/S0025315411001147

Zalmon, I. R., Novelli, R., Gomes, M. P., & Faria, V. V. (2002). Experimental results of an artificial reef programme on the Brazilian coast north of Rio de Janeiro. *ICES Journal of Marine Science, 59*(March), S83–S87. doi:10.1006/jmsc.2002.1273

Zalmon, I. R., Sá, F. S., Dornellas Neto, E., Rezende, C. E., Mota, P. M., & Almeida, T. C. M. (2014). Impacts of artificial reef spatial configuration on infaunal community structure— Southeastern Brazil. *Journal of Experimental Marine Biology and Ecology*, *454*, 9–17. doi:10.1016/j.jembe.2014.01.015

Chapter 6
An Overview of the First National Artificial Reef Project of Turkey

Altan Lök
Ege University, Turkey

Aytaç Özgül
Ege University, Turkey

Vahdet Ünal
Ege University, Turkey

ABSTRACT

Although artificial reef applications in Turkey started in the 1980s, the acceptance of its use as a tool in fisheries management ended in the 2000s. The first legal regulation on artificial reefs entered into force in 2001, and the National Artificial Reef Program (NARP) was prepared in 2009. The program was aimed to have two main purposes: (1) create new fishing grounds to promote small-scale fisheries and (2) conserve biodiversity in the littoral zone. The first pilot project within the framework of the NARP was completed in February 2013 in Edremit Bay. With the completion of the deployment of the ARs, a 5-year monitoring program was initiated. In this chapter, the historical development of AR studies in Turkey, the NARP, and the evaluation of the first project within the framework of the program are discussed.

INTRODUCTION

There is a global awareness today that the earth's capacity to absorb man's loading is limited, so that we must immediately reduce the loading that we have created. Unless the current consuming and destructive attitude of mankind is reversed, the collapse of all ecosystems on earth is inevitable. In the face of this reality, we urgently need to radically change our consumption habits on the one hand, and protect and improve our ecosystems on the other (Nakamura & Otake, 2015).

DOI: 10.4018/978-1-6684-2344-8.ch006

Fisheries enhancement is a rapidly growing field that aims to increase fishery productivity and likely to become increasingly important in the future (Becker et al., 2018). Artificial reefs (ARs) have long been an important tool in the hands of fisheries management in protecting and enhancing marine ecosystems around the world (Stone, et al., 1991; Jensen et al., 2000; Seaman et al., 2011). Last fifty years AR building has become popular and scientists from many countries are studying the use of ARs in ecosystem conservation and fishery management. Indisputably, the most fundamental positive contribution of artificial reefs to the ecosystem is that they create a carrying capacity in the area where they are placed (Bortone et al., 1998).

AR studies in Europe are concentrated on the Mediterranean coast (Jensen et al., 2000) and the first applications were made in Italy in the early 1970s (Bombace et al., 2000). In these early periods, protecting *Posidonia oceanica* meadows from illegal bottom trawling and improving traditional fisheries were the main objectives of artificial reef projects in the Mediterranean (Bombace, 1989). Italy was followed by France, Monaco, Spain and Portugal, while Turkey took its place in the Mediterranean AR school at the end of the 1980s.

The first applications were small-scale, based on individual efforts, and later they attracted the attention of universities, local governments and non-governmental organizations. In the early periods, AR projects were carried out to support small-scale commercial and recreational fisheries. Later, new approaches were added to these goals, such as protecting biodiversity and developing diving tourism. As a result of these developments, the state initiated the work of the National Artificial Reef Program (NARP). In this section, the historical development of AR studies in Turkey, the NARP and the evaluation of the first project within the framework of the program are discussed. In addition, the problems encountered in this project and suggestions for the future are presented.

BACKGROUND

Historical Development of ARs in Turkey

AR applications in Turkey started in the 1980s. In 1983, for the first time by the Ege University Hydrobiology Research Center, pipe-shaped AR made of concrete material were placed in the vicinity of Taş Island in İzmir Bay. This practice was followed by the deployment of terra cotta, concrete and rubber into the sea as reefs by Beykoz Fisheries Industry Vocational High School and Istanbul Underwater Clubs (Cirik & Neşer, 1999). However, there is no written record of these practices.

Izmir Municipality realized an AR project in order to prevent illegal trawling and to support sportive angling in 1989. In this project, ten old trolleybuses that were scrapped by the Izmir Municipality were sunk in the middle bay of Izmir in pairs (Fig 1a). Since a preliminary study was not carried out before the project, the site selection was not made correctly and the reefs were buried after one year due to excessive sedimentation (Lök & Tokaç, 2000). Nevertheless, this Project has made positive contributions to the development of reefs in Turkey. The most important contribution of this unsuccessful study is to increase awareness on the subject and pave the way for scientific studies.

Figure 1. Newspaper clipping on the placement of trolleybuses(a) and first planned AR project in Turkey (b)

The first scientific study on ARs was carried out by Ege University Faculty of Fisheries between 1991-1993 within the framework of a doctoral thesis (Lök, 1995). In the study carried out on the coast of Hekim Island, located in the middle bay of İzmir, 30 artificial reef blocks constructed of reinforced concrete in hollow cubic design were used (Fig 1b). After the results of this study were published, especially the coastal local governments showed great interest in the subject. New cooperative projects were initiated among local governments, fishery cooperatives and universities. Thus, a joint movement emerged in the support and development of ARs and local governments provided financial support, universities provided scientific and technical knowledge and fisheries cooperatives provided practical information on local conditions (Lök & Tokaç, 2000; Lök et al., 2011).

Legislative Framework

Although AR applications started, no legal regulation existed until 1999. The Ministry of Agriculture and Rural Affairs, as it was called at that time, started to work to eliminate this deficiency. At the end of the meetings organized with the participation of the stakeholders, a "Project Guide for Artificial Reef Applications" was prepared. This guideline are still in use today and consist of six parts; 1-description, 2-procedures, 3-objectives, 4-site-selection criteria, 5- construction materials and 6- design aspects (Lök et al., 2011). Although still far from fully covering all biological and engineering aspects, the guideline represents an important starting point in Turkey (Lök et al., 2002)

Institutions or organizations that want to make an AR project in Turkey must obtain permission from the Ministry of Agriculture and Forestry, General Directorate of Fisheries and Aquaculture. The stages in the AR project process are progressing as shown in Figure 2. For this permit application, a research institution is required to prepare a Preliminary Study Report containing all the details of the project. After this report has been prepared within the framework of the Project Guide for Artificial Reef Applications, an application can be made.

Figure 2. AR project implementation process

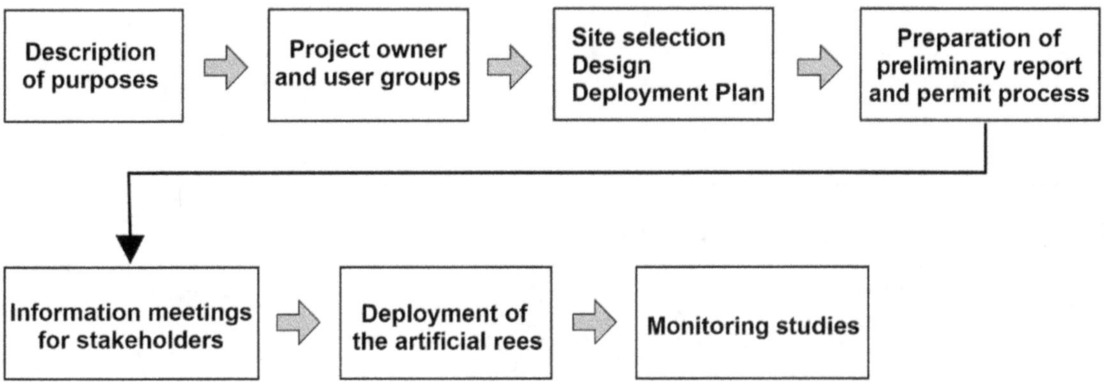

National Artificial Reef Program

The Ministry has decided to use ARs as a tool to protect coastal ecosystems that are important for sustainability. To this end, it started to work on preparing the NARP in 2008. In the meetings attended by decision makers and experts from universities, priority was given to determining the aims of the program and potential AR sites.

The Program was aimed to have two main purposes:

1. Create new fishing grounds to promote small-scale fisheries
2. Conserve biodiversity in the littoral zone

Three provinces with a coast to the Black Sea and all provinces with a coast to the Aegean Sea and the Mediterranean Sea were identified as potential AR areas.

As a first step, studies were carried out to determine the fisheries structure and oceanographic conditions in the identified potential provinces between September 2008 and March 2009. At the end of this phase, it was decided to deploy ARs on the coast of 8 provinces in total (Fig 3). In the next period, the AR project to be carried out in each province was planned to be carried out in three phases, pre-construction phase, construction phase and post-construction phase, as suggested in many guides (for example Fabi et al., 2015).

Figure 3. Provinces where AR projects will be implemented within the framework of the NARP

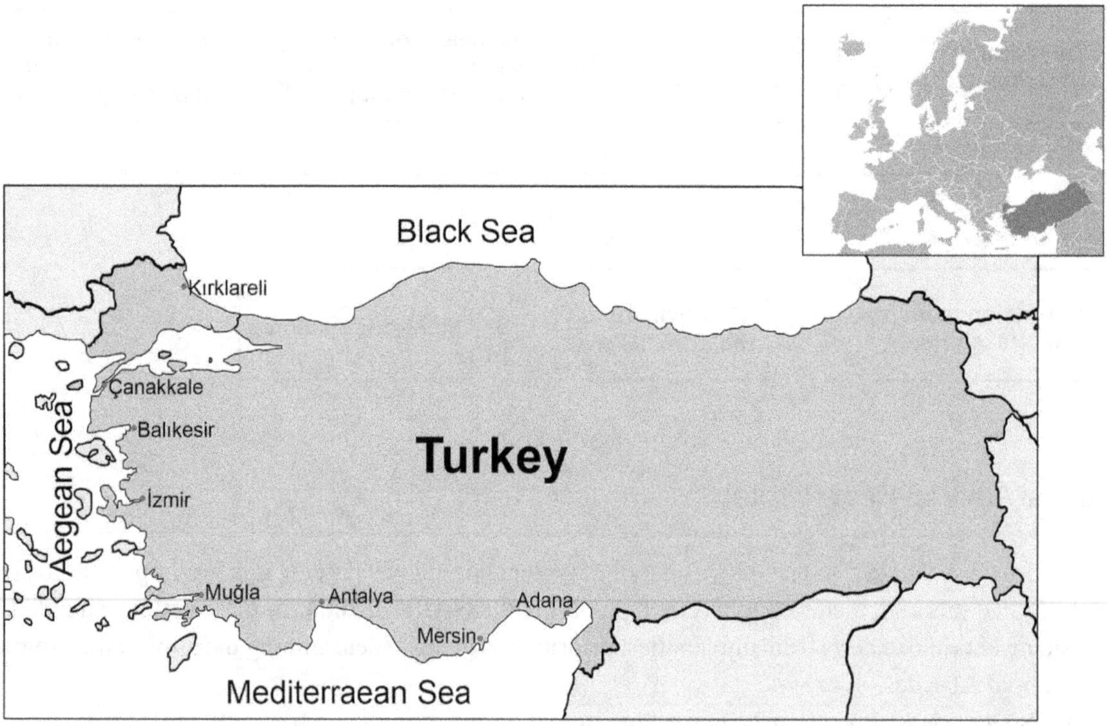

The year 2020 was targeted for the completion of the NARP in all designated areas. The pros and cons of the NARP will be discussed in the following sections.

FIRST NATIONAL AR PROJECT

Edremit Bay, located on the coast of the northern Aegean Sea of Turkey, was chosen as the first national project area. Many factors played a role in this choice. Edremit Bay has a high habitat diversity with its sea meadows, natural reefs, sandy and flat mud areas. Therefore, it is an ecologically and economically important bay that supports species diversity and productivity. Seagrass meadows cover the depths of the entire bay from 3m to 20m and provide shelter and feeding environment for many species. Edremit Bay, which is a spawning ground for the sardine species, is also an important habitat for other small pelagic species such as anchovy and horse mackerel. Rich populations of small pelagic fish species, predatory fish such as bluefish and marine mammals such as dolphins also attract to the Bay. In addition, its rich bottom habitat structure causes it to support important demersal species such as flatfish, shrimp, whiting and red mullet.

Rich habitat and species diversity in Edremit Bay supports local fisheries. Although large commercial fisheries such as purse seine and trawler operate in the region, small-scale fishing is quite common (Ceyhan et al., 2006).

In addition to all these, excessive construction on the Edremit Bay coasts, population growth especially in summer months, and the inoperability of wastewater treatment systems increase eutrophication and pollution on nearshore habitats. Another important problem in the region is the violation of the closed fishing areas by large-scale fishers and illegal fishing under the name of recreational fishing. Undoubtedly, all these factors mentioned above were effective in the selection of Edremit Bay as the first national AR area.

Pre-Construction Phase

From this point on, the one-year pre-construction phase determined at the planning stage had to begin. However, due to various administrative reasons, this process had to be completed within a month. Due to the short duration, studies on ecosystem and environmental factors, socio-economic effects, management plan could not be carried out.

In this process, the fishery structure was determined by holding meetings with the fishermen's cooperatives in the region, the local personnel of the ministry and the coast guard units. It was aimed to reveal the basic features of the ecosystem by making oceanographic measurements and underwater observations. Productive areas for fishing, natural reef areas, areas where seagrass is distributed, river mouths, military zones were excluded. Ultimately, seven rectangular areas where parallel to the shore were designated for the deployment of artificial reefs (Fig. 5). Fishermen took an active part in making site selection decisions.

It was decided to use reinforced concrete in the construction of ARs. Two different artificial reef designs were used for production and protection purposes (Fig. 4).

Figure 4. AR designs a) production reefs and b) protection reefs (Units are mm)

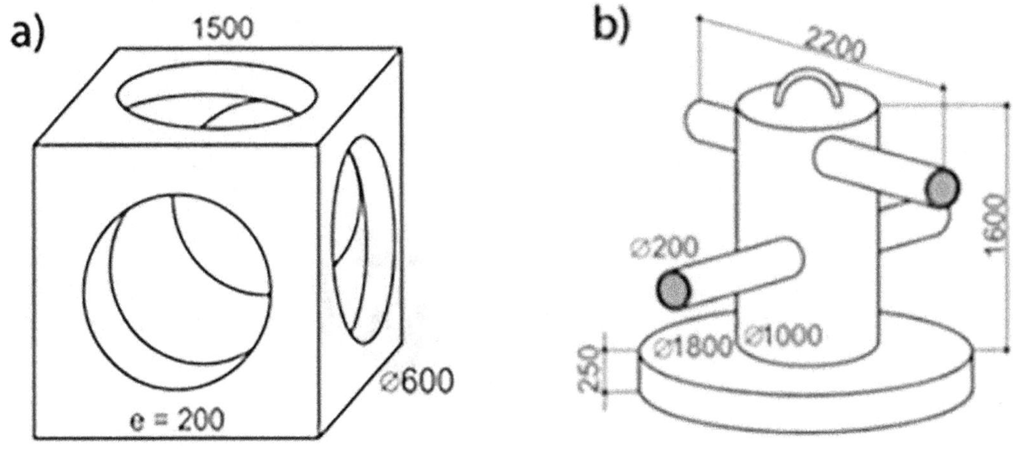

Construction Phase

The deployment plan of the ARs was made by placing protection-oriented designs on the outer line of the determined rectangular areas and protection-purposed designs on the inner part (Fig 5). AR units for production were placed in clusters. Each cluster consists of 30 units. An AR group was created from 3 clusters. Ten AR groups and their surrounding 35 protection reefs formed a rectangular AR system.

Figure 5. Deployment plan an AR system

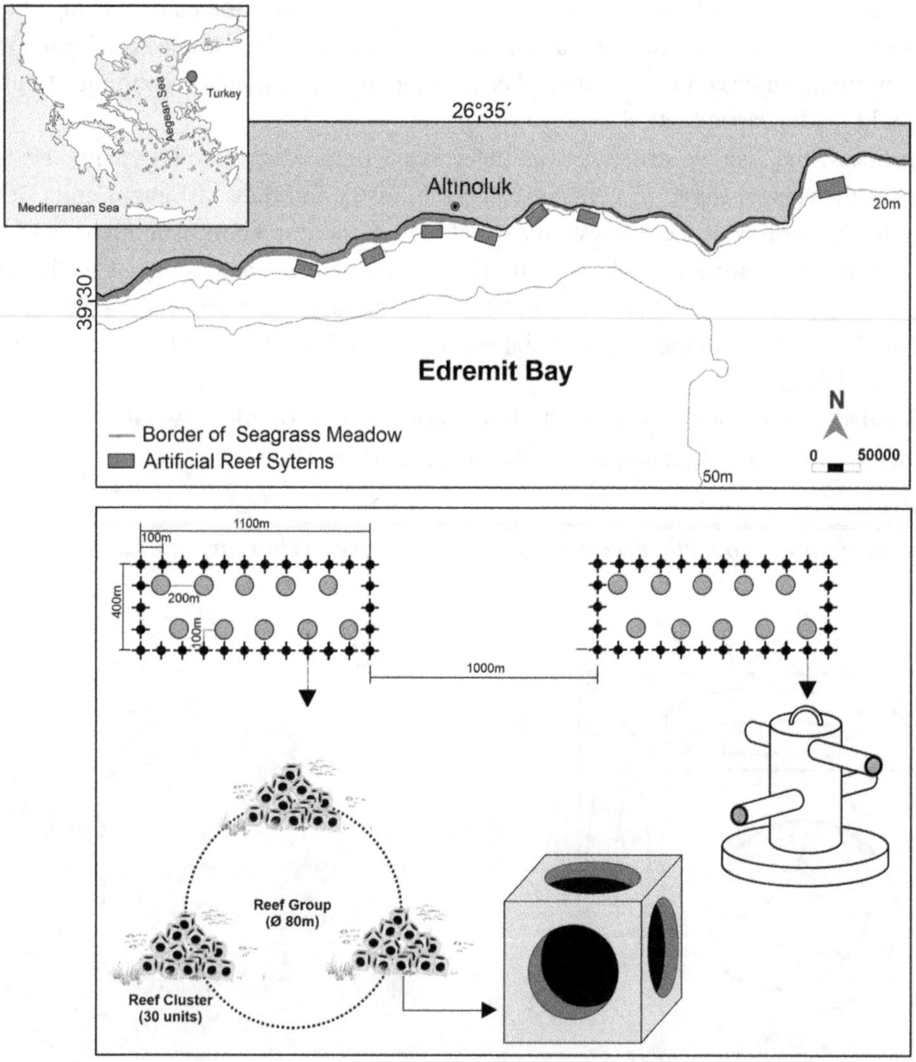

The AR units were constructed close to the seashore and transported to the deployment site by a pontoon. The location of each AR cluster was pre-determined by GPS and marked with buoys. AR units were placed at the points marked with buoys by free fall technique. In the project, a total of 7 AR systems consist of 6750 production and 250 protection reef units were planned, built and placed between 2011 and 2013 (Fig 6). Total bulk volume of the project is 25 000m^3.

Figure 6. AR units a) production b) protection units built and ready to ship

The corners of each rectangular AR system are marked with beacon buoys for navigational safety and management purposes. The final location of the AR clusters, border corners of each AR system and features of beacon buoys were reported to Turkish Naval Forces Office of Navigation, Hydrography and Oceanography.

Post-Construction Phase

While the planning phase of ARs is particularly important, it is not the end. ARs are planned, deployed, monitored and managed.

Deployment

The deployment of AR started in September-2011 and ended in February-2013. AR units were deployed at the coordinates determined in the pre-construction phase, in clusters containing 30 artificial reef units, using the free fall technique (Fig. 7). The details of the deployment plan are shown in the Fig. 5.

Figure 7. Deployment of artificial reef units

Monitoring

In the process after ARs are deployed, sustainable management and use of ARs and determination of success criteria are especially important. Therefore, a five-year monitoring study was planned to see if the national AR project was serving the purposes set out in the NARP.

In this section, the general results of the two work packages included in the monitoring project, determination of fish community structure and socio-economic impacts will be given. There are two other work packages carried out within the scope of the monitoring study but not included in this section. The first of these is the study of the biology of fish species around ARs, and the other is the abundance and distribution of ichthyoplankton.

Fish Community Structure

Studies to determine fish community structure were carried out seasonally for 5 years. Non-destructive underwater visual counting method was used to determine fish community structure. Because artificial reef clusters are large in volume, two divers performed the count at the same time. One diver counted around the base of the cluster and the other in the upper zone for 5 minutes. Samples were made at 6 stations and from 3 different AR clusters randomly selected at each station. In total, 18 artificial reef clusters from 6 stations were visited in one seasonal sampling.

A total of 51 fish species belonging to 20 families were determined. This result has higher values than other studies conducted in the Mediterranean (Table 1). The five most dominant species in terms of number of individuals are *D. vulgaris, C. chromis, S. umbra, S. maena* and *D. sargus* (Fig. 8). In terms of biomass, the five most dominant species are *D. vulgaris, S. umbra, D. sargus, D. puntazzo* and *D. dentex.* Sparidae, Labridae, Serranidae, Scorpaenidae and Mullidae were determined as the most dominant families.

Figure 8. Dominant fish species; D. vulgaris, C. chromis and S. umbra, respectively

A study was conducted using small scale fishing methods (Trammel net, long-line and line fishing) in a nationally supported project (TÜBİTAK-112O383) in the artificial reef area at the same time as the current monitoring study. When the results of this study carried out by fishing around AR and underwater visual census results are combined, a total of 72 fish species are reached. This figure covers 14% of Turkey's marine fish and 16% of fish in the Aegean Sea. This shows us that Edremit Bay ARs perform very well in terms of fish species diversity.

Table 1. Comparison of fish species and family numbers with other studies

Study area	Reef age	No Family	No species	Publication
Edremit Bay	5	20	51	Monitoring study
Hekim Island	8	15	29	Lök and Gül, 2005
Gümüldür-Ürkmez	7	15	39	Lök et al., 2011
Dalyanköy	10	13	35	Lök et al., 2011
Laona, Italy	12	16	44	Relini et al., 1999
Cote Bleue, France	8	13	36	Charbonnel et al., 2000

Sixty-one percent of fish species detected in AR areas have economic value. At the same time, economic species are dominant in terms of number of individuals and biomass. This result shows us that AR have the potential to support small-scale fisheries in the region (Fig. 9).

Figure 9. Fish species of high economic value caught from the AR site; from left to right: Dentex, lobster and blue fish

According to the report published by the IUCN in 2009 (Vié et al., 2009), the populations of *Dentex dentex* and *Sciaena umbra* species in the Mediterranean are classified as Vulnerable. *D. dentex* and *S. umbra* species, which are among the most important fish species in the Mediterranean in terms of ecological and economic aspects, are distributed throughout the AR system. This shows us that ARs provide suitable shelter and feeding environment for these two endangered species.

Socio-Economic İmpacts

With the completion of the deployment of the ARs (2013-February), a 5-year monitoring program was initiated. Monitoring of socio-economic impacts was achieved using original data from face-to-face surveys with stakeholders, particularly small-scale commercial fishers and recreational fishers, field observations, stakeholder meetings and cooperative records for the time period of 2013 and 2017. In addition to fishers, surveys were also conducted with the shop owners selling fishing gears and equipment, diving clubs and fishery cooperatives. However, in this chapter, only remarkable findings of commercial and recreational fishers are presented, and these findings are based on a survey of 598 fishers (295 commercial and 303 recreational) in total. According to general results of the 5-year monitoring program:

- A significant number of commercial fishers claimed that the ARs were not deployed as planned; they were thrown ashore more than necessary, which caused the narrowing of the fishing areas and the decrease in the amount of some species' catch.
- It was observed a positive change in the catch composition and catch quality, but a negative change in the catch amount.
- It has been observed that fishing equipment shop owners and dive center owners are generally satisfied with the reefs.
- Five years after the deployment of ARs, it has been observed that the reefs do not provide equal benefits to all stakeholders in the region, even to all fishers.
- Changes before and after ARs were compared in various aspects, both groups showed some satisfaction, but we know that recreational fishers are the most beneficiary group from ARs.
- Approximately 67.4% of recreational fishers stated that they are satisfied or absolutely satisfied with the ARs. Considering that 19.6% of the interviewed recreational fishers are moderately satisfied with the ARs, it is concluded that almost all recreational fishers are satisfied with ARs in some way. No doubt that this is desirable, but one of the main aims of the project was to support small-scale commercial fisheries. Therefore, what important was that satisfaction with ARs would also be high among commercial fishers rather than recreational fishers.
- One of the biggest problems with fisheries management and ARs is illegal fishing under the recreational umbrella. Many people who call themselves "recreational or amateur" perform this activity for commercial purposes. It has been observed that a large part of the "recreational fishers" in the region sell the fish they catch. This clearly creates unacceptable results in terms of unfair competition, black economy and fisheries management. Naturally, fishers in this group stand out as the most satisfied with the deployment of the reefs. It can be said that reefs constitute an alternative source of income for this group.

Lack of Management

More than a decade ago, Lök et al. (2011) reported that an artificial reef management plan (ARMP) was prepared by the ministry together with the scientists and will be open to users for discussion. Management options such as gear restrictions, no-take zones, spatial segregation, and "reef guard" personnel were accepted in the ARMP. According to the authors, harvest rotation and closed season regulations were under discussion. The main idea accepted by all bodies of the ARMP is that management options should be simple, and easily applicable. However, after years, there is still no AR in Turkey managed according to a plan. There is also no management advisory committee of any AR. Results of the monitoring program indicated that this situation was also not changed in the national AR project. However, in the monitoring reports prepared each year during the five-year monitoring program, recommendations were made to the Directorate General of Fisheries and Aquaculture regarding the establishment of an advisory committee of ARs management and the preparation of an AR management plan. Although it has been four years since the monitoring studies ended and the final report was submitted to official institutions, it was observed that there was no progress regarding reef management or the establishment of a management advisory committee in the pilot project area. This finding shows that the support given to artificial reef deployment by official institutions is not given to their management in Turkey.

GAPS AND CHALLENGES

There was not enough time allocated for the planning and preliminary studies of the Edremit Bay Artificial Reef Project, which was carried out within the framework of the NARP. Some of the consequences of this situation can be listed as follows:

- Failure to provide necessary information and exchange of ideas with all stakeholders,
- Project objectives are not defined very clearly and target-oriented.,
- Inability to determine the clear criteria to be used to evaluate the success of the project, since data to reveal the biological and environmental factors and the current situation of fisheries were not collected before the project,
- Due to the limited budget, monitoring can be done with a limited scope.

Like other similar applications in the aquatic environments, ARs may require post-installation management to make sure that they provide the desired outcomes for both biological resources and user groups (Fabi et al., 2015). In Edremit ARs, while a management plan was supposed to be developed before the ARs are deployed, it was prepared neither before nor after the deployment. One of the important problems caused by this situation is the rupture of the buoys indicating AR areas and the associated risks. The absence of marker buoys not only made it difficult to control the AR area, but also caused some fishing gear to get stuck on ARs and harmed the fishermen.

In multi-purpose AR applications, it is difficult to achieve success if there are no legal regulations determining site-specific fishing activities (Fabi et al., 2015). There are two restrictions that regulates fishing in AR areas. The first is fishing gear restriction. Only line and longline fishing is permitted in the AR area. The second is segregated area closure. The goal of this regulation is to ban four of the seven existing AR areas for fishing and to release three of them, and to replace them every two years. Apart from these regulations, for example, there are no regulations regarding user rights user groups. This lack of regulation can cause conflicts among user groups

DISCUSSION

There is neither a management plan, nor management objectives and criteria determined by stakeholders for measuring success have been defined for ARs in Edremit Bay. According to Fabi et al., (2011), fish stock enhancement and fishery management are the main goals for ARs in the Mediterranean Sea, while nature conservation & restoration, research, and recreation have been the main purposes in the other European regions to date. Although both conservation and fishery management were stated in the purpose of the national ARs project of Turkey, the results showed that none of these goals were reached entirely. However, the limited monitoring studies carried out in the region indicated that ARs created some support for biodiversity and small-scale fisheries including also recreational fisheries. On the other hand, the lack of a management plan for the ARs related areas is a major shortcoming. Such a deficiency causes undesirable side effects such as illegal fishing of artificial reefs and, conflicts between user groups. Fabi et al., (2011) pointed out the occurrence of the same problem in many countries. According to the authors, there is a noteworthy lack of management plans, in many countries, for the reefs after their deployment. Indeed, management measures aimed at regulating/limiting access to the reefs are essential to ensure the maintenance of reef resources over time and to avoid possible conflicts between the different users (Fabi et al., 2011). Therefore, ARs should be managed based on a plan. Monitoring, control, and surveillance, which are the most important components of the management, are of vital importance in order to provide sustainable benefits from the reef. So, we propose the creation of an advisory committee of artificial reef management, although not yet established and too late, accompanied by an adaptive artificial reef management plan. If one of the major objectives of the ARs Project was to support small-scale fisheries (as stated by the responsible government agency), then reef management can be considered within the fisheries management and some special regulations specific to the region should be introduced. A locally established fisheries management advisory committee or reef management advisory committee coordinated with the fisheries management authority may be helpful to develop and implement these specific regulations. It can be discussed within such committee that the reefs should be divided into parcels that recreational fishing should be subject to strict measures and regulations like commercial fishing, that a ban on fishing on certain days, and their monitoring should be followed. Establishing a unit under the ministry responsible for artificial reef control and inspection, which has examples in some coastal countries, and ensuring management is extremely important in terms of maximizing the benefits to be obtained from ARs.

According to Cardenas et al. (2011), ARs have the potential to increase fish landings or yields among artisanal fisheries, especially small-scale fishers in coastal areas. Similarly, Santos et al. (2011) reported that in southern Portugal, the use of artificial reefs was a response to problems related to fisheries management, as traditional management measures were insufficient in guaranteeing small-scale fisheries sustainability. However, we observed in the present study that the emergence of this positive potential of ARs is related to the successful planning and then effective management of ARs.

CONCLUSION

Edremit Bay AR project is facing difficulties and deficiencies of lack of management but on the other hand, this project created a particularly important laboratory area for future applications. The emergence of such a laboratory could be one of the most important benefits of the project. Therefore, ensuring that this laboratory will be used for scientific and educational purposes seems to be the future benefit of the project. In order to realize more successful AR applications in the future, lessons should be learned from the mistakes made in this project and scientific studies in this field should be supported.

ACKNOWLEDGMENT

We are grateful to many people for their help during the project. Thanks former and present managers and presidents of fishery cooperatives. A special thanks regional and central fisheries management authority for their support during the field works. This report would not have come out if these people had not cooperated. We thank all of them very much as well as all fishers who cooperated and answered the survey. We would also like to thank our colleagues, students and boat personnel who worked with devotion and made it possible for us to make underwater observations. But the biggest thanks to Halil Ataç who is the former head of Altınoluk Fishery Cooperative. His assistance always was extremely helpful.

REFERENCES

Becker, A., Taylor, M. D., Folpp, H., & Lowry, M. B. (2018). Managing the development of artificial reef systems: The need for quantitative goals. *Fish and Fisheries*, *19*(4), 740–752. doi:10.1111/faf.12288

Bombace, G. (1989). Artificial reefs in the Mediterranean Sea. *Bulletin of Marine Science*, *44*, 1023–1032.

Bombace, G., Fabi, G., & Fiorentini, L. (2000). Artificial reefs in the Adriatic Sea. In A. C. Jensen, K. J. Collins, & A. P. M. Lockwood (Eds.), *Artificial Reefs in European Seas* (pp. 31–63). Kluwer Academic Publishers. doi:10.1007/978-94-011-4215-1_3

Bortone, S. A., Cody, R. P., Turpin, R. K. & Bundrick, C. M. (1998). The impact of artificial-reef fish assemblages on their potential forage area. *Italian Journal of Zoology*, 265-267.

Burt, J., Bartholomew, A., & Firth, L. (2021). Policy and management considerations for artificial reefs in the Arabian Gulf. *Sheikh Saud bin Saqr Al Qasimi Foundation for Policy Research. Policy Paper*, *51*, 1–31.

Cardenas, E. R. B. (2011). Artificial Reefs in Artisanal Fisheries. In S. A. Bortone, F. P. Brandini, G. Fabi, & S. Otake (Eds.), *Artificial Reefs in Fisheries Management* (pp. 209–220). CRC Press Taylor & Francis Group.

Ceyhan, T., Akyol, O., & Ünal, V. (2006). Edremit Körfezi (Ege Denizi) Kıyı Balıkçılığı Üzerine Bir Araştırma [An investigation on the artisanal fisheries in Edremit Bay (Aegean Sea)]. *Ege University Journal of Fisheries & Aquatic Science*, *23* (1/3), 373-375.

Cirik, Ş., & Neşer, M. (1999, July). *Ülkemiz deniz teknolojisinde yeni bir uygulama alanı: yapay barınaklar* [A new field of application in marine technology in our country: artificial reefs]. *National Congress on Shipbuilding and Marine Technology*, İstanbul, Turkey.

Fabi, G., Scarcella, G., Spagnolo, A., Bortone, S. A., Charbonel, E., Goutayer, J. J., Haddad, N., Lök, A., & Trommelen, M. (2015). Practical Guidelines for the Use of Artificial Reefs in the Mediterranean and the Black Sea. FAO-GFCM, Studies and Reviews, No. 96.

Fabi, G., Spagnolo, A., Bellan-Santini, D., Charbonnel, E., Çiçek, B. A., García, J. J. G., Jensen, A. C., Kallianiotis, A., & Neves dos Santos, M. (2011). Overview on Artificial Reefs in Europe. *Brazilian Journal of Oceanography*, *59*(spe1), 155–166. doi:10.1590/S1679-87592011000500017

Jensen, A. C., Collins, K. J., & Lockwood, P. M. (2000). Current Issues Relating to Artificial Reefs in European Seas. In A. C. Jensen, K. J. Collins, & A. P. M. Lockwood (Eds.), *Artificial Reefs in European Seas* (pp. 489–499). Kluwer Academic Publishers. doi:10.1007/978-94-011-4215-1_29

Jensen, A. C., Collins, K. J., & Lockwood, P. M. (2000). Introduction and background to 'Artificial Reefs in European Seas'. In A.C. Jensen, K. J. Collins & A. P. M. Lockwood (Ed.), Artificial Reefs in European Seas (pp. ix-xii). Kluwer Academic Publishers.

Lök, A. (1995). *Yapay resiflerin uygulanabilirliği üzerine bir araştırma* [A study on the feasibility of artificial reefs] [Ph.D. dissertation]. Ege University, Science Institute, İzmir, Turkey.

Lök, A., Düzbastilar, F. O., Gül, B., Özgül, A., & Ulaş, A. (2011). The role of artificial reefs in fisheries management in Turkey. In S. A. Bortone, F. P. Brandini, G. Fabi, & S. Otake (Eds.), *Artificial Reefs in Fisheries Management* (pp. 155–166). CRC Press Taylor & Francis Group.

Lök, A., Metin, C., Ulaş, A., Düzbastılar, F. O., & Tokaç, A. (2002). Artificial reefs in Turkey. *ICES Journal of Marine Science, 59*, 192–195. doi:10.1006/jmsc.2002.1221

Lök, A., & Tokaç, A. (2000). Turkey: A new region for artificial habitats. In A.C. Jensen, K. J. Collins & A. P. M. Lockwood (Ed.), Artificial Reefs in European Seas (pp. 21-30). Kluwer Academic Publishers.

Nakamura, M., & Otake, S. (2015). Enhancement of marine habitats. In International Compendium of Coastal Engineering (pp. 375-396). World Scientific Publishing Co. Pte. Ltd. doi:10.1142/9789814449434_0011

Santos, M. G., Monteiro, C. C., & Leitão, F. (2011). The Role of Artificial Reefs in the Sustainability of Artisanal Fisheries. In S. A. Bortone, F. P. Brandini, G. Fabi, & S. Otake (Eds.), *Artificial Reefs in Fisheries Management* (pp. 221–238). CRC Press Taylor & Francis Group.

Seaman, W. Jr, Grove, R., Whitmarsh, D., Santos, M. N., Fabi, G., Kim, C. G., Relini, G., & Pitcher, T. (2011). Artificial reefs as unifying and energizing factors in future research and management of fisheries and ecosystems. In S. A. Bortone, F. P. Brandini, G. Fabi, & S. Otake (Eds.), *Artificial Reefs in Fisheries Management* (pp. 7–30). CRC Press Taylor & Francis Group. doi:10.1201/b10910-3

Stone, R. B., McGurrin, J. M., Sprague, L. M., & Seaman, W. Jr. (1991). Artificial Habitats of the World: Synopsis and Major Trends. In W. Seaman Jr & L. M. Sprague (Eds.), *Artificial Habitats for Marine and Freshwater Fisheries* (pp. 31–60). Academic Press Inc. doi:10.1016/B978-0-08-057117-1.50008-1

Vié, J.-C., Hilton-Taylor, C., & Stuart, S. N. (Eds.). (2009). Wildlife in a Changing World – An analysis of the 2008 IUCN Red List of Threatened Species™. IUCN.

KEY TERMS AND DEFINITIONS

Edremit Bay: It is one of the largest bays located in the north of Turkey's Aegean Sea coast.

Long-Line: It is a fishing method that involves setting out a large number of short lines carrying hooks, which are attached to a longer main line at regular intervals. The short lines are suspended horizontally at a predetermined depth with the help of surface floats.

Multi-Propose Artificial Reef: Multipurpose ARs can be designed to provide a range of functions and benefits to the community and environment. Multipurpose artificial reefs can provide shelter and feeding habitats for marine life, for example, while preventing illegal bottom trawling.

National Artificial Reef Program: It is a program prepared between 2008 and 2009 by the relevant ministry of the state in Turkey to carry out artificial reef applications within the framework of a specific plan.

Recreational Fisheries: Recreational fishing is defined as fishing of fish species that do not constitute the individual's primary resource to meet basic needs.

Small-Scale Fisheries: Small-scale fisheries can be broadly characterized as a dynamic and evolving sub-sector of fisheries employing labor intensive harvesting, processing and distribution technologies to exploit marine and inland water fishery resources.

Trammel Net: A trammel net consists of two or three layers of netting with a slack small mesh inner netting between two layers of large mesh netting within which fish will entangle. This fishing method use out Using this fishing method within the artificial reef area is prohibited. But it can be used outside the field boundaries.

Section 3

Dive Tourism Use

Chapter 7
Portuguese Artificial Reef Modules:
Are They Suitable for Dive Tourism?

Jorge H. P. Ramos
https://orcid.org/0000-0002-5115-7919
CinTurs, University of Algarve, Portugal

ABSTRACT

Artificial reefs (AR) for diving must have certain characteristics that are not activity-limiting. Likewise, these characteristics are generally fundamental for good diving experiences. In this aspect, AR must not be at a great depth, and they must have a diverse bio-colonization by several species of various taxa and attract diver interest. In Portugal, Algarve (South) and Nazaré (Centre) reef modules were deployed at depths that can be considered as potential for diving. In this chapter, an approach is made using the analytic hierarchy process (AHP) and fuzzy logic (a method called fuzzy analytic hierarchy process: FAHP) defining what is the likely interest for potential diving tourism activities in each of the locations presenting AR modules. The results obtained by using these methods allowed to further discover the importance of AR in Algarve and Nazaré.

INTRODUCTION

The concept of artificial reefs (AR) originates in Japan and dates to a few centuries ago (White et al., 1990). However, it was only in the 20th century that large-scale projects began to be developed (Sheehy, 1982; Thierry, 1988). The countries that developed reef programs were Japan for commercial fishing purposes and the US primarily for recreational activities (McGurrin et al., 1989; Polovina & Sakai, 1989; Bohnsack, 1991). Since the 1960s, AR have started to expand to other geographies and are currently dispersed all over the world (e.g., Bohnsack & Sutherland, 1985; Fabi & Spagnolo, 2011; Bortone & Otake, 2020).

AR are in coastal areas, lakes, or other bodies of water and serve different purposes depending on the entities involved in their construction and implementation (Frank, 2018). One of the fundamental

DOI: 10.4018/978-1-6684-2344-8.ch007

issues that enter the decision process is whether there is a need to implement AR (Gordon Jr. & Ditton, 1986). This decision depends on the expressed willingness of the stakeholders and the balance between demand and supply in the market (e.g., Sutton & Bushnell, 2007; Tunca et al., 2014).

AR are often used as protection tools, deployed on the seabed to divert fishing or recreational practices that may cause damage (Streich, 2017; Tynyakov, 2017). Another function of AR is to provide protective shelter for small marine organisms in transitional waters between nurseries and the ocean (Santos et al., 2005). Reef modules can also have another function, named as serving for recreational diving (e.g., Stolk et al., 2007; Tunca et al., 2012; Bideci & Cater, 2019).

In the case of diving, it is important to understand whether the tourism industry is looking for alternatives or complements to the places that are already visited (Oliveira et al., 2015); and knowing what kind of structures can be implemented and which locations are the most suitable for doing so (Kirkbride-Smith et al., 2013).

The use of AR for diving is particularly relevant in several respects. Specifically, when the structures are deployed to divert divers from places with natural reefs to places artificialized with these structures, but which in a way mimic the natural reefs and thus attract dive tourists (e.g., Perkol-Finkel et al., 2006; Salaün et al., 2020). There are many types of functional structures for certain purposes of protecting living organisms and mitigating potentially harmful practices in the marine environment (e.g., anti-trawling modules) (e.g., Ramos-Esplá et al., 2000), but not all reef structures are consistent to serve tourist diving (Ramos et al., 2006).

The first artificial reef modules in Portugal were supported by the fisheries and aquaculture research institute (then INIP) and were implemented in 1990 (Santos et al., 1995). The last AR were placed in 2010 and were backed by a municipality with the support of the fisheries and aquaculture research institute (then INIAP) (Ramos et al., 2019). The first purposes of these modules were to protect mollusks, crustaceans, and young fish of various species, to serve as a complement to the diversification of fisheries in small-scale commercial fisheries, and eventually to provide diverse ecosystem services and other functions. Among these other functions, we highlight the promotion of recreational activities, such as diving (Almeida et al., 2016).

The purpose of this chapter is to assess the AR or reef sites in Portuguese waters that can be used for diving activities and to know what criteria can influence the decision to choose these AR. For this, a simplified technique of the fuzzy analytic hierarchy process (FAHP) is used.

BACKGROUND

Artificial reefs for diving purposes must fulfill certain requirements to be viable for the operators and their clientele. These requirements are relevant criteria in decision-making for diving activity.

Diving Goals (Demand and Supply)

The diving industry has reached a considerable dimension, involving many players, thanks to the stimuli provided by great adventurers such as Commander Cousteau (Cousteau & Dumas, 1953). On an economic level, the diving market has a variety of peculiarities, either on the demand side or on the supply side (Lemke & Olech, 2011; Dimopoulos et al., 2019).

On the demand side, what instigates the diver to enjoy the hobby is the fact that in this activity, the practitioner can be free from the daily routine (Palau-Saumell et al., 2019). The practice of diving allows the diver to experience sensations with almost no gravity and move in all directions, going up and down. In addition, the diver can experience an aquatic habitat and have the feeling that is part of it. The diver tries to get the maximum satisfaction out of the activity, both in terms of the place visited and in terms of personal and interpersonal development goals (Cater, 2007).

On the supply side, there are several types of motivation for diving (e.g., recreational, commercial, scientific). There is an entire industry associated with diving that also includes host communities (Dimmock & Musa, 2015). In recreational diving, companies try to offer their customers a diverse range of possibilities. Generally, the diving supply has to do with the diver's experience (Kirkbride-Smith et al., 2013). For less experienced divers, shallow depth is a particularly crucial factor to consider, while for an experienced diver, diving in deeper waters allows for a larger number of places and access to other challenges (Dimmock, 2007). The dive operator always tries to provide the best possible experience for their clientele, giving them the best feasible options to practice the activity (Kirkbride-Smith et al., 2016). On the supply side, the existence of AR makes it possible to expand the diversification of dive sites and alleviate the pressure on natural reefs (Verkoeyen & Nepal, 2019).

Criteria Used in the Decision to Dive

Several studies indicate that diving AR must have certain attributes/criteria that are not activity-limiting themselves (MacCarthy et al., 2006; Kirkbride-Smith et al., 2013; Schuett et al., 2016). Likewise, these characteristics are generally fundamental to providing good diving experiences (Lök et al., 2002; Bideci & Cater, 2019).

Regarding biodiversity by organisms, there must be good colonization by several species of different trophic levels with good levels of diversity and generate the interest of the diver (Moura et al., 2006; Encarnação & Calado, 2018; Özgül et al., 2019). In the geographical criteria, dive sites must have good accessibility and the potential existence of conflicting activities fits into an important criterion to be considered and avoided. The most pleasant places for diving are eventually those where it is easier to practice the activity; places, where there are other activities or limited access (e.g., fishing or aquaculture production areas), are less desirable and access becomes more difficult (Pears & Williams, 2005; Shani et al., 2012). Bathymetry is a very limiting factor for a diver, where a more experienced person can go down to a depth of about 40 m (Loske 2013), while an inexperienced diver should not go deeper than 18 m (Hunt, 1995; Johansen, 2013). Regarding atmospheric criteria, currents can be limiting, as places with strong flows are dangerous as they often jeopardize the safety of divers, such as surface streams that can lead to seasickness while onboard, and deeper streams make it difficult for the diver to maneuver near the bottom (Newman, 2011). Visibility underwater is an especially key factor to take advantage of the satisfaction that a good dive can give because if visibility is poor, the dive is not as pleasant (Garrod, 2007; Thirumoorthi et al., 2013). Atmospheric conditions depend much on the chosen locations and the season of the year, because where/when there is harsh weather and more uncomfortable temperatures, they require greater training on the part of divers and become less sought after by less experienced practitioners or those with less propensity for this type of diving (Giddy & Rogerson, 2018). In terms of economic criteria, the clientele is essential because places that have much tourist demand are more likely to provide diving activities in the vicinity, tour operators that supply diving activities in these places try to get a good list of spots to visit including artificial reefs if they are nearby (Uyarra et al., 2009).

Options/Alternatives: The Reef Modules in Portugal

In Portugal, on the Algarve coast (South) and in Nazaré (Centre) reef modules were deployed. Off the Algarve, were deployed between 1990 and 2003 (Santos & Monteiro, 2007), while off Nazaré reefs were deployed in 2010 (Almeida et al., 2016).

Pilot AR – In the Algarve, the pilot project consisted of the Faro and Olhão AR whose objective was to allow the protection of juveniles of various species of mollusks, crustaceans, and fish in the transition from the sheltered inland waters of the Ria Formosa (where they are born) to the oceanic waters where they are more exposed to predators and more adverse conditions (in later stages of development and adults). There were post-deployment studies in which it was verified that the reef presence was an asset, both for fish aggregations (Santos & Monteiro, 1997) and fish catch or production (Santos & Monteiro, 1998). These sets had post-deployment follow-up and monitoring in several ways (e.g., Santos et al., 2005; Whitmarsh et al., 2008).

Large-Scale AR – As the pilot AR results were positive, for example at the level of biological colonization by marine species (mainly fish, but also other organisms), it was decided to move towards a larger-scale project. Thus, from 1998 to 2003, many modules were implemented in several reef sets in two phases (1st phase from 1998 to 2001 and the 2nd phase from 2002 to 2003). A few of these sets had post-deployment follow-up and monitoring – e.g., Faro-Ancão (Falcão et al., 2007; Lino et al., 2009; Moura et al., 2011). In others, follow-up was scarcer: Tavira and Cacela (Santos et al., 2011), Vilamoura (Ramos & Santos, 2015), Oura (Ramos et al., 2019), Alvor (Cancela da Fonseca et al., 1997).

Nazaré AR – In Nazaré pre-deployment studies were carried out in 2008 (Dias, 2011). The placement of the structures was done in 2010. Post-deployment studies took place in 2015 and covered several aspects, including perception and comparison of ecosystem services and other functions by potential and actual users (Ramos et al., 2019), and colonization assessment (i.e., biological colonization of structures, an abundance of organisms around, experimental fisheries, scientific diving), and socioeconomic impact (i.e., the importance of the AR to the local community) (Ramos et al., 2021).

AHP and Fuzzy Logic

Every day people face decision analysis problems, which are more complex the more factors they include. In cases where it is difficult to adequately quantify the preferences of certain criteria over the alternatives (choice options), the analytic hierarchy process (AHP) is a technique that can be used (e.g., Ananda & Herath, 2003). AHP was developed to facilitate decision analysis. AHP is an approach to decision-making that uses pairwise comparisons and allows decision-makers to select and make a better choice among several alternatives (Saaty, 2008). AHP can consider and cover both quantitative and qualitative aspects in decision-making. It allows synthetically reducing complex decisions, which eases the decision process to reach a certain goal (Smith & Tighe, 2006).

Thomas Saaty developed this technique in the 1970s (Saaty, 1977; Saaty & Erdener, 1979). It is a technique widely used internationally by different academics and practitioners (Maleki & Zahir, 2013). The potential of this technique is exceptionally large, and it is possible to use it in multi-criteria decisions (Huang et al., 2011), planning (e.g., Abdullah & Najib 2016), conflict analysis (Vargas & Zoffer, 2019), forecasting (Lawal et al., 2016). al., 2014), and other applications in many areas of knowledge (Vargas, 1990; Saaty, 2008). In the social sciences, AHP can be used to quantify and derive measures of intangibles (e.g., Valentino et al., 2021). Although AHP provides many analysis benefits, it has certain limitations (Ishizaka & Labib, 2009).

Fuzzy logic is generally used when there are data whose knowledge suffers from uncertainty and imprecision (Zadeh, 1965; Frikha & Moalla, 2015). Therefore, fuzzy logic is used in conjunction with AHP in decision-making processes when circumstances exist where the degree of uncertainty is relevant. In these cases, fuzzy AHP is used (Ho & Ma, 2018).

METHODOLOGY

Based on the literature analysis related to the most important attributes/criteria in the decision to dive in this study, an adaptation of Ramos et al. (2006) where 5 categories were used for the criteria in the dive decision – biological, geographic, atmospheric, economic, and incentives – (Figure 1).

Figure 1. Criteria used in the dive decision (demand and supply)

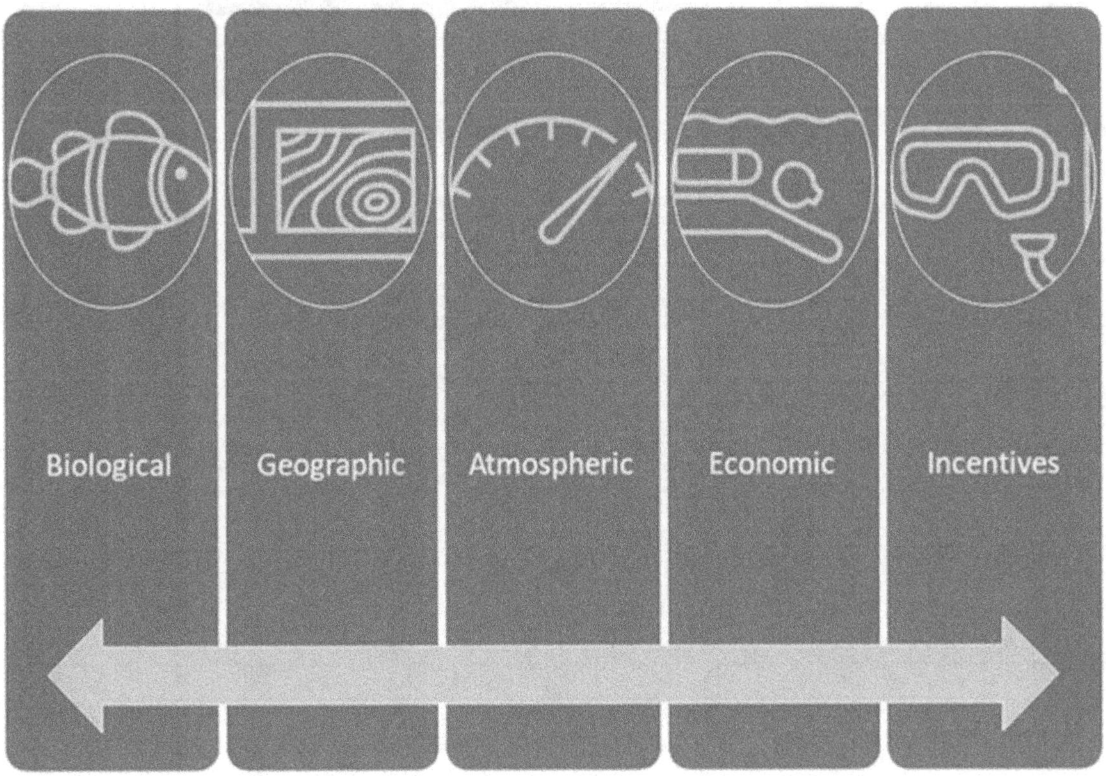

A tree structure (AHP) was designed. For the options/alternatives of reef modules implemented in Portugal, an allocation was made for each of the criteria. The expert assessment was carried out according to the characteristics of the AR depending on their geographic location, depth, and typology, and according to the results obtained in the analysis of published scientific literature. As there is uncertainty and imprecision in certain data, fuzzy logic was used in conjunction with AHP.

Goal

At the top of the hierarchy tree (AHP) comes the question that is of fundamental importance to answer: Are the AR modules implemented in Portugal suitable for diving? (Figure 2). The search for answers to this premise of the usefulness of reef modules for diving is the basis of the methodology of this study. This objective, which is based on the question asked, must consider the supply and demand market for diving and the importance of reef modules in that market.

Figure 2. The AHP tree used in this analysis: The goal at the top, the 5 decision criteria (and related 15 sub-criteria) in the middle, and the 8 divable alternatives/options at the bottom of the tree

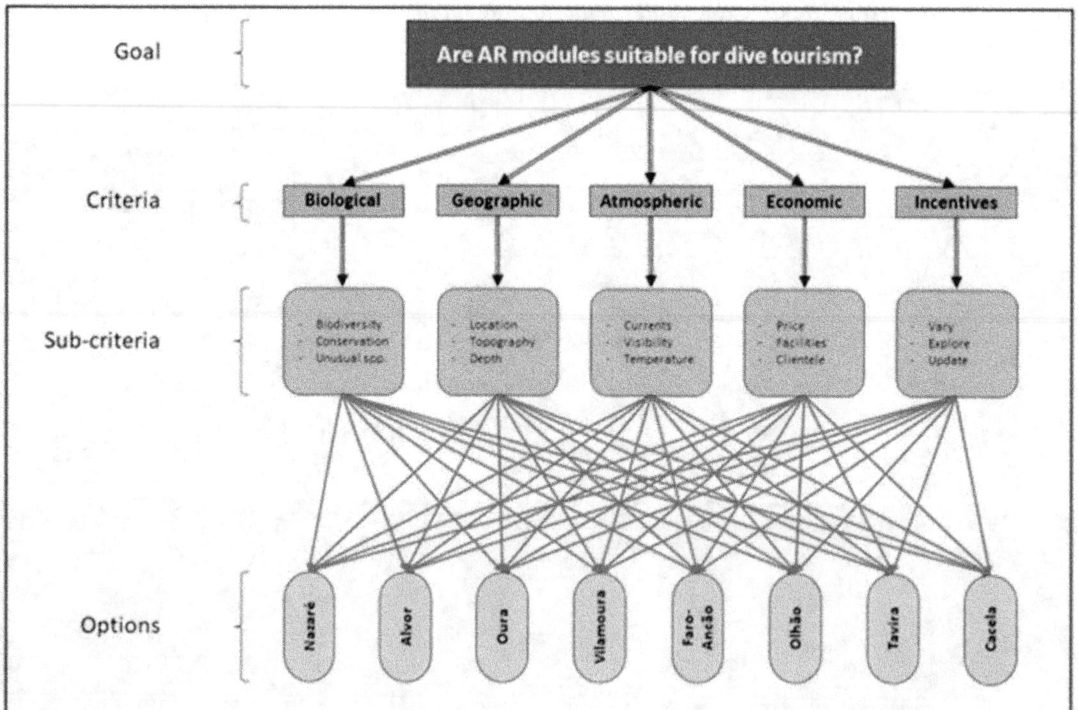

Criteria

People have several motivations to practice diving activities. These motivations derive from the cost-benefit analysis that practitioners take from diving. Companies that provide services to tourists and other diving enthusiasts tend to adjust their supply based on these criteria. The relative importance of each criterion depends on several factors and is sometimes also an individual choice. For simplicity, the criteria are considered equally important.

Biological

Biodiversity – The modules must have good colonization by several species of organisms from different trophic levels, presenting a good biodiversity spectrum, and generating the diver's interest.

Conservation – Spots with good preservation of the diving space, without marine litter and deterioration of the sites visited (underwater). It may also be important to consider the places on land in the vicinity that give access to diving.

Unusual Species – The presence of unusual species in dive sites can be a great attraction for visitors. These unusual species can be harmless, but they can also pose a danger to visitors.

Geographic

Location – The place where operators take their clientele should have good accessibility and should be away from potentially conflicting activities. The most pleasant places for diving are those where it is generally easier to practice the activity. Places where there are other activities or that have restrictions/limitations (e.g., aquaculture or fisheries production zones), must be avoided or eradicated.

Topography – The relief of the place where the dive takes place must be appealing to the practitioner or visitor. Generally, very rugged natural reefs with a very intricate relief are appealing to divers. Hence AR have limitations, especially if they were not designed from scratch for diving, this kind of structure is less appealing (e.g., modules with geometric shapes).

Bathymetry – This is a very limiting factor for divers, particularly if they still have little experience in the activity. Generally, a more experienced diver can go to a depth of around 40m, while an inexperienced diver should not go deeper than 18m.

Atmospheric or Other Elements of Nature

Currents – Places with strong currents have some danger as they often jeopardize the safety of divers, such as surface currents that can lead to seasickness while on the vessel, and deep currents hinder the diver's stability when maneuvering near the bottom.

Underwater Visibility – This is a particularly key factor to take advantage of the satisfaction that a good dive can give, if the visibility is bad, the dive is not so pleasant. The poor visibility of the water can also have to do with the atmospheric conditions, and they require more training on the part of divers and become less sought after by less experienced divers.

Temperature – The temperature underwater is also important for the duration and pleasure enthusiasts get from diving, particularly if a diver is not wearing a dry suit.

Economic

Price – The price of a visit or package of underwater visits is generally not a limiting factor for this type of recreational activity. However, the activity of diving is by nature a little expensive and it requires from the devotee financial capacity to practice it. Additionally, operators that provide higher quality diving services generally have higher prices. In the specific case of AR modules, the price charged may be higher because it costs the operator more (e.g., if AR location is further away from land, leading to more money being spent on fuel).

Facilities – The reception and support conditions, boat, trip to the dive site, as well as the equipment that the operator has for rent to its customers, are important to stimulate and attract customers.

Clientele – Places where there is a lot of tourist demand are more likely to provide diving activities nearby. Tour operators supplying diving activities in these locations try to provide a good list of underwater visitation options including AR if these are submerged nearby.

Incentive

Varying the Locations – The diversification of diving locations is one of the components that most attracts people to practice this activity. The experienced diver likes to have in their portfolio a panoply of places already known, and always desires to learn more about what is new. Diving on AR is one of those experiences that allow the diver to broaden the range.

Explore a Place – There is also the possibility that a dive that was interesting in a specific place at a certain time left open the possibility of knowing it better. Dive into that location comparatively (e.g., before and after reef deployment), with a certain frequency (e.g., annual visit), or routinely (e.g., in regular monitoring tasks).

Updating the Diver's Skills – In the evolution of the practice of diving, as in other sports or recreational activities, when a beginner is at their debut, they like to evolve, and when is more advanced, they like to take on new challenges that make them become an improved diver.

Alternatives

Location – For this study, the locations of the sunken AR modules in Portugal were defined, that is along the Algarve coast and the coast of Nazaré (Figure 3). For each reef set, the characteristics of its geographic setting were described, based on the factors mentioned in the literature and on the information and characteristics inherent to the reef sets themselves. The characteristics inherent to the reef sets were collected from institutional information, gray literature, or from the experience of professionals who have used the structures.

Figure 3. Location of the AR systems (red dots): A) Central and South regions of Portugal, B) Central AR system of Nazaré, and C) South AR systems scattered along the Algarve coast (from left to right: Alvor, Oura, Vilamoura, Faro-Ancão, Olhão, Tavira, and Cacela)

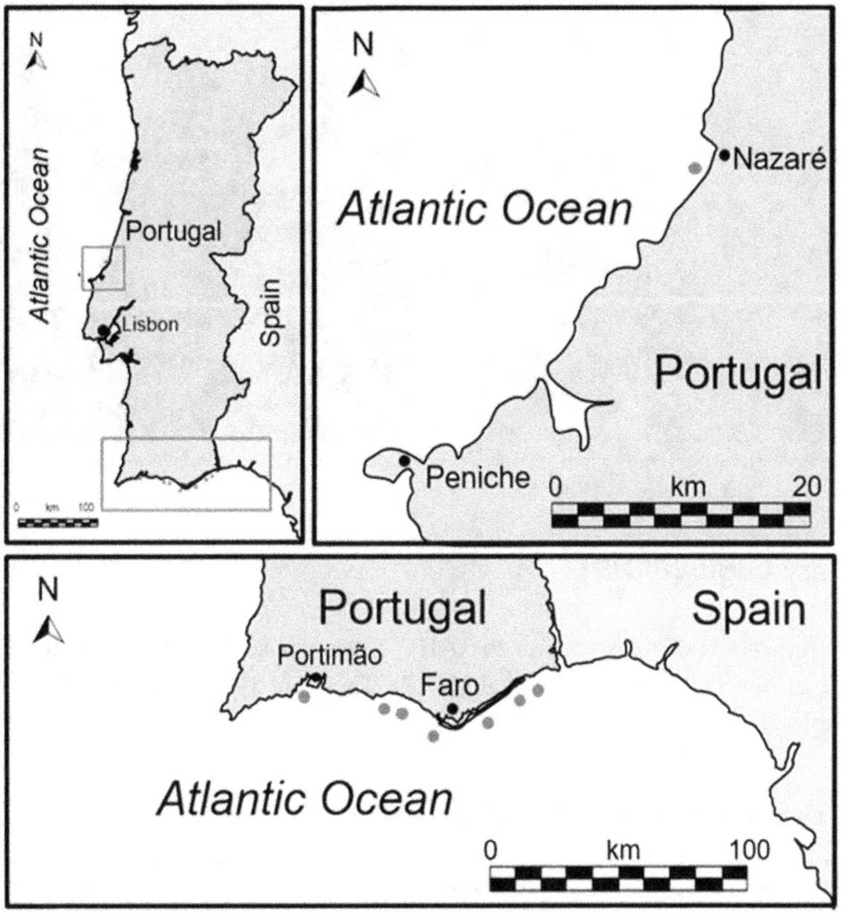

AR modules in Portugal were deployed on the south coast of the Algarve (either in the pilot or large-scale phase) and the west coast off Nazaré (a single-phase). The pilot AR were those of Olhão and Faro. The large-scale AR in the 1st phase were those of Cacela, Vilamoura, and Alvor, and in the 2nd phase were those of Tavira, Ancão, and Oura. The AR in central Portugal was deployed off the town of Nazaré, close to the submarine canyon. Concerning AR there are two types. The smaller ones are cubic (Figure 4 right), and the larger ones are octagonal (Figure 4 left). At shallower waters are the small-sized modules, while further out the larger ones. Nazaré AR only presents cubic modules.

Figure 4. The AR concrete modules; Left: cubic blocks (2.7m³/each); Right: octagonal blocks (174m³/each)
Photo credits: ProAquática (left) and IPMA (right)

AHP and Fuzzy Logic (FAHP)

To make measurements and judgments in the AHP, a 9-point scale is normally used and pairwise comparisons are made (Saaty, 1990; Kim et al., 2017). There are 17 possible possibilities for pairwise comparison (Table 1).

Table 1. The 9-point importance scale for AHP judgments

Intensity of importance	Definition	Explanation
1	Equally important	Two criteria contribute equally to the objective
3	Moderately more important of one over another	Experience and judgment strongly favor one criterion over another
5	Strongly more important	Experience and judgment strongly favor one criterion over another
7	Very strongly more important	A criterion is strongly favored, and its dominance demonstrated in practice
9	Extremely more important	The evidence favoring one criterion over another is of the highest possible order of affirmation
2, 4, 6, 8	Intermediate values between two adjacent judgments	When a compromise is needed
Reciprocals of above	Reciprocal values for inverse comparison	A reasonable assumption

Source: Saaty (1990)

The procedures used to carry out the analysis by the AHP are those that are normally found in the literature (Saaty, 2004). To make the judgments, a comparison matrix is used, which consists of a list of criteria in which this is present, either in the columns or in the rows. Comparisons are made in terms of relative importance.

The number of judgments has to do with the number of criteria used. In the present case, ten judgments are made, because there are five criteria used. AHP allows for a little inconsistency, but there is a measure of inconsistency for each set of judgments. Judgment consistency can be calculated by the consistency ratio (CR), which is defined as follows:

$$CR = CI / RI \tag{1}$$

Where CI is the consistency index, and RI is the random index (Table 2).

Table 2. Mean consistency of random matrices

n	1	2	3	4	5	6	7	8	9	10
RI	0.00	0.00	0.52	0.89	1.11	1.25	1.35	1.40	1.45	1.49

Source: Saaty (2000)

In each judgment matrix there is a little inconsistency that can be computed as a function for its maximum eigenvalue (λmax) and considering the order n of the matrix. For this, the following expression is used:

$$CI = \lambda_{max} - n / n - 1 \tag{2}$$

Inconsistencies of less than 10% are generally accepted. In case the values are higher, the judgments must not be viable and must be elicited again.

Next, pairwise comparisons are made for the options/alternatives, regarding each of the criteria. The relative importance of the judgments and the ranking of priorities derived therefrom for the alternatives should be shown in a final table. Priority vectors are the large main vectors of pairwise comparison matrices and must be normalized according to Saaty (2004: 13).

In the fuzzy logic part, the open-source software FisPro was used (Guillaume & Charnomordic, 2011). This software allows the user to create systems to make inferences in fuzzy logic.

In this chapter, an approach is made using a combination of the analytic hierarchy method with fuzzy logic (fuzzy analytic hierarchy process – FAHP). It is identified the potential interest for a dive by tourism in each of the AR module locations. Selected important criteria for the development of diving activities are listed and verified how likely each AR locations suit the criteria through an expert evaluation. Based on this assessment, it is possible to verify which are the most likely places for recreational diving, specifically regarding tourism (Figure 5).

In FAHP, the considered criteria (and their subdivisions) are used. Regarding the sites with modular artificial reefs, there are seven different sites off the Algarve and one off Nazaré town. There are still other experimental sites, but they will not be considered in this study.

Figure 5. Fuzzy logic: membership functions (Note: For the Inputs is given one example only, as it was used the same membership functions for all the five criteria)

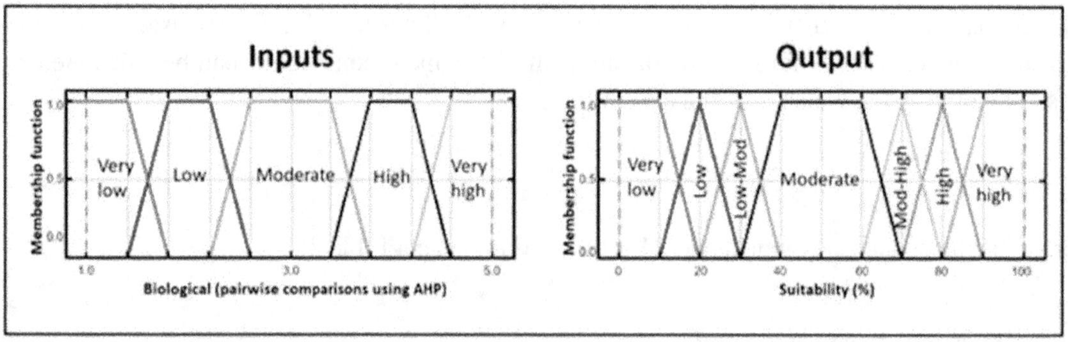

RESULTS

Through the characterization of the various AR regarding the criteria – via an expert evaluation –, the following table is obtained (Table 3). There is a great similarity between many reef sets. However, there are differences that go into the FAHP process as distinctive. These main differences are as follows.

In the AR of Nazaré there are sets of reefs at low depth and close to the coastline, and they were placed with the purpose of diving. The water temperature on this coast is lower than the AR in the Algarve.

In the Alvor AR there are reef sets at low depth and relatively close to the coastline. This area is immensely popular for nautical tourism activities.

The Oura and Vilamoura AR are relatively far from the coast and have very much pressure from fishing. Despite being in areas with a great supply and demand for nautical tourism, their demand for diving is limited due to fishing pressure and possible externalities (e.g., fishing congestion, lost gear involving various modules).

The Faro-Ancão AR is the reef system that, due to its larger dimension, allows diving operators and their clientele to have more incentives to practice the activity. This means that divers can vary the locations and explore more places.

The Olhão AR is the reef system whose location is within an exclusive area for offshore aquaculture since 2008 (DR 2008). Therefore, there is a limitation for diving, which is only allowed for the activities of companies operating within the area (APPA).

The Tavira AR is in the eastern part of the Algarve. This is an area where the water is milder due to its proximity to the Mediterranean Sea.

The Cacela AR is also found where the water is milder. However, in this area there is little offer of nautical activities. There are also testimonies of reef congestion due to lost fishing gear around the modules.

Table 3. Expert weights to construct fuzzy logic rules

Alternatives → / Criteria ↓	Nazaré	Alvor	Oura	Vilamou	F-A	Olhão	Tavira	Cacela
Biological	4.3	4.3	3.7	3.7	4.3	4.3	4.3	3.7
Biodiversity	High	High	High	High	High	High	High	High
Conservation	High	High	Average	Average	High	High	High	Average
Unusual spp	Average	Average	Average	Average	Average	Average	Average	Average
Geographic	3	3	2.3	2.3	2.3	1.7	2.3	2.3
Location	Average	Average	Average	Average	Average	Low	Average	Average
Topography	Low	Low	Low	Low	Low	Low	Low	Low
Depth	Low	Low	Average	Average	Average	Average	Average	Average
Atmospheric	2.3	3	3	3	3	3.7	3.7	3.7
Currents	Average	Average	Average	Average	Average	Average	Average	Average
Visibility	Average	Average	Average	Average	Average	Average	Average	Average
Temperature	Low	Average	Average	Average	Average	High	High	High
Economic	3	3.7	3.7	3.7	3	2.3	3	2.3
Price	Average	Average	Average	Average	Average	Average	Average	Average
Facilities	Average	Average	Average	Average	Average	Low	Average	Low
Clientele/Tourism	Average	High	High	High	Average	Average	Average	Average
Incentives	1.7	1.7	1.7	1.7	3	1	1.7	1.7
Vary	Low	Low	Low	Low	Average	Low	Low	Low
Explore	Low	Low	Low	Low	Average	Low	Low	Low
Update	Average	Average	Average	Average	Average	Low	Average	Average

Note: Low = 1; Average = 3; High = 5 (except "Depth" and "Price" which are the other way around).

Based on the FAHP analysis, the results obtained lead to believe that the AR modules with more potential for the diving activity are those from the bay of Alvor, Faro-Ancão, and a little less for those from Nazaré (Figure 6). The ones from Alvor are close to the coast and in places with high tourist demand, and which have other artificial complements such as the sunken ships of the Ocean Revival Project (Nunes & Roeder, 2014). In the case of the Faro-Ancão AR, because it is the largest continuous reef set and because it offers other diving alternatives, such as a sunken trawler for diving purposes (Ramos et al., 2006). In the case of the Nazaré AR, it has certain potential for a dive as well, as it is remarkably close to the coast at a shallow depth, but the complete set is relatively small.

Figure 6. FAHP results (outputs) for each AR diving option/alternative; it was used the open-source software FisPro (Guillaume & Charnomordic, 2011)

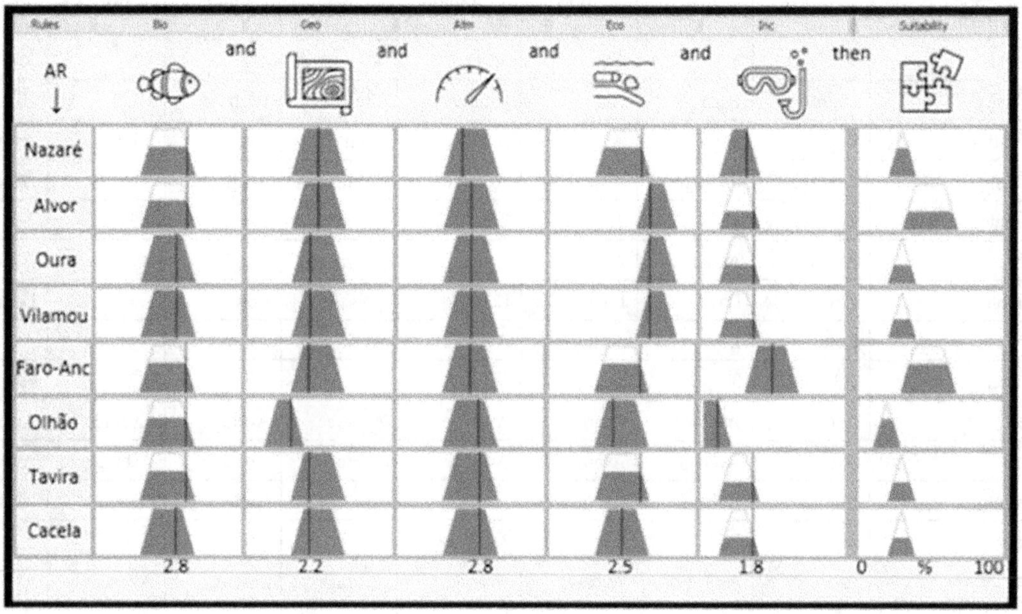

DISCUSSION

The first modules of artificial reefs in Portugal were not built with the main purpose of supporting recreational activities, such as for example diving (Santos & Monteiro, 1997, 1998). Their main functions were of an ecological and socioeconomic nature. In the ecological part, these first AR were implemented to provide protection to small organisms – e.g., mollusks, crustaceans, and fish – in their transition from inland waters to the open ocean. The other purpose of the AR in the socio-economic part would be as a support tool in the management of local small-scale fisheries.

Contrary to the first ones, the last sets of AR were deployed considering not only the ecological and socioeconomic part as before, but also already including modules deployed closer to the coast and which had recreational diving as their main objective (Ramos et al., 2021). However, most reef sets that have been implemented – both the oldest and the most recent – are divable.

Although the factors that go into the decision to dive in AR may come from different criteria, there are two sub-criteria framed in the geographic factors that stand out as more limiting in the decision. The sub-criteria are the depth and location of the AR (MacCarthy et al., 2006; Schuett et al., 2016).

The reefs deployed off the Algarve have two types of modules: small and large. On the one hand, the cubic modules are smaller (2.7m^3) and deployed in shallower bathymetric up to about 20m. On the other hand, octagonal modules are larger (174 m^3) and deployed in bathymetric depths up to about 40 m. In the pilot projects, a few parallelepipeds with a volume slightly smaller than the octagonal AR were also deployed. The depth here may have influence on the choice of the AR typology. Less experienced divers can dive the cubic modules that are closer to the shore at a shallower depth (smaller modules), while more experienced divers are able to dive in the modules that are deeper (larger modules).

Despite the depth being a limiting factor and which is in line with the diver's experience, there are other constraints that appear more due to the difficulty of access. For example, AR that are in areas delimited for offshore aquaculture, are no longer available to diving operators, unless there are agreements between the entities that work in the aquaculture areas with the diving operators. In the Algarve, in the case of the Olhão AR, the modules are in the middle of an offshore aquaculture restricted area.

The reflections generated from the results obtained using this methodological analysis allow us to determine how important each AR is for tourist diving operators that intend to develop activities in these spots (Figure 7).

Figure 7. Simplified conceptual framework towards decision-making processes of AR for human use, such as diving

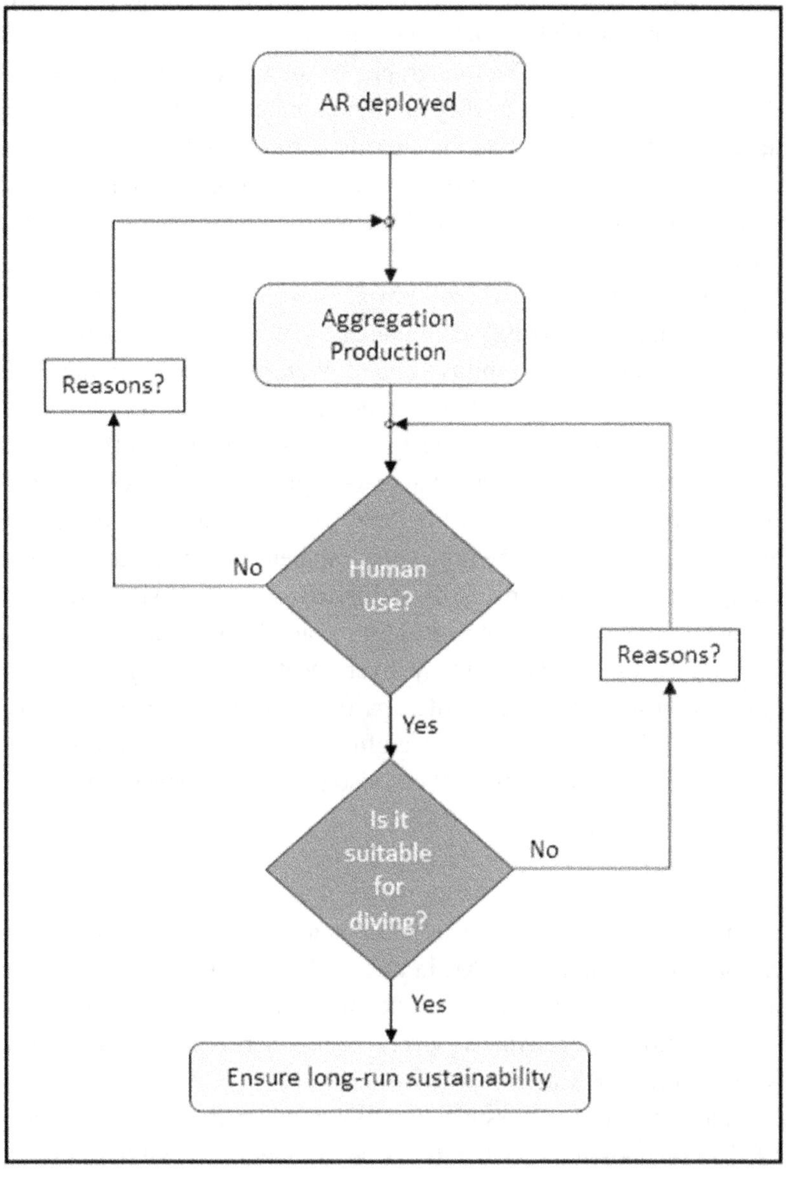

Criteria

Concerning biological criteria, the biodiversity of the reef area is reached after a few years, mimicking other sites with reef characteristics, namely the natural reefs. The existence of a great diversity of marine life is an attraction for tourist or leisure diving. The presence of AR improves what was previously largely devoid of marine life, and somehow approaches or even reaches the equivalent of a natural reef. The conservation or degradation of sites with AR is highly dependent on several factors, both environmental and human. In the case of environmental factors, there may be reef modules covered with sand or sediment due to bottom currents, but they may also be subject to anthropogenic factors and several modules may be covered with nets or other gear derived from intensive fishing on site. The probability of encountering peculiar or unusual species is a factor that generates particular interest in diving.

The geographic location of AR is of particular importance in choosing a dive site. Sites with privileged access for diving are helpful – e.g., tacit rules as in Nazaré or future marine protected areas (MPAs) in specific Algarve areas –, whereas places where there is an eventual competition for space with other activities, diving is not recommended. The topography of the places is an attraction for diving, in this case, the reef modules are not appealing as they are monotonous structures (cubes and octagons). The depth at which the reef structures are located is a particularly crucial factor when it comes to diving, as for an experienced diver the depth should be around 30m (and never more than 40m), while for an inexperienced diver this depth should be less than 18m. Most of the submerged reef modules in Portugal are found at these depths.

Regarding atmospheric factors, places where there is a prevalence of both surface and bottom currents are factors that interfere with a good diving performance. In addition, it is especially important for good diving satisfaction that the underwater visibility is good. Waters with a lot of turbidities are unsatisfactory for leisure and tourist activities. The water temperature can also be a factor that somehow affects diving satisfaction, even though there is equipment on the market that can offset this negative factor.

Regarding economic factors, it is important that there is provided a service supply by local operators or in the vicinity of AR and that they can enable these services at a fair price since diving has many inherent costs related to tours. Likewise, the range of equipment and the ability to meet the needs of the clientele are especially important to be able to diversify and sustain a business related to diving.

As for diving incentives, the possibility of being able to vary the dive sites becomes a key factor for the clientele, thus, the reef modules do not offer many diversification options, just being a different option. And the possibility of exploring different dive sites, particularly in an area with reef modules, can be an incentive for a diver to visit this type of place more often. Finally, the possibility of allowing improvements to be made in the skills of divers in the AR can also serve as an incentive to dive into these places.

Options/Alternatives

Although the sites with reef modules all have similarities to each other, at a structural level they differ in relation to several factors. Currently, all AR in this study are mature and were colonized over time reaching good levels of biodiversity, which can obviously be a good indicator to promote diving. However, there are reefs that stand out negatively as several user fishers report that there are nets doing ghost fishing (e.g., a few reefs in the AR Cacela).

There is a reef set that is no longer accessible due to the implementation of an aquaculture production zone in the AR of Olhão (Ramos et al., 2017). In terms of topography, all have the same composition, i.e., 105 blocks spread in sets of 3, just varying the number of groups. In almost all reef systems there are groups that are closer to land and thus more suitable for less experienced divers. However, the AR modules in Nazaré are the ones that are closer to the coast and were specifically created just for diving (Almeida et al., 2016).

Regarding atmospheric criteria, the Tavira modules suffer from undercurrents (Júnior et al., 2021), while those of Nazaré are in an area where the water tends to be cooler (Baptista et al., 2019). Visibility is highly variable and depends more on weather conditions and the time of year than on the AR itself.

The reefs in the Algarve, especially those in the west and central areas, are the ones with the greatest tourist potential, as this is where there are more active diving operators and more tourists (e.g., AR of Alvor, Oura, and Vilamoura). There are dive operators that have been in the market for many years and that have developed enough conditions to practice this type of leisure (Rangel et al., 2014).

AR modules are usually not attractive to divers, sunken ships are more appealing. The larger reef systems allow offering variations for divers to explore (e.g., Faro-Ancão, Vilamoura, and Oura). The deeper reefs allow more experienced divers to also make their evolutions and upgrade their skills.

In general, the AR on the Portuguese coast have played a role that essentially serves to promote an adequate habitat for several marine species and, at the same time, help in the management of fisheries, namely in limiting the gears. In terms of policy aspects, the present results indicate that AR closer to the coast could be more easily used in recreational/tourist diving, as they are potentially easier to manage because they involve less fishing activities and because they are potentially more favorable and interesting for diving activities. This selection – based on less distance from the coast and promotion of habitat with greater biodiversity – can have a positive effect with respect to the promotion of recreational/tourist diving activities.

CONCLUSIONS

In Portugal, AR modules were created essentially to protect marine species in an environment of transition from the waters in the inland lagoons (i.e., *rias* of Alvor and Formosa) to the sea. The AR also have a component that consists of helping small-scale fisheries to achieve more "stable" fisheries and allowing for higher sustainability in fishing. Reef modules deployed closer to the coast were intended to be used by dive operators, however, this feature was only better defined in a more recent AR (i.e., Nazaré).

On the positive side, it can be said that the benefits derived from the reef modules can be their use by tour operators to diversify the dive sites and alleviate the pressure on natural reefs. On the negative side is the low attractiveness of structures with reef modules. Enthusiasts dive once to try them out and eventually tend not to repeat the dive.

ACKNOWLEDGMENT

This paper is financed by Portuguese National Funds provided by FCT - Foundation for Science and Technology through project UIDB/SOC/04020/2020. J. Ramos is supported by Scientific Employment Stimulus from FCT (2021.02180.CEECIND).

REFERENCES

Abdullah, L., & Najib, L. (2016). Sustainable energy planning decision using the intuitionistic fuzzy analytic hierarchy process: Choosing energy technology in Malaysia. *International Journal of Sustainable Energy*, *35*(4), 360–377. doi:10.1080/14786451.2014.907292

Almeida, M., Raposo, A. C., Guerra, M. T., Gaudêncio, M., Maranhão, P., Rodrigues, N. V., & Leandro, S. (2016). Macrobenthic Assemblages in an Artificial Reef in Central-West Portugal: Preliminary Results. *Front. Mar. Sci. Conference Abstract: XIX Iberian Symposium on Marine Biology Studies.*

Ananda, J., & Herath, G. (2003). The use of Analytic Hierarchy Process to incorporate stakeholder preferences into regional forest planning. *Forest Policy and Economics*, *5*(1), 13–26. doi:10.1016/S1389-9341(02)00043-6

Baptista, J., Martinho, F., Martins, R., Carneiro, M., Azevedo, M., Vieira, A. R., Gomes, P., & Pardal, M. A. (2019). Water temperature gradient shapes the structure and composition of nearshore marine fish communities in southern Europe. *Journal of Sea Research*, *154*, 101807. doi:10.1016/j.seares.2019.101807

Bideci, C., & Cater, C. (2019). In search of underwater atmosphere: A new diving world on artificial reefs. In *Atmospheric turn in culture and tourism: Place, design and process impacts on customer behaviour, marketing and branding*. Emerald Publishing Limited. doi:10.1108/S1871-317320190000016023

Bohnsack, J. A. (1991). Habitat structure and the design of artificial reefs. In *Habitat Structure* (pp. 412–426). Springer. doi:10.1007/978-94-011-3076-9_20

Bohnsack, J. A., & Sutherland, D. L. (1985). Artificial reef research: A review with recommendations for future priorities. *Bulletin of Marine Science*, *37*(1), 11–39.

Bortone, S. A., & Otake, S. (Eds.). (2020). *Modern Fisheries Engineering: Realizing a Healthy and Sustainable Marine Ecosystem*. CRC Press. doi:10.1201/9780429328039

Cancela da Fonseca, L., Boaventura, D., Ré, P., Pereira, P., & Santos, M. N. (2007). Caracterização das Comunidades de Macroinvertebrados Bentónicos no Sistema Recifal do Alvor (Costa Sul do Algarve). *13º Congresso do Algarve*, 403-413.

Cater, C. (2007). Perceptions of and interactions with marine environments: Diving attractions from great whites to pygmy seahorses. In *New frontiers in marine tourism* (pp. 71–86). Routledge.

Cousteau, J. Y., & Dumas, F. (1953). The silent world. Academic Press.

Dias, D. A. C. M. (2011). *Estudo da valorização de cinzas de biomassa na produção de materiais para a construção de recifes artificiais* (Doctoral dissertation). Faculdade de Ciências e Tecnologia.

Dimmock, K. (2007). Scuba diving, snorkeling, and free diving. In *Water-based tourism, sport, leisure, and recreation experiences* (pp. 146–166). Routledge. doi:10.1016/B978-0-7506-6181-2.50013-0

Dimmock, K., & Musa, G. (2015). Scuba diving tourism system: A framework for collaborative management and sustainability. *Marine Policy*, *54*, 52–58. doi:10.1016/j.marpol.2014.12.008

Dimopoulos, D., Queiros, D., & van Zyl, C. (2019). Sinking deeper: The most significant risks impacting the dive tourism industry in the East African Marine Ecoregion. *Ocean and Coastal Management, 181*, 104897. doi:10.1016/j.ocecoaman.2019.104897

DR (2008). Decreto Regulamentar n.º 9/2008 de 18 de março. Ministério da Agricultura, do Desenvolvimento Rural e das Pescas. Diário da República, 1ª série, N.º 55 de 18 de março de 2008.

Encarnação, J., & Calado, G. (2018). Effects of recreational diving on early colonization stages of an artificial reef in North-East Atlantic. *Journal of Coastal Conservation, 22*(6), 1209–1216. doi:10.100711852-018-0630-0

Fabi, G., & Spagnolo, A. (2011). *Artificial reefs in the management of Mediterranean Sea fisheries*. CRC Press.

Falcão, M., Santos, M. N., Vicente, M., & Monteiro, C. C. (2007). Biogeochemical processes and nutrient cycling within an artificial reef off Southern Portugal. *Marine Environmental Research, 63*(5), 429–444. doi:10.1016/j.marenvres.2006.12.001 PMID:17239434

Frank, M. (2018). *Artificial reefs: marine and freshwater applications*. CRC Press.

Frikha, A., & Moalla, H. (2015). Analytic hierarchy process for multi-sensor data fusion based on belief function theory. *European Journal of Operational Research, 241*(1), 133–147. doi:10.1016/j.ejor.2014.08.024

Garrod, B. (2007). Market segments and tourist typologies for diving tourism. In *New Frontiers in Marine Tourism* (pp. 53–70). Routledge. doi:10.4324/9780080551081-13

Giddy, J. K., & Rogerson, C. M. (2018). Tracking SCUBA diving adventure tourism in South Africa. *EuroEconomica, 37*(2), 47–62.

Gordon, W. R. Jr, & Ditton, R. B. (1986). A user-resource planning framework for offshore recreational artificial reefs. *Coastal Management, 13*(3-4), 369–395.

Guillaume, S., & Charnomordic, B. (2011). Learning interpretable fuzzy inference systems with FisPro. *Information Sciences, 181*(20), 4409–4427. doi:10.1016/j.ins.2011.03.025

Ho, W., & Ma, X. (2018). The state-of-the-art integrations and applications of the analytic hierarchy process. *European Journal of Operational Research, 267*(2), 399–414. doi:10.1016/j.ejor.2017.09.007

Huang, I. B., Keisler, J., & Linkov, I. (2011). Multi-criteria decision analysis in environmental sciences: Ten years of applications and trends. *The Science of the Total Environment, 409*(19), 3578–3594. doi:10.1016/j.scitotenv.2011.06.022 PMID:21764422

Hunt, J. C. (1995). Divers' accounts of normal risk. *Symbolic Interaction, 18*(4), 439–462. doi:10.1525i.1995.18.4.439

Ishizaka, A., & Labib, A. (2009). Analytic hierarchy process and expert choice: Benefits and limitations. *Or Insight, 22*(4), 201-220.

Johansen, K. (2013). Scuba diving education and training. In *Scuba diving tourism* (pp. 99–114). Routledge.

Júnior, L. D. O., Garel, E., & Relvas, P. (2021). The structure of incipient coastal counter currents in South Portugal as indicator of their forcing agents. *Journal of Marine Systems, 214*, 103486. doi:10.1016/j.jmarsys.2020.103486

Kim, N., Park, J., & Choi, J. J. (2017). Perceptual differences in core competencies between tourism industry practitioners and students using Analytic Hierarchy Process (AHP). *Journal of Hospitality, Leisure, Sport and Tourism Education, 20*, 76–86. doi:10.1016/j.jhlste.2017.04.003

Kirkbride-Smith, A. E., Wheeler, P. M., & Johnson, M. L. (2013). The relationship between diver experience levels and perceptions of attractiveness of artificial reefs-examination of a potential management tool. *PLoS One, 8*(7), e68899. doi:10.1371/journal.pone.0068899 PMID:23894372

Kirkbride-Smith, A. E., Wheeler, P. M., & Johnson, M. L. (2016). Artificial reefs and marine protected areas: A study in willingness to pay to access Folkestone Marine Reserve, Barbados, West Indies. *PeerJ, 4*, e2175. doi:10.7717/peerj.2175 PMID:27547521

Lawal, D. U., Yusof, K. W., Hashim, M. A., & Balogun, A. L. (2014, June). Spatial analytic hierarchy process model for flood forecasting: An integrated approach. [). IOP Publishing.]. *IOP Conference Series. Earth and Environmental Science, 20*(1), 012029.

Lemke, L., & Olech, L. (2011). Dive tourism. In The Long Tail of Tourism: Holiday Niches and Their Impact on Mainstream Tourism (pp. 105-114). Gabler Research. doi:10.1007/978-3-8349-6231-7_12

Lök, A., Metin, C., Ulaş, A., Düzbastilar, F. O., & Tokaç, A. (2002). Artificial reefs in Turkey. *ICES Journal of Marine Science, 59*(suppl), S192–S195. doi:10.1006/jmsc.2002.1221

Loske, A. (2013). Fundamentals of SCUBA-diving physics. *International Journal of Sports Science, 3*(2), 37–45.

MacCarthy, M., O'neill, M., & Williams, P. (2006). Customer satisfaction and scuba-diving: Some insights from the deep. *Service Industries Journal, 26*(5), 537–555. doi:10.1080/02642060600722841

Maleki, H., & Zahir, S. (2013). A comprehensive literature review of the rank reversal phenomenon in the analytic hierarchy process. *Journal of Multi-Criteria Decision Analysis, 20*(3-4), 141–155. doi:10.1002/mcda.1479

McGurrin, J. M., Stone, R. B., & Sousa, R. J. (1989). Profiling United States artificial reef development. *Bulletin of Marine Science, 44*(2), 1004–1013.

Moura, A., Boaventura, D., Cúrdia, J., Santos, M. N., & Monteiro, C. C. (2006). Biomass production of early macrobenthic communities at the Faro/Ancão artificial reef (Portugal): Effect of depth and reef layer. *Bulletin of Marine Science, 78*(1), 83–92.

Moura, A., Fonseca, L. C. D., Boaventura, D., Santos, M. N., & Monteiro, C. C. (2011). Estimation of secondary production of the Faro/Ancão artificial reefs. *Brazilian Journal of Oceanography, 59*(spe1), 91–94. doi:10.1590/S1679-87592011000500011

Newman, J. (2011). *Scuba Diving and Snorkeling for Dummies*. John Wiley & Sons.

Nunes, C. H., & Roeder, F. (2014). Tourism and Algarve's underwater heritage valorisation. A case study. *Journal for Maritime Research, 11*(1), 35–44.

Oliveira, M. T., Ramos, J., & Santos, M. N. (2015). An approach to the economic value of diving sites: Artificial versus natural reefs off Sal Island, Cape Verde. *Journal of Applied Ichthyology, 31*, 86–95. doi:10.1111/jai.12953

Özgül, A., Lök, A., Tanrıkul, T. T., & Alos, J. (2019). Home range and residency of Scorpaena porcus and Scorpaena scrofa in artificial reefs revealed by fine-scale acoustic tracking. *Fisheries Research, 210*, 22–30. doi:10.1016/j.fishres.2018.10.008

Palau-Saumell, R., Forgas-Coll, S., Sánchez-García, J., & Prats, L. (2019). Motivation and attachment to a diving destination: The case of Medes Islands (Catalonia, Spain). *Journal of Vacation Marketing, 25*(3), 301–319. doi:10.1177/1356766718778867

Pears, R. J., & Williams, D. M. (2005). *Potential effects of artificial reefs on the Great Barrier Reef: background paper*. CRC Reef Research Centre.

Perkol-Finkel, S., Shashar, N., & Benayahu, Y. (2006). Can artificial reefs mimic natural reef communities? The roles of structural features and age. *Marine Environmental Research, 61*(2), 121–135. doi:10.1016/j.marenvres.2005.08.001 PMID:16198411

Polovina, J. J., & Sakai, I. (1989). Impacts of artificial reefs on fishery production in Shimamaki, Japan. *Bulletin of Marine Science, 44*(2), 997–1003.

Ramos, J., Caetano, M., Himes-Cornell, A., & dos Santos, M. N. (2017). Stakeholders' conceptualization of offshore aquaculture and small-scale fisheries interactions using a Bayesian approach. *Ocean and Coastal Management, 138*, 70–82. doi:10.1016/j.ocecoaman.2017.01.008

Ramos, J., Lino, P. G., Himes-Cornell, A., & Santos, M. N. (2019). Local fishermen's perceptions of the usefulness of artificial reef ecosystem services in Portugal. *PeerJ, 6*, e6206. doi:10.7717/peerj.6206 PMID:30647998

Ramos, J., & Santos, M. N. (2015). Facilitating fishing decisions in an artificial reef area off southern Portugal: A case study using generalized additive models. *Journal of Applied Ichthyology, 31*, 24–34. doi:10.1111/jai.12948

Ramos, J., Santos, M. N., Whitmarsh, D., & Monteiro, C. C. (2006). The usefulness of the analytic hierarchy process for understanding reef diving choices: A case study. *Bulletin of Marine Science, 78*(1), 213–219.

Ramos, J., Tuaty-Guerra, M., Almeida, M., Raposo, A. C., Gaudêncio, M. J., Silva, A. D., Rodrigues, N., Leandro, S. M., & Caetano, M. (2021). An artificial reef at the edge of the deep: An interdisciplinary case study. *Ocean and Coastal Management, 210*, 105729. doi:10.1016/j.ocecoaman.2021.105729

Ramos-Esplá, A. A., Guillén, J. E., Bayle, J. T., & Sánchez-Jérez, P. (2000). Artificial anti-trawling reefs off Alicante, South-Eastern Iberian Peninsula: Evolution of reef block and set designs. In *Artificial reefs in European seas* (pp. 195–218). Springer. doi:10.1007/978-94-011-4215-1_12

Rangel, M. O., Pita, C. B., Gonçalves, J. M. S., Oliveira, F., Costa, C., & Erzini, K. (2014). Developing self-guided scuba dive routes in the Algarve (Portugal) and analysing visitors' perceptions. *Marine Policy, 45*, 194–203. doi:10.1016/j.marpol.2013.10.011

Saaty, T. L. (1977). A scaling method for priorities in hierarchical structures. *Journal of Mathematical Psychology, 15*(3), 234–281. doi:10.1016/0022-2496(77)90033-5

Saaty, T. L. (1990). How to make a decision: The analytic hierarchy process. *European Journal of Operational Research, 48*(1), 9–26. doi:10.1016/0377-2217(90)90057-I

Saaty, T. L. (2000). *Fundamentals of decision making and priority theory with the analytic hierarchy process* (Vol. 6). RWS publications.

Saaty, T. L. (2004). Decision making—The analytic hierarchy and network processes (AHP/ANP). *Journal of Systems Science and Systems Engineering, 13*(1), 1–35. doi:10.100711518-006-0151-5

Saaty, T. L. (2008). Decision making with the analytic hierarchy process. *International Journal of Services Sciences, 1*(1), 83–98. doi:10.1504/IJSSCI.2008.017590

Saaty, T. L., & Erdener, E. R. E. N. (1979). A new approach to performance measurement the analytic hierarchy process. *Design Methods and Theories, 13*(2), 62–68.

Salaün, J., Pioch, S., & Dauvin, J. C. (2020). Description of a theoretical social-ecological approach to manage artificial reefs. *Vie et Milieu/Life & Environment, 70.*

Santos, M. N., & Monteiro, C. C. (1997). The Olhao artificial reef system (south Portugal): Fish assemblages and fishing yield. *Fisheries Research, 30*(1-2), 33–41. doi:10.1016/S0165-7836(96)00567-X

Santos, M. N., & Monteiro, C. C. (1998). Comparison of the catch and fishing yield from an artificial reef system and neighbouring areas off Faro (Algarve, south Portugal). *Fisheries Research, 39*(1), 55–65. doi:10.1016/S0165-7836(98)00169-6

Santos, M. N., & Monteiro, C. C. (2007). A fourteen-year overview of the fish assemblages and yield of the two oldest Algarve artificial reefs (southern Portugal). In *Biodiversity in Enclosed Seas and Artificial Marine Habitats* (pp. 225–231). Springer. doi:10.1007/978-1-4020-6156-1_19

Santos, M. N., Monteiro, C. C., & Lasserre, G. (1995). A four-year overview of the fish assemblages and yield on two artificial reef systems off Algarve (south Portugal). *Proceedings of the 30th European Marine Biological Symposium*, 345-352.

Santos, M. N., Monteiro, C. C., & Lasserre, G. (2005). Observations and trends on the intra-annual variation of the fish assemblages on two artificial reefs in Algarve coastal waters (southern Portugal). *Scientia Marina, 69*(3), 415–426. doi:10.3989cimar.2005.69n3415

Santos, M. N., Monteiro, C. C., & Leitão, F. (2011). The role of artificial reefs in the sustainability of artisanal fisheries. *Artificial reefs in Fisheries Management*, 221-237.

Schuett, M. A., Ding, C., Kyle, G., & Shively, J. D. (2016). Examining the behavior, management preferences, and sociodemographics of artificial reef users in the Gulf of Mexico offshore from Texas. *North American Journal of Fisheries Management, 36*(2), 321–328. doi:10.1080/02755947.2015.1123204

Shani, A., Polak, O., & Shashar, N. (2012). Artificial reefs and mass marine ecotourism. *Tourism Geographies, 14*(3), 361–382. doi:10.1080/14616688.2011.610350

Sheehy, D. J. (1982). The use of designed and prefabricated artificial reefs in the United States. *Marine Fisheries Review, 44*(6), 4–15.

Smith, J. T., & Tighe, S. L. (2006). Analytic hierarchy process as a tool for infrastructure management. *Transportation Research Record: Journal of the Transportation Research Board, 1974*(1), 2–9. doi:10.1177/0361198106197400101

Stolk, P., Markwell, K., & Jenkins, J. M. (2007). Artificial reefs as recreational scuba diving resources: A critical review of research. *Journal of Sustainable Tourism, 15*(4), 331–350. doi:10.2167/jost651.0

Streich, M. K., Ajemian, M. J., Wetz, J. J., & Stunz, G. W. (2017). A comparison of fish community structure at mesophotic artificial reefs and natural banks in the western Gulf of Mexico. *Marine and Coastal Fisheries, 9*(1), 170–189. doi:10.1080/19425120.2017.1282897

Sutton, S. G., & Bushnell, S. L. (2007). Socio-economic aspects of artificial reefs: Considerations for the Great Barrier Reef Marine Park. *Ocean and Coastal Management, 50*(10), 829–846. doi:10.1016/j.ocecoaman.2007.01.003

Thierry, J. M. (1988). Artificial reefs in Japan—A general outline. *Aquacultural Engineering, 7*(5), 321–348. doi:10.1016/0144-8609(88)90014-3

Thirumoorthi, T., Wong, K. M., & Musa, G. (2013). Scuba diving satisfaction. In *Scuba Diving Tourism* (pp. 170–176). Routledge.

Tunca, S., Miran, B., & Ünal, V. (2012). Decisions of stakeholders for the proposed artificial reef deployment: Analytic hierarchy process approach. *Ege Journal of Fisheries and Aquatic Sciences, 29*(1), 21–29.

Tunca, S., Miran, B., & Ünal, V. (2014). Perception and demand for artificial reef by relevant local groups in Altınoluk (Turkey). *Ege Journal of Fisheries and Aquatic Sciences, 31*(1), 5–10. doi:10.12714/egejfas.2014.31.1.02

Tynyakov, J., Rousseau, M., Chen, M., Figus, O., Belhassen, Y., & Shashar, N. (2017). Artificial reefs as a means of spreading diving pressure in a coral reef environment. *Ocean and Coastal Management, 149*, 159–164. doi:10.1016/j.ocecoaman.2017.10.008

Uyarra, M. C., Watkinson, A. R., & Cote, I. M. (2009). Managing dive tourism for the sustainable use of coral reefs: Validating diver perceptions of attractive site features. *Environmental Management, 43*(1), 1–16. doi:10.100700267-008-9198-z PMID:18810525

Valentino, V. H., Setiawan, H. S., Saputra, A., Haryanto, Y., & Putra, A. S. (2021). Decision Support System for Thesis Session Pass Recommendation Using AHP (Analytic Hierarchy Process) Method. *International Journal of Educational Research & Social Sciences, 2*(1), 215-221.

Vargas, L. G. (1990). An overview of the analytic hierarchy process and its applications. *European Journal of Operational Research, 48*(1), 2–8. doi:10.1016/0377-2217(90)90056-H

Vargas, L. G., & Zoffer, H. J. (2019). Applying AHP in conflict resolution. *International Journal of the Analytic Hierarchy Process, 11*(1), 143–147. doi:10.13033/ijahp.v11i1.649

Verkoeyen, S., & Nepal, S. K. (2019). Understanding scuba divers' response to coral bleaching: An application of Protection Motivation Theory. *Journal of Environmental Management*, *231*, 869–877. doi:10.1016/j.jenvman.2018.10.030 PMID:30419442

White, A. T., Chou, L. M., De Silva, M. W. R. N., & Guarin, F. Y. (1990). Artificial reefs for marine habitat enhancement in Southeast Asia. *ICLARM Education Series, 11*, 45.

Whitmarsh, D., Santos, M. N., Ramos, J., & Monteiro, C. C. (2008). Marine habitat modification through artificial reefs off the Algarve (southern Portugal): An economic analysis of the fisheries and the prospects for management. *Ocean and Coastal Management*, *51*(6), 463–468. doi:10.1016/j.ocecoaman.2008.04.004

Zadeh, L. (1965). Fuzzy sets. *Information and Control*, *8*(3), 338–353. doi:10.1016/S0019-9958(65)90241-X

Chapter 8
Artificial Reef Applications for Diving Tourism:
Artificial Wreck Reefs in Turkey

Aytaç Özgül
Faculty of Fisheries, Ege University, Turkey

Altan Lök
Faculty of Fisheries, Ege University, Turkey

ABSTRACT

Diving tourism has been very popular in the world and Turkey. In addition to reasons such as seeking hobbies and socializing, marine biodiversity, natural reefs, wrecks, and corals cause more and more people to dive every day. In many regions in Turkey, which has many underwater riches, diving is prohibited for reasons such as archaeological and military zones and natural reserves. Therefore, alternative diving points have been created for diving tourism by using artificial reefs in recent years. Within the scope of 35 artificial wreck reefs, in which diving tourism has been carried out in Turkey, 3000 amphoras, many thematic sculptures, 22 ships, 9 planes, and 1 tank were launched. While decommissioned vehicles were used in the first years, underwater parks with different themes have been created in recent years. In this study, artificial wreck reefs in Turkey were examined, and the permit process, the structures, and the interests of divers were evaluated. In addition, mistakes made in material and location selection and interactions with other sectors are also discussed.

INTRODUCTION

Diving tourism is a type of tourism that includes promotional, sportive and educational dives for the purpose of seeing, photographing, and sport fishing the flora, fauna and archaeological cultural assets of the underwater world, and accommodation and hospitality services provided on the shore (Garrod and Gössling, 2008; Yaşar, 2011). It is developing day by day and spreading to wider masses due to the increase in people's interest in the marine environment, as well as the ease of transportation, the

DOI: 10.4018/978-1-6684-2344-8.ch008

development of underwater technology, the widespread use of underwater photography and social media (Dimmock, 2007; Musa and Dimmock, 2013). This development in diving tourism is also revealed by the statistics of institutions providing scuba diving education. According to the Professional Association of Diving Instructors (PADI), there are more than 6,600 dive centers in 186 countries and, since 1967 there are more than 28 million certified divers worldwide (PADI, 2021). The USA, Italy, Germany, France and England are demarcated as the most important source markets for diving tourism, which is defined as an important and growing component of the international tourism market (Tabata, 1992; CBI, 2019). Today, marine-based tourism employs 2.4 million people with USD 59 Billion in revenue in USD 124 Billion GDP (NOAA, 2019).

Diving tourism attractions are dominated by those in the area of coral reefs, located in tropical latitudes, the most known of which are the Great Barrier Reef in Australia and the Red Sea in the Middle East (Bideci and Cater, 2019). In addition to coral reefs, they are focal points for diving tourism in areas where there are objects such as naturally or artificially sunk ships, planes, tanks (Stolk et al., 2007; Vantreeck and Schuhmacher, 1998). Today, destinations and tourism businesses working to serve diving tourism are increasing day by day and scuba diving tourists travel throughout the world visiting popular diving points in 91 countries (Johansen, 2013). The Great Barrier Reef, the Red Sea, the Caribbean Sea, Indonesia, Thailand, Philippines and Malaysia are destinations where scuba diving is an important tourism industry. These dive spots are seen as 'must dive' places that are widely promoted in social and other media (Lew et al., 2013; Salim et al., 2013). Scuba diving tourism has grown to become a multi-billion-dollar industry, drawing millions of people to explore the underwater world (De Groot and Bush, 2010; Wongthong and Harvey, 2014). The boom in scuba diving has led to significant investments in various products, from retail to education and dives, encouraging the emergence and growth in the number of scuba diving schools, diving equipment shops, and scuba diving charter businesses (Dimmock et al., 2013).

Diving tourism is very popular among alternative tourism options also in Turkey. Many diving centers and hotels on the Mediterranean Sea and the Aegean Sea coasts are investing in diving tourism. Antalya, Kaş, Ayvalık, Bodrum, Saros Bay and Gökçeada are quite popular regions for diving tourism on the Turkish coasts (as recognized by the Ministry of Culture and Tourism in 2020). According to the data of Istanbul and Marmara, Aegean, Mediterranean, Black Sea Regions (IMEAK) Chamber of Shipping (2020), there are 279 diving centers in Turkey and, 7.2 billion dollars of revenue were obtained from sea tourism in 2019. In many regions in Turkey, which has many underwater richness both ecologically and archaeologically; scuba diving is prohibited for reasons such as archaeological sites, military zones and natural reserves. Therefore, with the rapid growth in the diving tourism sector, the search for alternative diving spots is increasing.

In this chapter, artificial wreck reefs in Turkey were examined, the process of permission, the materials and structures used, and the interest of divers in these areas were evaluated. In addition, mistakes made in material and site selection and interactions with other sectors are also discussed. The aim of this section is to examine the artificial wreck reefs in Turkey, to evaluate the legislation, the materials and structures used, and the interest of divers in these areas. In addition, problems made in artificial wrecks projects and interactions with other sectors were also discussed and a general evaluation was made.

THE EFFECT OF DIVE TOURISM ON THE MARINE ENVIRONMENT

As in all tourism sectors, scuba diving is discussed under three main headings: sustainability, environmental, social and economic (Haddock-Fraser and Hampton, 2012; Wongthong and Harvey, 2014). This is the great potential that the diving tourism sector has to drive and promote conservation, attract tourism, generate income, improve people's quality of life and promote community pride (De Groot and Bush, 2010; Mota and Frausto, 2014). Diving tourism is one of the most popular recreational activities of the marine environment. According to Dearden et al. (2007) this activity "provides an important source of income for certain coastal communities and has a high potential for raising awareness of marine conservation concerns". Factors such as underwater visibility, water temperature, depth, current, distance from the diving point to the shore and accommodation facilities also affect the sustainability of diving tourism (De Groot and Bush, 2010). However, a damaged reef and an ecosystem with reduced biodiversity quickly lose their appeal.

Since divers prefer sites that are in the best possible natural condition, such as well-preserved coral reefs, diving-based tourism provides an economic incentive as well as a financial means for reef preservation (Parsons and Thur 2008). Whereas reef degradation presents a direct economic loss (Wielgus et al., 2003; Kragt et al., 2009). Diving tourism can negatively impact underwater ecosystems if not conducted in a proper manner (Davis and Tisdell, 1996; Dimmock and Musa, 2015). Therefore, more precautions for protection are taken in geographies where the diving tourism sector develops compared to other regions. However, diving into specific diving points, where there is natural diversity and biological resources – for instance, well-preserved coral reefs and well-organized varied artificial reefs like shipwrecks – is always a better option for divers (Bideci and Carter, 2019). This preference helps to create an economic market in marine-based tourism, which aids reef conservation and promotes more diving areas (Spalding et al., 2017; Tynyakov et al., 2017; Ramos et al., 2021).

Natural reefs provide a diverse and stimulating setting for diving tourism, as well as other marine-based activities. Although, the constant visitation of natural reefs or wrecks by divers might have adverse effects. Anthropogenic intrusions, such as fishing, chemical discharge, oil pollution and tourist activity, threaten reefs and lead to the loss of their biodiversity (Wilkinson 2000). Sensitive ecosystems such as coral reefs damaged by natural or anthropological cases take many years to regenerate. In addition, it is another negative situation that the creatures in the reef area leave the area completely or temporarily due to stress (Hawkins and Roberts, 1992; Harriott et al., 1997). An increase in the number of dives in an area result in more frequent physical contact between the marine habitat and divers, often occurring in habitat damage due to the divers' inexperience and ignorance (Davis and Tisdell, 1995). However, balancing the needs of local economies while protecting natural reefs is a major challenge for managers and policymakers.

CAN ARTIFICIAL REEFS BE USED FOR DIVING TOURISM?

The growing pressure on marine resources, such as coral reefs and natural wrecks due to the presence of scuba divers, also requires new strategies to divert some divers from the fragile natural reefs (Davis and Tisdell, 1995; Epstein et al., 1999). Utilizing artificial reefs to reduce diver pressure or to create new dive spots is one of the most common solutions (Hsui and Wang, 2013). Redirecting the underwater activities from natural reefs to artificial reefs can be an ideal substitution, using adequate technology to protect the environment while also meeting the demand for scuba-diving (Van Treeck and Schumacher, 1998).

Types of Artificial Reefs used for Diving

Before we explore the potential of artificial reefs it is necessary to define the term "artificial reef". It is defined as any structure placed on the seabed or water column to protect sensitive ecosystems, increase marine biodiversity, prevent illegal fishing, develop diving tourism and fisheries, or conduct scientific studies (Bohnsack and Sutherland, 1985). According to this definition, one of the uses of artificial reefs is to create suitable areas for diving tourism. It is used extensively as both conservational tools and tourism attractions in many regions in the world, including North America, Europe, Australia, the Pacific and Southeast Asia, the Caribbean Islands and the Mediterranean (Jensen, 2002). In the USA, for example, there are over 500 approved artificial reef projects in coastal waters, primarily for recreational use, with 350 of them in Florida alone (Pears and Williams, 2005).

The materials used in the design of artificial reefs vary across the globe. Concrete is the predominantly used material in European seas (Bombace et al., 1994). Concrete structures in conjunction with steel and fiberglass, are used exclusively in Japan. Rubber is used in Australia, Jamaica and The Philippines (Branden et al., 1994; Pickering, et al., 1998). In the USA the use of "materials of opportunity" is widespread (Stone et al., 1991). These include various cleansed objects, including oil and gas platforms in the Gulf of Mexico (Wilson and Van Sickle, 1987).

In countries such as the USA, Canada and Australia, objects such as decommissioned ships, aeroplanes, etc. are used to create a diving spot, while artificial coral reefs are used to create alternative diving points in regions such as the Red Sea, Australia and Southeast Asian countries such as Indonesia, Malaysia, Philippines (Harriott et al., 1997; Jakšić et al., 2013). Whilst artificial reefs are not viewed by some scientists as "perfect" substitutes for natural coral reefs, there is valid evidence that they are valued by scuba divers (Oliveira et al., 2015). Nonetheless of significance, artificial reefs have been proven to alleviate user pressure on nearby natural reef habitats and to contribute substantially to localhost economies. The experiences provided by recreating on and around artificial reefs are valued by many divers (Stolk et al., 2007). Positive feedback from divers, the demand for more artificial reef sites, and the opportunities that arise from related tourism development are some of the factors that have contributed to the growing interest in the deliberate creation of artificial reefs around the world (Harriott et al., 1997; Jakšić et al., 2013). Furthermore, it is also defined as *Diver Aggregation Device*, which allows for better control of the divers' impact on the sites (Van Treeck and Schumacher, 1998).

Coral Transplantation

Coral reefs are biogenic, three-dimensional marine habitats composed of carbonate structures that are deposited by hermatypic scleractinian corals and are generally found in areas where water temperature does not fall below 18°C for extended periods (Achituv and Dubinsky, 1990). With the development of diving tourism in recent years, coral reefs are under intense pressure due to recreational and sports underwater activities (Green and Shirley, 1999). In regions where coral reefs exist, artificial reefs are used in two ways to reduce diver pressure from diving tourism (Pickering et al., 1998; Abelson, 2006; Polak and Shashar, 2012). Firstly, 3D structures such as ships, planes, etc. are used to create an alternative diving spot to coral reefs. New dive points with artificial reefs are included in diving tours and the pressure on coral reefs is reduced. Coral has not been transplanted to these structures but over time the corals develop naturally and the artificial reef becomes more attractive to divers. Secondly, the artificial coral reefs are created by transplanting coral to artificial reefs deployed in areas far from the coral reefs. It attracts divers in a shorter time as it is supported by coral transplantation (Soong and Chen, 2003; Rinkevich, 2000). It is available when both applications are used together (Epstein et al., 2003).

Artificial Wreck Reefs

The word sunken generally describes structures such as ships, aircraft, tanks that are partially or completely sunk to the seabed by accident and/or because of war. The difference of artificial wrecks reef from this definition is that they are deliberately sunk at a point determined in line with a project, after going through the cleaning processes (Fabi et al., 2015). With the increasing popularity of artificial wreck reefs, structures such as decommissioned ships, petrol platforms, aircraft are sunk as artificial reefs to reduce the pressure of divers and/or to create an alternative diving point for diving tourism (Leeworthy et al., 2006; Edney and Spennemann, 2015). Creating new dive points using artificial wreck reefs is very popular in oceanic countries such as the USA, Canada, and Australia (Stolk et al., 2005; Shani et al., 2012). To date, more than 700 ships have been sunk as artificial reefs off the continental US coastline (Pendleton, 2005). However, such practices have begun to be seen in countries such as Malta, Greece, Cyprus, Israel and Turkey, which are located in an important ecosystem such as the Mediterranean Sea (Love et al., 2006; Jimenez et al., 2016; Şensurat-Genç et al., 2017).

Legislation to Create an Artificial Reef in the World

In the use of decommissioned vehicles such as aircraft, ships and tanks as artificial wreck reefs, guides are used for the process of cleaning and sinking these vehicles. These guidelines are usually prepared by that country's Maritime Administration Environmental Protection Agency (EPA). Also in the following international agreements, there are regulations for the process of deployment of artificial reefs. The London Convention is one of the first worldwide conventions relating to the protection of the marine environment from human activities and has been in force since 1975. In 2008, specific guidelines for the placement of artificial reefs were developed within the context of The London Convention and Protocol (London Convention and Protocol/UNEP, 2009). The Convention for the Protection of the North-East Atlantic Marine Environment (OSPAR), which came into force in 1998, unlike others, guides materials, design, location, monitoring and testing, among others, as well as the work that needs to be done before artificial reefs can be approved (OSPAR, 2009).

ARTIFICIAL WRECK REEFS FOR DIVING TOURISM IN TURKEY

The artificial reef phenomenon that started in the 1980s in Turkey aimed to protect/increase marine biodiversity and support small-scale fisheries. Similar to other Mediterranean countries, while artificial reefs are used to protect sensitive ecosystems, prevent illegal fishing and support small-scale fisheries in Turkey, there has been an increase in reef projects for the development of diving tourism after 2006 (Şensurat-Genç et al., 2017). Within the scope of 35 artificial wreck reefs projects carried out to date, 13,000 amphorae, many thematic sculptures, 22 ships, 9 planes and 1 tank have been deployed to create alternative diving spots. Although artificial reef wrecks for diving tourism are found on all coasts of Turkey, the Southern Aegean and Mediterranean coasts are home to mostly artificial wreck reefs. However, with the pressure of diving schools and local governments, applications are increasing for the creation of artificial wrecks for the development of diving tourism even in regions such as the Marmara Sea and the Black Sea where the diving season is shorter (Fig.1; Table.1)

The conditions required for scuba diving, such as water temperature and visibility in mostly EU countries, cannot be met throughout the year cause the diving tourism industry to shift to new markets such as Turkey. Countries such as Turkey, Malta and Cyprus have become more attractive to divers by using artificial wreck reefs in recent years as well as their natural and historical richness (Albayrak et al., 2021). Unlike EU countries, it is relatively easy to create artificial wreck reefs in the eastern Mediterranean, especially due to the legal situation. Therefore, there is a flow to countries such as Turkey and Malta in the European diving tourism market.

Figure 1. Artificial reef wrecks on the coast of Turkey

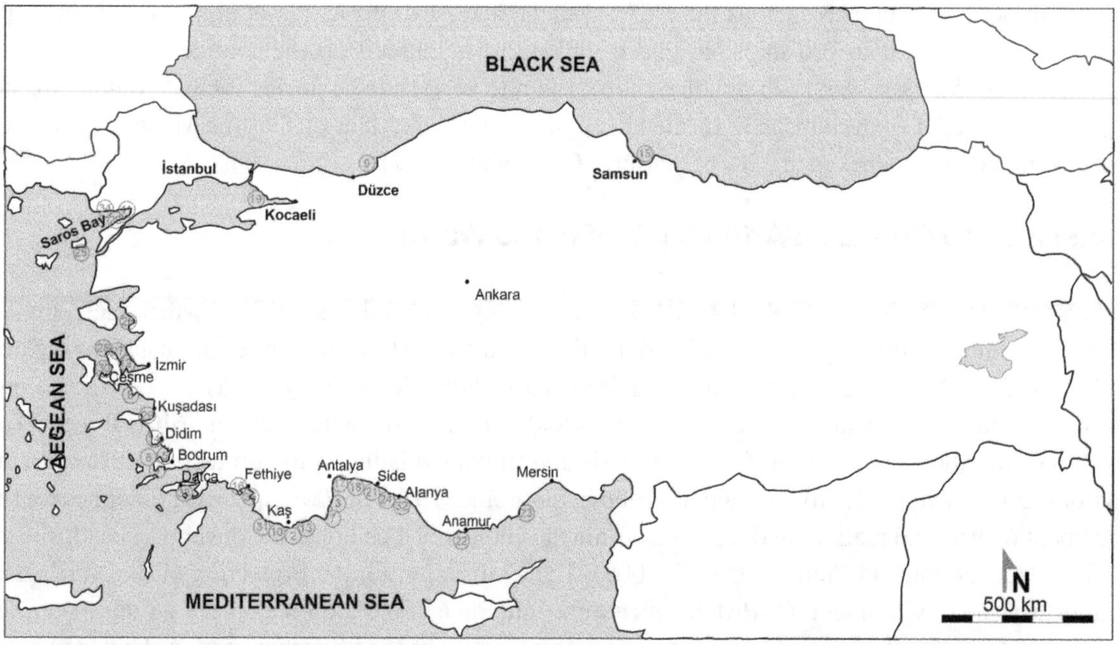

Table 1 The locations, depths and characteristics of artificial reef wrecks on the Turkish coast

No	Location	Material	Depth (m)	Date	Location	Wrecks Type
1	Alanya - Damlataş	Boat (TCSG-118)	26	2006	Mediterranean Sea	
2	Antalya - Kaş	Replica Wooden Boat	22	2006	Mediterranean Sea	
3	Antalya - Kemer	Boat (TCSG-120)	29	2007	Mediterranean Sea	
4	Bodrum Karada	Boat (Pınar-1)	30	2007	Aegean Sea	
5	Bodrum Karada	Boat (TCSG-115)	33	2007	Aegean Sea	
6	Bodrum-Karada	Airplane (C-47)	28	2008	Aegean Sea	
7	Antalya-Tekirova	Boat (TCSG-120)	32	2008	Mediterranean Sea	
8	İzmir - Sığacık	Boat (TCSG-114)	30	2010	Aegean Sea	
9	Düzce - Akçakoca	Airplane (C-47)	29	2009	Black Sea	
10	Antalya - Kaş	Airplane (C-47)	22	2009	Mediterranean Sea	
11	Saros Bay	Underwater Museum	16-20	2010	Aegean Sea	
12	İzmir - Mordoğan	Airplane (C-47)	18	2011	Aegean Sea	
13	Antalya-Çakırbağ	Boat (TCSG-119)	21	2011	Mediterranean Sea	
14	Aydın Didim	Boat (TCSG-118)	30	2011	Aegean Sea	
15	Samsun-Atakum	Boat (TCSG-125)	20	2012	Black Sea	

Table 2. (continued)

16	Fethiye	Boat (TCSG-121)	28	2012	Aegean Sea	
17	Antalya	Airplane (Dornier Do28)	25	2012	Mediterranean Sea	
18	Antalya- Üç Adalar	Airplane (C-47)	28	2013	Mediterranean Sea	
19	Kocaeli - Karamürsel	Boat (İnönü)	20	2013	Marmara Sea	
20	Saros Bay	Boat (TCSG-122)	30	2014	Aegean Sea	
21	Antalya - Side	Boat	27	2014	Mediterranean Sea	
22	Mersin - Anamur	Airplane	20	2014	Mediterranean Sea	
23	Mersin - Erdemli	Boat (TCSG-113)	25	2015	Mediterranean Sea	
24	Antalya - Side	Underwater Museum	11-25	2015	Mediterranean Sea	
25	Gökçeada	Boat (TCSG-132)	24	2016	Aegean Sea	
26	İzmir - Dikili	Boat (TCSG-51)	35	2016	Aegean Sea	
27	İzmir - Karaburun	Ferry (9 Eylül)	34	2016	Aegean Sea	
28	İzmir - Karaburun	Ferry (Alaybey)	36	2016	Aegean Sea	
29	Aydın - Kuşadası	Airbus A300	25	2016	Aegean Sea	
30	İzmir - Çeşme	Boat (TCSG 68)	30	2017	Aegean Sea	
31	Antalya - Kaş	Tank	20	2017	Mediterranean Sea	

Table 3. (continued)

32	Alanya	Boat (TCSG-127)	27	2017	Mediterranean Sea	
33	Datça	Boat (TCSG-52)	20	2018	Aegean Sea	
34	Saros Bay	Airplane (A330-300)	30	2019	Aegean Sea	
35*	Fethiye	Underwater Museum	30	2022	Aegean Sea	

* It will be created in 2022.

Legislation to Create an Artificial Reef in Turkey

Until 1999, there was no legal regulation on artificial reef studies in Turkey. With the increasing interest in artificial reefs, "The Artificial Reef Applications Project Planning Guide" was prepared by the Ministry of Food, Agriculture and Livestock. At present, the permission of artificial reef projects is subject to the permission of the Ministry of Agriculture and Forestry, together with this guide. Institutions and organizations that want to realize an artificial reef project should apply to the Department of Resource Management and Fisheries Structures of the General Directorate of Fisheries and Aquaculture, with a preliminary study report to be prepared by universities (Lök et al., 2011). This form must include purposes of the project, type of vessel, location of sinking site, sinking method, fisheries condition of the site, user groups and possible benefits etc. Although there is no roadmap to be followed in sinking vehicles such as ships and aircraft to be used as artificial wreck reefs in this guide, it is also insufficient on how to prepare these decommissioned vehicles before they are sunk. Although there is no legal obligation, the above-mentioned US guide has been tried to be followed in cleaning the ships and planes that will be used as sunken ships. In this context, vehicles are covered with concrete after all materials are removed and the bilge line is washed until the shell and body remain. Concrete coating is not applied to all vehicles, only the bilge line of some vehicles is washed and holes are drilled to accelerate the sinking. In these preparation processes, paint cleaning is not performed on the outer surfaces due to the high cost. Additionally, the coordinates of the place where the artificial wreck are informed to the General Directorate of Navigation, Hydrography and Oceanography (ONHO).

Antalya: When diving tourism in Turkey is mentioned, Antalya comes to mind first. Kaş, Kemer, Üç Adalar and Side within the borders of Antalya are the most important diving spots. In addition to the natural reefs, 7 boats, 3 planes and 1 tank were artificially sunk for diving tourism in Antalya. Otherwise, Side underwater museum, which is the second-largest museum in the world in terms of the number of sculptures, was opened in 2015. In the museum, 117 sculptures in 5 different themes are located between 11-25 m deep. The cost of diving for a day in the Side underwater museum varies between 25-40 dollars (Fig.2).

Figure 2. Photographs of artificial wreck reefs in Antalya; Artificial Wreck Reefs

Bodrum Karaada: The artificial reef area in Bodrum, one of the most important tourism centers of Turkey, was created in 2007. With the joint efforts of local dive centers, 2 ships and 1 aircraft were placed at an average depth of 30m after cleaning. The artificial wreck reefs are visited by 150 thousand divers per year (Fig.3).

Figure 3. Photographs of artificial wreck reefs in Bodrum Artificial Wreck Reefs

Saros Bay Artificial Wreck Reefs: The artificial reef area, established in 2010 in Saros Bay, which is one of the important diving points of the North Aegean Sea, draws attention with its different applications. In the project, besides decommissioned vehicles such as ships, aircraft and tanks, artificial reef blocks were also used to increase the biodiversity in the region. In addition, there is an underwater museum with the theme of the Çanakkale war to evaluate the potential of the Çanakkale region, which receives thousands of visitors every year (Fig.4).

Figure 4. Photographs of artificial wreck reefs in Saros Artificial Reefs Area

Karaburun Artificial Wreck Reefs: To develop diving tourism and create an alternative to natural reefs in the small town of Izmir, Karaburun, two ferries were placed in the sea after being cleaned in 2016. Passenger ferries named 9 Eylül and Alaybey, which have cultural significance for İzmir, were used as artificial wreck reefs. With a length of 46.7 m and a width of 7.5 m, the depth in the area where the Ferries sit on the seafloor is approximately 37 meters. In Karaburun, where the tourism season is generally limited to summer months, diving tourism is developing rapidly with the effect of artificial wreck reefs. Today, 4 diving centers operate in the Region. Approximately 12 500 dives were made in artificial reef wrecks in 2019. Within the scope of a project completed last year, unlike other artificial sunken reef areas, the effects of shipwrecks in this region on the marine environment and diving tourism in the region were examined (Şensurat-Genç et al., 2021). In particular, it is expected that such scientific studies will be increased and provide data to decision-making institutions.

Figure 5. Photographs of artificial wreck reefs in Karaburun Artificial Wreck Reefs

CONCLUSION

Decommissioned military-civilian ships, planes and tanks have been sunk as artificial reefs in Turkey seas since 2006. The main goal of sinking these vehicles is to contribute to diving tourism, as in some other countries (United States, Canada, Australia, etc.). In addition, these artificial reefs also serve to increase marine biodiversity and create new hunting grounds for sportive and commercial fisheries in the region. The diving tourism sector is growing rapidly, especially with the support of artificial wreck reefs at many points in the southern Aegean Sea and Mediterranean coasts. This increasing interest has spread to other regions as well, and efforts to create artificial wreck reefs are increasing even in regions where the diving season is limited such as the Black Sea and Marmara coasts, especially in the northern Aegean Sea.

Many diving points in Turkey are closed to diving due to military zones, archaeological or natural protection areas. With the development of the diving tourism sector, natural reefs can't meet the current demand due to environmental pollution, overfishing or other anthropogenic effects. Although making use of artificial reef wrecks seems like a solution, for now, there are very few scientific studies on the effects of these structures on the marine environment and the level of attracting divers in a sensitive ecosystem like the Mediterranean. The number of scientific studies on artificial submerged reefs in Turkey is quite limited. In these studies, the fish community structure in the shipwrecks was examined (Lök and Özgül, 2010; Acarlı et al., 2020). Along with the fish community structure, the number of scientific studies that evaluate the effects of artificial shipwrecks on the marine ecosystem and diving tourism should be increased. There is no doubt that such studies will contribute to the management of artificial wreck reefs.

Artificial wreck reefs are much more advantageous for wreck dive training and skill development than natural shipwrecks. Since these wrecks are in shallower and more controlled areas, they have a significant potential for both diving safety and reducing diver effects on historical wrecks. In addition, these alternative diving spots are preferred for underwater photography, technical and deep diving training. Similar to the examples in the world, thematic underwater parks or museums have begun to be established in Turkey. In these applications, which are purely for diving tourism, different cultural or historical objects are used instead of decommissioned vehicles. This practice, which was carried out for the first time in Turkey in Saros Bay, was followed by the Side underwater museum. In the next year, an underwater thematic museum will be established in Fethiye to try to take a share of the underwater tourism market. On the other hand, it is to bring the shipwrecks from the 1st World War in the Dardanelles to tourism. Under the name of Gallipoli Historical Underwater Park, 105-year-old historical wrecks are opened to diving tourism. After the wreck of HMS Majestic, more than 22 shipwrecks are expected to be brought into diving tourism.

Artificial wreck reefs applications for diving tourism are increasing in Turkey. In the face of the demands of dive centers to create new artificial wreck sites, a commission consisting of universities, fishermen, NGOs and diving centers should be established under the relevant ministries, and an active role should be taken in the evaluation and management of artificial reef projects. In addition, the guide used in the design of artificial reef projects in Turkey needs to be updated. In particular, there are important deficiencies in the cleaning and immersion of vehicles such as ships and aircraft, which are used as artificial reefs. The cleaning of the buildings to be used as artificial wrecks, site selection, deployment and who will supervise all these processes should be handled meticulously. In addition, the creation of uncontrolled artificial reef areas by diving centers or similar institutions should be prevented. Enforcement should be increased by the Ministry and the number and regions of artificial reefs to be used for diving tourism should be determined.

REFERENCES

Abelson, A. (2006). Artificial reefs vs coral transplantation as restoration tools for mitigating coral reef deterioration: Benefits, concerns, and proposed guidelines. *Bulletin of Marine Science*, *78*(1), 151–159.

Acarlı, D., Semih, K., & Kocabaş, S. (2020). TCSG-132 Gemi Batığı Yapay Resifinin (Gökçeada, Kuzey Ege Denizi) Biyoçeşitliliği. Acta Aquatica Turcica, 16(3), 313–329. doi:10.22392/actaquatr.677175 doi:10.22392/actaquatr.677175

Achituv, Y., & Dubinsky, Z. (1990). Evolution and zoogeography of coral reefs. In Ecoystems of the World, Volume 25 — Coral Reefs. Elsevier Science.

Albayrak, T., Caber, M., & Cater, C. (2021). Mass tourism underwater: A segmentation approach to motivations of scuba diving holiday tourists. *Tourism Geographies*, *23*(5-6), 5–6, 985–1000. doi:10.1080/14616688.2019.1696884

Bideci, C., & Cater, C. (2019). Dive and Novelty Seeking in Experimental Artificial Reefs. *Journal On Tourism & Sustainability*, *3*(1), 22–31.

Bohnsack, J. A., & Sutherland, D. L. (1985). Artificial Reef Research: A review with Recommendations for Future Priorities. *Bulletin of Marine Science, 37*(1), 11–39.

Bombace, G., Fabi, G., Fiorentini, L., & Speranza, S. (1994). Analysis of the efficacy of artificial reefs located in five different areas of the Adriatic Sea. *Bulletin of Marine Science, 55*, 559–580.

Branden, K. L., Pollard, D. A., & Reimers, H. A. (1994). A review of recent artificial reef developments in Australia. *Bulletin of Marine Science, 55*(2-3), 982–994.

CBI. (2019). *Dive tourism from Europe. Centre for the Promotion of Imports from Developing Countries.* Retrieved from https://www.cbi.eu/market-information/tourism/dive-tourism/europe/

Chamber of Shipping of Turkey. (2020). https://www.denizticaretodasi.org.tr/media/SharedDocuments/sektorraporu/MaritimeSectorReport2021.pdf

Davis, D., & Tisdell, C. (1995). Recreational scuba-diving and carrying capacity in marine protected areas. *Ocean and Coastal Management, 26*(1), 19–40. doi:10.1016/0964-5691(95)00004-L

Davis, D., & Tisdell, C. (1996). Economic Management of Recreational Scuba Diving and the Environment. *Journal of Environmental Management*, (February), 229–248.

De Groot, J., & Bush, S. R. (2010). The potential for dive tourism led entrepreneurial marine protected areas in Curacao. *Marine Policy, 34*(5), 1051–1059. doi:10.1016/j.marpol.2010.03.004

Dearden, P., Bennett, M., & Rollins, R. (2007). Perceptions of diving impacts and implications for reef conservation. *Coastal Management, 35*(2-3), 305–317. doi:10.1080/08920750601169584

Dimmock, K. (2007). Scuba diving, snorkeling and free diving. In G. Jenning (Ed.), *Water-based tourism, sport, leisure, and recreation experiences* (pp. 128–148). Butterworth-Heinemann. doi:10.1016/B978-0-7506-6181-2.50013-0

Dimmock, K., & Cummins, T. (2013). History of scuba diving tourism. In G. Musa & K. Dimmock (Eds.), *Scuba Diving Tourism* (pp. 14–28). Routledge.

Dimmock, K., & Musa, G. (2015). Scuba Diving Tourism System: A framework for collaborative management and sustainability. *Marine Policy, 54*, 52–58. doi:10.1016/j.marpol.2014.12.008

Edney, J., & Spennemann, D. H. (2015). Can artificial reef wrecks reduce diver impacts on shipwrecks? The management dimension. *Journal of Maritime Archaeology, 10*(2), 141–157. doi:10.100711457-015-9140-5

Epstein, N., Bak, R. P. M., & Rinkevich, B. (1999). Implementation of a small-scale "no-use zone" policy in a reef ecosystem: Eilat's reef-lagoon six years later. *Coral Reefs, 18*(4), 327–332. doi:10.1007003380050207

Epstein, N., Bak, R. P. M., & Rinkevich, B. (2003). Applying forest restoration principles to coral reef rehabilitation. *Aquatic Conservation, 13*(5), 387–395. doi:10.1002/aqc.558

Garrod, B., & Gössling, S. (2008). Introduction. In B. Garrod & S. Gössling (Eds.), *New Frontiers in Marine Tourism: Diving experiences, Sustainability, Management* (pp. 8–29). Elsevier Press. doi:10.1016/B978-0-08-045357-6.50004-8

Green, E. P., & Shirley, F. (1999). The global trade in coral. WCMC, World Conservation Press.

Haddock-Fraser, J., & Hampton, M. P. (2012). Multistakeholder values on the sustainability of dive tourism: Case studies of Sipadan and Perhentian Islands, Malaysia. *Tourism Analysis*, *17*(1), 27–41. doi:10.3727/108354212X13330406124016

Harriott, V. J., Davis, D., & Banks, S. A. (1997). Recreational diving and its impact in marine protected areas in eastern Australia. *Ambio*, *26*(3), 173–179.

Hawkins, J. P., & Roberts, C. M. (1992). Effects of recreational SCUBA diving on fore-reef slope communities of coral reefs. *Biological Conservation*, *62*(3), 171–178. doi:10.1016/0006-3207(92)91045-T

Hsui, C. Y., & Wang, C. C. (2013). Synergy between fractal dimension and lacunarity index in design of artificial habitat for alternative SCUBA diving site. *Ecological Engineering*, *53*, 6–14. doi:10.1016/j.ecoleng.2013.01.014

Jakšić, S., Stamenković, I., & Đorđević, J. (2013). Impacts of artificial reefs and diving tourism. *Turizam*, *17*(4), 155–165. doi:10.5937/Turizam1304155J

Jensen, A. (2002). Artificial reefs of Europe: Perspective and future. *ICES Journal of Marine Science*, *59*(1), 3–13. doi:10.1006/jmsc.2002.1298

Jimenez, C., Hadjioannou, L., Petrou, A., Andreou, V., & Georojou, A. (2016). Fouling communities of two accidental artificial reefs (modern shipwrecks) in Cyprus (Levantine Sea). *Water (Basel)*, *9*(1), 2–11. doi:10.3390/w9010011

Johansen, K. (2013). Scuba diving education and training. In G. Musa & K. Dimmock (Eds.), *Scuba diving tourism: contemporary geographies of leisure, tourism and mobility* (pp. 81–96). Routledge.

Kragt, M. E., Roebeling, P. C., & Ruijs, A. (2009). Effects of Great Barrier Reef degradation on recreational reef-trip demand: A contingent behaviour approach. *The Australian Journal of Agricultural and Resource Economics*, *53*(2), 213–229. doi:10.1111/j.1467-8489.2007.00444.x

Leeworthy, V. R., Maher, T., & Stone, E. A. (2006). Can artificial reefs alter user pressure on adjacent natural reefs? *Bulletin of Marine Science*, *78*(1), 29–37.

Lew, A. A. (2013). A world geography of recreational scuba diving. In G. Musa & K. Dimmock (Eds.), *Scuba Diving Tourism* (pp. 29–51). Routledge.

Lök, A., Düzbastılar, F. O., Gül, B., Özgül, A., & Ulaş, A. (2011). The role of artificial reefs in fisheries management in Turkey. In S. A. Bortone, F. P. Brandini, G. Fabi, & S. Otake (Eds.), *Artificial Reefs in Fisheries Management* (pp. 155–166). CRC Press.

Lök, A., & Özgül, A. (2010). Fish assemblages around fish aggregation devices in Aegean Sea coast of Turkey. *Rapports et Procès Verbaux des Réunions - Commission Internationale pour l'Exploration Scientifique de la Mer Méditerranée*, *39*, 573.

London Convention and Protocol/UNEP. (2009). London Convention and Protocol/UNEP Guidelines for the Placement of Artificial Reefs.

Love, M. S., Schroeder, D. M., Lenarz, W., MacCall, A., Bull, A. S., & Thorsteinson, L. (2006). Potential use of offshore marine structures in rebuilding an overfished rockfish species, bocaccio (*Sebastes paucispinis*). *Fish Bulletin*, *104*, 383–390.

Mota, L., & Frausto, O. (2014). The Use of Scuba Diving Tourism for Marine Protected Area Management. *International Journal of Social, Behavioural, Educational, Economic, Business and Industrial Engineering*, *8*(10), 3159–3164.

Musa, G., & Dimmock, K. (2013). Introduction: scuba diving tourism. In G. Musa & K. Dimmock (Eds.), *Scuba Diving Tourism* (pp. 1–13). Routledge.

National Oceanic and Atmospheric Administration (NOAA). (2019). *Tourism and recreation*. Retrieved from https://coast.noaa.gov/states/fast-facts/tourism-and-recreation.html

Oliveira, M. T., Ramos, J., & Santos, M. N. (2015). An approach to the economic value of diving sites: Artificial versus natural reefs off Sal Island, Cape Verde. *Journal of Applied Ichthyology*, *31*, 86–95. doi:10.1111/jai.12953

OSPAR Commission. (2009). Assessment of construction or placement of artificial reefs. London: Biodiversity Series, publ. no. 438/2009.

PADI Worldwide Certification. (2021). *Statistics*. https://www.padi.com/corporate/company-info

Parsons, G. R., & Thur, S. M. (2008). Valuing Changes in the Quality of Coral Reef Ecosystems: A Stated Preference Study of SCUBA Diving in the Bonaire National Marine Park. *Environmental and Resource Economics*, *40*(4), 593–608. doi:10.100710640-007-9171-y

Pears, R. J., & Williams, D. M. (2005). *Potential effects of artificial reefs on the Great Barrier Reef: Background paper.* CRC Reef Research Centre Technical Report No. 60. CRC Reef Research Centre.

Pendleton, L. H. (2005). Understanding the potential economic impacts of sinking ships for scuba recreation. *Marine Technology Society Journal*, *39*(2), 47–52. doi:10.4031/002533205787443944

Pickering, H., Whitmarsh, D., & Jensen, A. (1998). Artificial reefs as a tool to aid rehabilitation of coastal ecosystems: Investigating the potential. *Marine Pollution Bulletin*, *37*(8–12), 505–514.

Polak, O., & Shashar, N. (2012). Can a small artificial reef reduce diving pressure from a natural coral reef? Lessons learned from Eilat, Red Sea. *Ocean and Coastal Management*, *55*, 94–100. doi:10.1016/j.ocecoaman.2011.10.006

Ramos, J., Tuaty-Guerra, M., Almeida, M., Raposo, A. C., Gaudêncio, M. J., Silva, A. D., Rodrigues, N., Leandro, S. M., & Caetano, M. (2021). An artificial reef at the edge of the deep: An interdisciplinary case study. *Ocean and Coastal Management*, *210*, 105729. doi:10.1016/j.ocecoaman.2021.105729

Rinkevich, B. (2000). Steps towards the evaluation of coral reef restoration by using small branch fragments. *Marine Biology*, *136*(5), 807–812. doi:10.1007002270000293

Salim, N., Bahauddin, A., & Mohamed, B. (2013). Influence of SCUBA divers' specialization on their underwater behavior. *Worldwide Hospitality and Tourism Themes*, *5*(4), 388–397. doi:10.1108/WHATT-03-2013-0015

Şensurat-Genç, T., Lök, A., Özgül, A., Irmak, E., & Özkan, E. Y. (2021). Determination of Contribution to Diving Tourism and Environmental Effects of Shipwrecks Used as Artificial Reefs. TUBITAK 117Y0383project report.

Şensurat-Genç, T., Özgül, A., & Lök, A. (2017). The Use of Artificial Reefs for Recreational Diving. *Turkish Journal of Maritime and Marine Sciences*, *3*(1), 27–33.

Shani, A., Polak, O., & Shashar, N. (2011). Artificial reefs and mass marine ecotourism. *Tourism Geographies*, *1*, 1–22.

Soong, K., & Chen, T. A. (2003). Coral transplantation: Regeneration and growth Acrophore Fragments in a nursery. *Restoration Ecology*, *11*(1), 62–71. doi:10.1046/j.1526-100X.2003.00100.x

Spalding, M., Burkeb, L., Wood, S. A., Ashpolee, J., Hutchisone, J., & Zu Ermgassene, P. (2017). Mapping the global value and distribution of coral reef tourism. *Marine Policy*, *82*, 104–113. doi:10.1016/j.marpol.2017.05.014

Stolk, P., Markwell, K., & Jenkins, J. (2005). Perceptions of artificial reefs as scuba diving resources: A study of Australian recreational scuba divers. *Annals of Leisure Research*, *8*(2&3), 153–173. doi:10.1080/11745398.2005.10600968

Stolk, P., Markwell, K., & Jenkins, J. (2007). Artificial reefs as recreational scuba diving resources: A critical review of research. *Journal of Sustainable Tourism*, *15*(4), 331–350. doi:10.2167/jost651.0

Stone, R. B., Mc Gurrin, J. M., Sprague, L. M., & Seaman, W. (1991). Artificial habitats of the world: synopsis and major trends. In J. R. W. Seaman & L. Sprague (Eds.), *Artificial Habitats for Marine and Freshwater Fisheries*. Academic Press Inc. doi:10.1016/B978-0-08-057117-1.50008-1

Tabata, R. S. (1992). Scuba diving holidays. In B. Weiler & C. M. Hall (Eds.), *Special Interest Tourism* (pp. 171–184). Belhaven Press.

Treeck, V. P., & Schuhmacher, H. (1998). Mass Diving Tourism – A New Dimension Calls for New Management Approaches. *Marine Pollution Bulletin*, *37*(8-12), 499–504. doi:10.1016/S0025-326X(99)00077-6

Tynyakov, J., Rousseau, M., Chen, M., Figus, O., Belhassen, Y., & Shashar, N. (2017). Artificial reefs as a means of spreading diving pressure in a coral reef environment. *Ocean and Coastal Management*, *149*, 159–164. doi:10.1016/j.ocecoaman.2017.10.008

Wielgus, J., Glassom, D., & Fishelson, L. (2003). Long term persistence of low coral cover and abundance on a disturbed reef flat in the northern Red Sea. *Journal of Experimental Marine Biology and Ecology*, *297*(1), 31–41. doi:10.1016/S0022-0981(03)00354-X

Wilkinson, C. R. (2000). Executive summary. In C. R. Wikinson (Ed.), *Status of Coral Reefs of the World* (pp. 7–19). Australian Institute of Marine Science.

Wilson, C. A., & Van Sickle, V. R. (1987). Development of the Louisiana artificial reef pro-gramme. *Bulletin of Marine Science*, *44*(2), 1071–1072.

Wongthong, P., & Harvey, N. (2014). Integrated coastal management and sustainable tourism: A case study of the reef-based SCUBA dive industry from Thailand. *Ocean and Coastal Management*, *95*, 138–146. doi:10.1016/j.ocecoaman.2014.04.004

Yaşar, O. (2011). Underwater Diving Tourism on The Coasts of The Gulf of Saros. *Zeitschrift für die Welt der Türken/Journal of World of Turks ZFWT*, *3*(1), 33-55.

Chapter 9
Human Resources in an Artificial Reef Deployment:
A Case Study in Damas Beach, Trenggalek, Indonesia

Andik Isdianto
University of Brawijaya, Indonesia

Oktiyas Muzaky Luthfi
https://orcid.org/0000-0002-9550-9381
University of Brawijaya, Indonesia

Muhammad Arif Asadi
University of Brawijaya, Indonesia

Muchamad Fairuz Haykal
University of Brawijaya, Indonesia

Berlania Mahardika Putri
University of Brawijaya, Indonesia

Candra Adi Intyas
University of Brawijaya, Indonesia

Mochammad Fattah
University of Brawijaya, Indonesia

Pudji Purwanti
University of Brawijaya, Indonesia

Edi Susilo
University of Brawijaya, Indonesia

ABSTRACT

Artificial reefs are structures that can be natural or artificial materials that are intentionally placed on the seabed, placed at one time or over a long period of time, and ideally will work as part of a natural ecosystem. In artificial reefs deployment, it is necessary to pay attention to several factors so that the artificial reef itself functions optimally. Therefore, competent human resources are needed through the correct technical understanding of artificial reef deployment. The less-than-optimal deployment of artificial reefs in Damas waters is caused by a lack of understanding of human resources, so that when monitoring was carried out, it was found that 22 of the 25 artificial reefs found were mostly in poor condition.

DOI: 10.4018/978-1-6684-2344-8.ch009

INTRODUCTION

Damas Beach is an unspoiled beach located in Karanggandu Village, Trenggalek Regency (Indonesia). The condition of coral reefs at Damas Beach is categorized as bad because there are much rubble caused by fishing nets stuck in this area. Based on the problems in the region, a solution is needed to submerge the artificial reefs. The process of sinking artificial reefs was carried out in September 2017 with a total of 25 artificial reefs sunk in the shape of a cube in the waters of Damas Beach. The condition of artificial reefs that have been lowered in the waters of Damas Beach within a period of 2.5 years must be monitored with the aim of knowing the physical characteristics of the artificial reefs. However, in practice, the drowning process still lacks good technical attention, such as not properly identifying the surrounding environmental conditions. Therefore, in 2019, a monitoring team was formed, and dives were conducted to look for the presence of the artificial reef.

The diving activity was conducted for 4 days, and the previous team succeeded in finding the sinking location of the artificial reef. The condition of artificial reefs is documented in the waters. One month later, monitoring activities were conducted again to identify the artificial reefs and to collect data from various parameters, namely hydro-oceanography, physics and chemistry, detailed conditions of artificial reefs, sedimentation rate, depth of artificial reefs found, and biofouling colonies attached to and another biota such as fish and other invertebrates.

BACKGROUND

The Artificial Reef Structures

Artificial reefs are structures that are intentionally placed on the seabed to achieve the biological but also socio-economic or engineering goals of one or more stakeholders. The structure can be of natural or artificial material, placed at one time or over a long period, and will ideally serve as a habitat that functions as part of a natural ecosystem (Lindberg & Seaman, 2011). The forms of artificial reefs certainly have advantages and disadvantages depending on the type of base material.

These objects are expected to be a place for new coral animals to attach (Sirait et al., 2021; Thaeraniza et al., 2020). The submerged artificial reef model is a hollow concrete cube model. The surface area of the artificial reef was measured directly underwater using a sewing meter. Each rib is 60 cm long and 8 cm thick. The dimensions are presented in Figure 1.

Figure 1. Artificial Reef Size
Source: (Sirait et al., 2021)

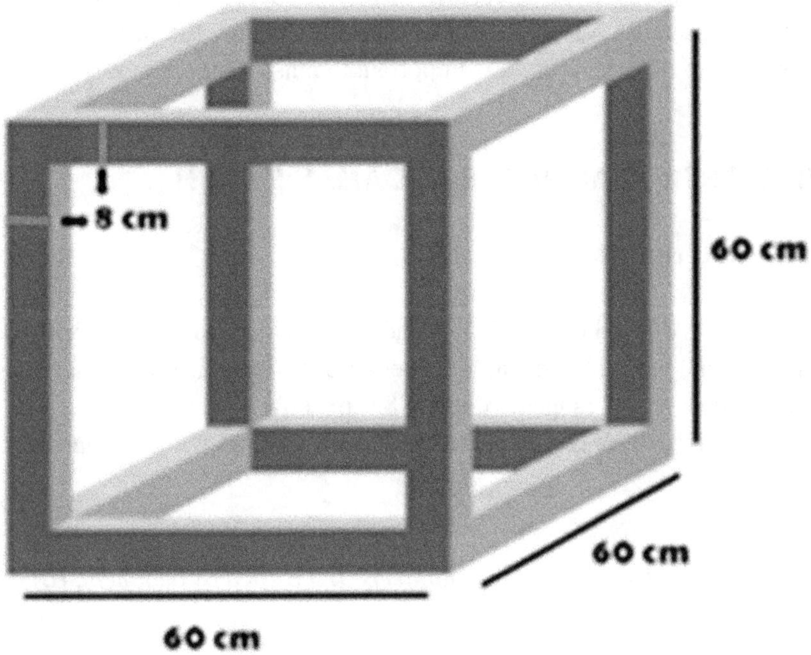

Management and Organizational Concerns

The sinking of artificial reefs at Damas Beach is part of the community service program in 2017 which was initiated by a lecturer from Universitas Brawijaya, namely Dr. Edi Susilo, MS, Dr. Ir. Pudji Purwanti, MP and Mochammad Fattah, S.Pi., M.Si in collaboration with local community groups. The purpose of sinking artificial reefs is to improve the condition of ecosystems that have been damaged due to fishing gear that are not environmentally friendly. In addition, the sinking of this artificial reef is in the hope of creating underwater beauty as a tourist attraction and the emergence of reef fish for fishing activities. However, in practice and supervision is not appropriate so that the results of this activity are not optimal. This failure does not escape the unclear distribution of the roles of the community and other stakeholders involved.

Underwater beauty in water can be used as ecotourism. Natural ecotourism in conservation areas aims to preserve the biodiversity of its ecosystems and generate income for the benefit of the area, local communities, local governments, and managers. The law on the financial balance between the central government and regional governments, especially in planning development activities independently, is expected to be able to optimize every available resource for sustainable and environmentally sound development. Based on this, in the management of resources, especially natural resources that are environmentally sound in the form of the development of natural tourism and ecotourism based on strengthening the role of the region and the community (Latupapua, 2008).

According to Suwantoro (2004), several more criteria to be considered for selecting ecotourism products, namely: 1) Education and information aspects; 2) Socio-cultural aspects of tourist destinations; 3) Environmental aspects; 4) Aesthetic aspects; 5) Ethical and reputational aspects. The community as a crucial element requires direct involvement in the arrangement of tourist areas. The role of the community is needed in providing quality services for tourists and preserving the surrounding environment so that tourism can continue to run, therefore it is important to make the community a tourism-aware society.

MONITORING ARTIFICIAL REEFS AT DAMAS BEACH

Physical Characteristics

The diving activity was carried out for 4 days and the previous team succeeded in finding the sinking location of the artificial reef. The results of observations of physical characteristics obtained that the depth of artificial reefs found ranged from 5.0-8.9 m. Physical characteristics of the artificial reefs found; artificial reefs buried by sediment; artificial reefs covered with nets and inclined artificial reefs (Figures 2 and 3).

Figure 2. Physical Characteristics of Artificial Reefs; (A) Buried by Sediment, (B) Covered by Nets and (C) Sloping Position
Source: (Sirait et al., 2021)

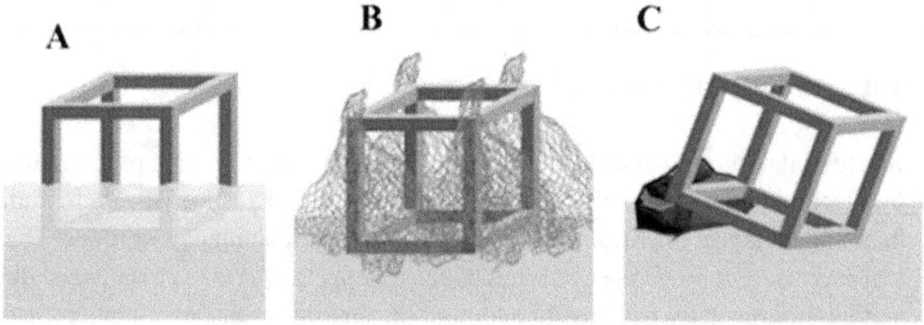

The value of sediment stockpile for each AR ranges from 10-21 cm. The results were obtained because the AR placement substrate is a muddy sand substrate. The physical characteristics of the AR covered by the net found only 1 AR out of 22 AR. The net coverage area on the sediment-covered surface is 15 AR from 22 AR. The closure of artificial reefs by nets is caused by net waste from floating net cages belonging to the community which is right on the surface of the waters. The existence of floating net cages causes a decrease in water quality and the presence of domestic waste in the bottom of the water. The domestic waste is in the form of aquaculture fish feed waste and net waste found at the bottom of the waters.

Figure 3. Physical Characteristics of Artificial Reefs; (A) Good, (B) Buried by Sediment, (C) Covered with Nets
Source: (Sirait et al., 2021)

The results of subsequent monitoring were found that AR had a slope which was obtained as much as 3 AR from 22 AR. The AR slope values found ranged from 5-33°. The slope of the artificial reef is caused by the process of laying artificial reefs using only ropes from the surface of the waters and not being directed by divers so that the AR structure is above the rubble. The process of lowering artificial reefs that experience a slope because it is not in accordance with the procedure so that the position of the artificial reef that is lowered is tilted. The procedure that must be observed is to bring the artificial reef structure by boat and be lowered with direct guidance by divers. Topography of the bottom of the waters with a maximum slope of 30° and wide, so that the artificial reef grip is strong. Physical characteristics of artificial reefs are described in Table 1.

Table 1. Data on the physical characteristics of artificial reefs

Physical Characteristics of Artificial Reefs							
AR	d (m)	Buried by Sediment (cm)	Category	Covered with Nets (%)	Category	Slope (°)	Category
1	5.0	-	-	-	-	33	Heavy
2	5.2	-	-	-	-	-	-
3	5.5	-	-	-	-	-	-
4	7.0	10	Slight	-	-	-	-
5	6.9	13	Medium	-	-	-	-
6	6.5	12	Medium	-	-	-	-
7	7.5	19	Medium	-	-	-	-
8	7.1	21	Heavy	75%	Heavy	-	-
9	8.6	13	Medium	-	-	-	-
10	8.1	20	Heavy	-	-	-	-
11	7.3	-	-	-	-	-	-
12	7.6	15	Medium	-	-	-	-
13	7.3	10	Slight	-	-	-	-
14	6.5	-	-	-	-	10	Slight
15	6.5	-	-	-	-	5	Slight
16	7.2	11	Medium	-	-	-	-
17	6.9	-	-	-	-	-	-
18	7.7	14	Medium	-	-	-	-
19	8.9	12	Medium	-	-	-	-
20	7.9	10	Slight	-	-	-	-
21	7.7	10	Slight	-	-	-	-
22	6.6	10	Slight	-	-	-	-

Source: (Sirait et al., 2021)

Artificial Reef Surface Area

Characteristics of AR that have been placed for 2.5 years at the study site, found 15 AR buried in sediment, 1 AR covered with nets, and 3 AR found in an inclined position. The use of AR as a medium to reduce damage to coral reef ecosystems is the right choice. But location selection must be considered before placing AR. AR surface area data are presented in Table 2.

Table 2. Data of artificial reef surface area

Artificial Reef	Physical Characteristics	Artificial Reef Surface Area (cm^2)
1	Sloping Position 33°	18.176
2	Good Condition	18.176
3	Good Condition	18.176
4	Buried by Sediment 10 cm	13.056
5	Buried by Sediment 13 cm	12.672
6	Buried by Sediment 12 cm	12.800
7	Buried by Sediment 19 cm	11.904
8	Buried by Sediment 21 cm & Covered with Nets 75%	11.648
9	Buried by Sediment 13 cm	12.672
10	Buried by Sediment 20 cm	11.776
11	Good Condition	18.176
12	Buried by Sediment 15 cm	12.416
13	Buried by Sediment 10 cm	13.056
14	Sloping Position 10°	18.176
15	Sloping Position 5°	18.176
16	Buried by Sediment 11 cm	12.928
17	Good Condition	18.176
18	Buried by Sediment 14 cm	12.544
19	Buried by Sediment 12 cm	12.800
20	Buried by Sediment 10 cm	13.056
21	Buried by Sediment 10 cm	13.056
22	Buried by Sediment 10 cm	13.056
Average		**14.394**

Source: (Sirait et al., 2021)

For the placement of AR within the 2.5-year period of laying it cannot be described specifically whether the placement of AR in the waters of Damas Beach will form a coral reef ecosystem due to the overall condition of the AR structure which is in poor condition.

WATER PARAMETERS AT DAMAS BEACH

Physics Oceanography

Depth

The results of depth measurements on Damas beach range from 0-45 meters (Fikri et al., 2021b). Most corals grow well at depths of less than 25 m. A depth of 5-10 meters is the optimal depth because it will facilitate the intensity of sunlight in helping the development of adherent biota (Purnawati & Hartati, 2007). At depths exceeding 70 m, coral reefs are difficult to find. Coral reefs can also live at depths of up to 50 m with truly clear water. This is related to the process of light penetration into the waters for photosynthesis (As- Syakur & Wiyanto, 2016). A useful criterion established by the Philippine Agency for Fisheries and Aquatic Resources for the placement of artificial reefs is to be at a depth of 5-20 meters, protected from waves, but easily accessible by local fishers. Based on research (Suryanti et al., 2011), depth has no effect on coral morphology, this is because at different depths the types of coral-based on the morphology found there are relatively no significant differences, only the number that dominates is different. Hermatypic corals are limited to a depth of about 70 meters from the water surface in clear seas with temperatures above 20° C throughout the year (Restu, 2016).

Brightness

Brightness is the level of intensity of sunlight that penetrates water, so it is strongly influenced by turbidity (Salim et al., 2017). The brightness of the water of water is a key factor both for the life of biota in the seawater column and for tourism objects in the sea. The brightness value is in accordance with the established criteria, which is >5 meters for coral reefs (Rizqika et al., 2018). Light plays a role in the coral calcification process. The amount of light available for daily coral growth is not only determined by the light intensity but also by the photoperiod or light duration (Schutter et al., 2012). Good brightness for coral growth is optimal light penetration for photosynthesis. The brightness of reef waters can exceed 50 meters on open ocean reef areas and can be less than one meter for example after a storm occurs around fringing reefs. The range of water brightness determines the depth of coral growth, different species have different tolerances to light levels, both maximum and minimum, so this also greatly determines the variation of the reef community (Restu, 2016). The results of the measurement of brightness on the Damas beach range from 2-6 meters (Fikri et al., 2021a).

Temperature

Temperature is an important indicator to determine the subsequent effect on the value of other water parameters, such as accelerating the occurrence of chemical reactions, reducing the solubility of gases in water, or increasing the smell or taste. The natural temperature for tropical waters suitable for living organisms ranges from 23-32°C (Siburian et al., 2017). The optimal temperature for coral growth ranges from 27-29°C, where the maximum temperature at which coral reefs are still alive is 36°C, and the lowest temperature is 15°C, but their development, metabolism, and calcification are disturbed (Giyanto *et al.*, 2017; Salim *et al.*, 2017). According to Sumarno & Muryanto (2014), the most optimal development of coral reefs occurs at an annual temperature of 23-25 °C, thus coral reefs can tolerate temperatures that reach 36-40 °C. The results of the measurement of the temperature of the waters on the Damas coast range from 25.18-29.59°C, when compared with the sources above, this value is still within the optimum reasonable limits. (Fikri et al., 2021a).

Turbidity

The optical properties of water which are determined based on the amount of light absorbed and emitted by the materials contained in the water are turbidity parameters. Turbidity is caused by organic and inorganic materials both suspended and dissolved such as mud, sand, organic materials such as plankton and other microorganisms (Salim et al., 2017). Turbidity can prevent light penetration into the water, which is feared to interfere with the photosynthesis process of zooxanthella (Andaris et al., 2015). The overall average results obtained that the highest turbidity in Damas waters is at station 5, a value of 4.32 NTU is obtained, it is possible because the mixing of sand which is the main substrate of this station The high turbidity in the waters will certainly affect the penetration of sunlight into seawater so that it will reduce the rate of photosynthesis. coral. In general, too much sediment can cause the death of coral biota because coral polyps will be covered by particles so that they cannot catch plankton in the waters (Fikri et al., 2021a).

Current

Ocean currents are the movement of water masses from one place (position) to another (Azis, 2006). The slow speed can cause the growth of microalgae which can threaten coral growth so that zooxanthella become stressed and leave the coral body (Tanamal et al., 2019). Current is needed in the process of coral growth in terms of supplying food in the form of microplankton. Coral growth in a place that is always stirred by currents and waves, is better than in calm and protected waters (Restu, 2016). The current also plays a role in the cleaning process of deposits and plays a role in the sedimentation process (Fikri et al., 2021b). According to Andaris et. al (2015), there is a strong correlation between speed and coral growth rate. The results of the current velocity measurement range from 0.1 to 0.4 m/s which is classified as low to moderate, because it is still below 0.5 m/s which is an indicator that the current is strong.

Ocean Wave

The phenomenon of ocean waves describes the transmission of energy and momentum. Ocean waves always cause a swing of water that moves incessantly on the sea surface layer and is rarely completely stationary. Ocean waves are generally caused by the influence of the wind, although there are other factors that can cause waves in the sea such as seismic activity on the seabed (earthquakes), volcanic eruptions, ship movements, the attraction of celestial bodies (the moon and sun) (Azis, 2006). According to Suryanti et. al (2011), corals that live in protected areas from waves (leeward zones) have slender and elongated branches, in contrast to strong waves (windward zones) which tend to grow in the form of short, strong, or creeping branches. The average wave heights of Damas Beach waters in the morning, afternoon, and evening are 5.6 cm, 8.14 cm, and 6.2 cm. Coral growth in current areas is better when compared to calm waters. Coral reefs are more developed in areas with large waves, because it is immensely helpful in preventing sedimentation in coral colonies. On the other hand, very strong waves, such as tsunami waves, can physically destroy corals (Fikri et al., 2021b).

Total Suspended Solid

Total Suspended Solid (TSS) is a solid material in suspended waters in the form of organic and inorganic substances (Fathiyah et al., 2017). Suspended solid material or Total Suspended Solid (TSS) is a place where heterogeneous reactions take place, which functions as the earliest sediment-forming material and can hinder the ability to produce organic substances in water (Jiyah, 2016). In waters that have a high concentration of TSS tend to experience high sedimentation. TSS concentrations affect corals directly or indirectly. Directly deposited sediments will clog coral polyps, while indirectly suspended sediments can block the penetration of sunlight needed for the photosynthesis of zooxanthellae (Rizka et al., 2020). The measurement results show that using Landsat 8 OLI/TIRS C1 Level-1 imagery acquired on March 16, 2020, with Path 199 and Row 66, the TSS value in Damas Beach waters, Trenggalek Regency, is around 35.6-351 mg/L (Fikri et al., 2021b). The TSS level in Damas waters is quite high when compared to the TSS quality standard level based on remote sensing. The distribution of TSS changes every season, especially in the rainy season which tends to experience a high increase in TSS levels, especially in areas close to the shoreline and the mouth of river mouths. Currents on Damas Beach are obtained with an average of 0.15 m/s which has a positive correlation or can affect the distribution of TSS (Yonar et al., 2021).

Chemical Oceanography

Salinity

Salinity is the degree of concentration of salt dissolved in water (Salim et al., 2017). The ideal salinity ranges from 30-36 ppm, where low salinity affects coral resistance to high temperatures. Meanwhile, according to (NOAA, 2021), most reef-building corals also require very salty (salty) water ranging from 32 to 42 parts ppm. Results of processing salinity data In September 2019, the range of salinity values is between 33.5-34.6 ppt, while in November the range of salinity obtained is between 31.5-34 ppt. Generally, the natural range of salinity in tropical waters such as Indonesian waters ranges from 32-35%o or less than 35%o for the highest salinity (Yusuf et al., 2012). Based on this explanation, the salinity level at the study site is still in the good category for coral survival (Moira et al., 2020). The effect of salinity varies greatly depending on the conditions of water and the resistance of each type of coral to different salinities (Andaris et al., 2015).

pH

Power of Hydrogen (pH) is the number of hydrogen ions in seawater expressed in hydrogen activity (Rizka et al., 2020). Marine waters have a pH ranging from 6.5 to 9.0. Changes in the pH value cause changes to the balance of carbon dioxide, bicarbonate, and carbonate content in the water (Siburian et al., 2017). The results of pH measurements in Damas waters have a value range of 6.9-8.4. According to Barus et al. (2018), In general, the measurement results of pH 7-8.5 are a common condition in Indonesian waters (tropical). Based on Corvianawatie & Abrar (2018), the pH value of the waters in all observation stations is still appropriate in levels between 7.0-8.5, which is classified as good.

Dissolved Oxygen

Dissolved oxygen is a chemical variable that has a particularly key role in the life of aquatic biota as well as a limiting factor for the life of biota. The solubility of oxygen can be reduced due to increasing water temperature and increasing salinity. Dissolved oxygen concentration is influenced by the respiration process of aquatic biota and the decomposition of organic matter by microbes (Patty et al., 2015; Siburian et al., 2017). Dissolved oxygen levels obtained at the time of data collection is associated with the statement (Rizka et al., 2020) included in good condition because the oxygen content in the water quality standard is more than 5 mg/L. So the DO value in the waters of the Damas Beach area is good or meets the seawater quality standards and for coral life (Moira et al., 2020). Corals can grow in DO conditions with levels above 3.5 mg/L (Sumarno & Muryanto, 2014).

Nitrate

Nitrate is the main form of nitrogen in water and is the main nutrient needed for plant and algae growth (Makatita et al., 2014). Nitrate levels are also influenced by salinity, that nitrate concentrations will decrease with increasing salinity (Utami et al., 2016). Normal nitrate levels in marine waters generally range from 0.10-0.50 µg.at/l or equivalent to 0.001-0.007 mg/l, where nitrate levels are still good for coral growth (Patty et al., 2015). Mentioned in the Ministry of Environment of the Republic of Indonesia (Ministry of Environment, 2014) that the standard quality of nitrate concentration suitable for the life of deep-sea biota is 0.008 mg/l. Based on observations at all stations for two months (September and November), nitrate levels contained in normal waters in marine waters generally ranged from 0.001-0.007 mg/l. Coral reefs are actually able to live in an environment that is low in nutrients (nitrate, phosphate, water) because of the ability of corals to produce their own nutrients, so it can be assumed that the largest contribution to aquatic productivity is coral organisms themselves (Moira et al., 2020).

Phosphate

Phosphate is a form of phosphorus that can be utilized by plants and is an essential element for higher plants and algae so that it can affect the level of water productivity (Makatita et al., 2014). Phosphorus compounds bound in sediments decompose with the help of bacteria through an abiotic process resulting in dissolved phosphate compounds that can diffuse back into the water column (Utami et al., 2016). According to Ministry of Environment of the Republic of Indonesia (Ministry of Environments, 2014) states that the phosphate level for marine biota is 0.015 mg/l, while (Ilyas et al., 2017) in his research suggested that phosphate levels for the chemical parameters of coral reef waters ranged from 0.27-5.51 mg/L. Damas Beach waters are categorized as quite fertile and still good for the growth and development of coral reefs. The increase in phosphate concentration can be caused by waste disposal found in the area around the waters carried by water currents (Moira et al., 2020).

ARTIFICIAL REEF BIOTA

Biofouling

Artificial reefs are also known to function to repair or restore the biological functions of ecosystems. The underwater surface of the sea is strongly influenced by the attachment of fouling organisms, especially barnacles (Figure 4). The process of forming a biofouling community through several processes in which colonization on a new surface occurs as the result of several stages. Initially, a biochemical layer is formed on a clean surface, this event is then followed by microbial attachment or microfouling (colonization of bacteria and diatoms) and the final stage is macrofouling (colonization of macroalgae and invertebrates).

The results of observations of invertebrate organisms showed that the total number of invertebrate organisms that had been found on all artificial reefs on the Damas coast was 9,293 biota. There were ten organisms found, including *tubeworm, barnacle, hermit crab, bryozoan, green algae, tunicate, hydroid, brown algae, sponge*, and *red algae*. (Thaeraniza et al., 2020).

Figure 4. Macrofouling Found Attached to Artificial Reef
Source: (Thaeraniza et al., 2020).

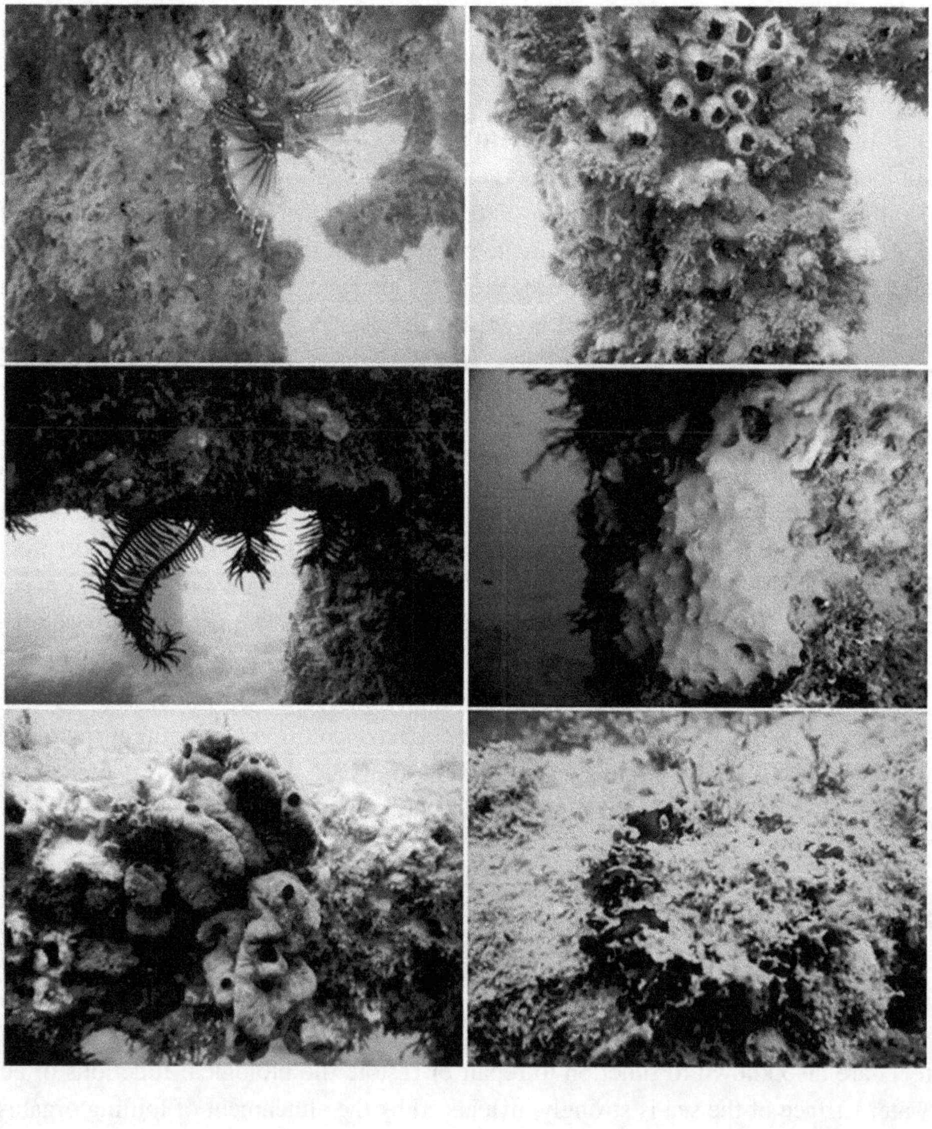

Reef Fish

The reef fish community is an important part of the coral reef ecosystem, not only for the fish themselves who make the coral reef ecosystem their vital habitat, namely as a spawning ground, nursery ground, and feeding ground. The abundance of reef fish is related to the condition of coral reef cover, the better the condition of the coral reefs in the area, the more coral fish in the area. According to Sale (1991) stated that an estimated 4000 species of fish live in coral reef areas and are associated with this Indo Pacific coral reef habitat or about 18% of the total fish available, namely: Acanthuridae, Apogonidae, Blenniidae, Chaetodontidae, Diodontidae, Fistulaidae, Labridae, Nemipteridae, Ostraciidae, Pinguipedidae, Scorpaenidae, Solenostomidae, Tetradontidae, Tetrarogidae and Pomacentterygiidae, and Pomacentterygiidae. Records of some fish species can be found in artificial reefs off Trenggalek Damas Beach (Figure 5).

Figure 5. Some Fish Known to Be Around Artificial Reefs
Source:(Sirait et al., 2021)

SOLUTIONS AND RECOMMENDATIONS

Management and Technical Implementation of Artificial Reef Project

In addition to monitoring, the team also communicated with the community and received information that there was no socialization and education regarding the technicalities and benefits of sinking artificial reefs when it was about to do artificial reef sinking. In addition, the implementation of the conservation zone in the sinking area was not conducted properly, and the monitoring activities conducted by the previous team were not sustainable, so the first thing the team did was to monitor the sinking location for 2 years.

Activities Before and After Sinking of Artificial Reefs

Before sinking an artificial reef, you must pay attention to several things such as the site selection correct sinking technique, the technology to be applied, and the artificial reef material used, because these three things are some of the factors that can make an artificial reef a success or a failure. Description of the ideal artificial reef according to (NHRCP, n.d.) are: (1) Stable in normal to major storms; (2) Made of durable, solid, and non-toxic material; (3) Designed to have a high surface complexity (texture) for recruitment of corals, sponges, and other organisms; (4) Designed to provide high structural complexity for fish and other animals; (5) Designed to blend into natural reefs, or designed to stand out and convey a message (sculpture and art). As for the construction, it is endeavored to make the surface of the artificial coral reefs rough, textured or wavy, in order to facilitate the attachment of coral larvae for recruitment.

Based on Kambey et al. (2017) in the waters of Malalayang, 15 species of demersal fish were found that were attracted to artificial reefs made of bamboo, indicating that the artificial reefs were good enough to be developed. The use of bamboo as a natural base material that does not damage and is cheap and easy to obtain. Compared to using other materials such as the use of concrete, it can affect the quality of the surrounding waters, because the cement mixture in concrete containing calcium peroxide has a fairly high pH range (Alwi et al., 2020). In the research in Damas, using concrete materials, therefore it is necessary to review the coral reef materials that will be used.

Beside materials, or the performance of an artificial reef to be successful, it is important to know several criteria like: (1) Location close to fishing settlements; (2) Separated from natural reefs; (3) Depth based on distance from coastal waters; (4) Water conditions according to the optimum value of coral growth; (5) The location is far from the main fishing grounds; (6) The condition of the substrate is solid and flat to prevent artificial reefs from sinking to the bottom; (7) Orientation with fish migration patterns; (9) Does not interfere with shipping navigation; (10) The location is close to tourism and provides easy access for diving activities (Dhiecha et al., 2004; Salleh et al., 2018).

For restoration activities, artificial reefs must be considered natural in terms of aesthetics and appearance when overgrown with coral or other reef biotas. According to Gomez et al. (2007), the main roles of artificial reefs in coral reef restoration activities include: (1) stabilizing and restoring the complexity of areas filled with coral fractures such as those caused by fish bombs, and attracting fish and corals in areas with low probability of recovery; (2) support tourism or education activities and public awareness that requires easy and safe access to reefs; (3) reduce pressure from divers on natural reefs in densely populated areas. Based on several criteria from various sources, the appropriate placement of artificial reefs is in a limited use zone.

After sinking artificial reefs, according to Ampou et al. (2019), stated several things that need to be done, namely: (1) It is necessary to carry out routine monitoring activities related to the status of corals and their associations on artificial reefs every 2-6 months; (2) Cleaning the artificial reef media at least once every 1 – 3 months; (3) Take and test seawater quality samples, at least representing the season to determine the quality standards of seawater around artificial reef sites, (4) Scientific studies of growth, health, disease, modeling of ocean currents and other related matters are very important needed as supporting data. In research in Damas waters, monitoring is not carried out regularly after the sinking and is only conducted 2 years later.

STAKEHOLDER ROLE

Mindset and Public Perception

Several applications of artificial reefs have had success, such as in Bangsring, Banyuwangi. In its implementation, what needs to be emphasized is the mindset of the surrounding community on the importance of maintaining coral reef ecosystems. The stakeholders involved are fishers, environmental NGOs, and academic partners. Several stages were carried out, namely, (1) Survey of the location and condition of the Coast; (2) Socialization of the Village Development Work Program; (3) Counseling on the Conservation of Marine Biota and Coral Reefs; (4) Manufacture of artificial reefs; (5) Sinking of artificial reefs; (6) Monitoring and evaluation of success indicators; (7) Publication; (8) Final reporting (Erwanto & Masluha, 2019).

Apart from Bangsring, the coral reef conservation project in Pemuteran, Northwest Bali, is the result of positive perceptions and high community participation that strongly correlated with demographic background, level of community interaction, and their perceived personal benefits from the project. Positive perceptions are built with a personal approach, namely using essential local beliefs in this case related to Balinese Hindu beliefs which encourage adherents to protect nature because God's soul. This positive perception leads to a high level of community participation to support the project (Trialfhianty & Suadi, 2017). Based on these two examples, it can be concluded that the participation of the local community's mindset strongly supports the success of conservation activities, one of which is the sinking of artificial reefs.

Policy by the Government

The success of the application of other artificial reefs can also be seen in the policies of the Polewali Mandar local government. Government support through socialization and education has had a major impact on fishers, including no longer using explosives to find fish. Despite the 1.02% increase in fishery production from 2009, there has been a better chance in the Polewali Mandar fishery sector. The decrease in the use of explosives and poisons is due to patrol activities as a collaborative effort of the Marine Fisheries Service, Pokmaswas, Satpolair, and the Polewali Mandar District Forestry Service. In addition, the empowerment of coastal communities is carried out by the local government by forming a Floating Net Cage Fisherman Group (Pratiwi et al., 2014).

Based on several examples of the successful application of artificial coral reefs, conclusions can be drawn that effective communication between the community and researchers is needed as well as the implementation of government policies in the artificial reef sinking area so that it can become an area that can support sustainability or improve the condition of the waters around Damas beach. In addition, if there will be another artificial reef sinking activity, the sinking location must have accurate coordinates and there are number markings on each artificial reef so that in the subsequent monitoring and evaluation process it will be more effective and efficient.

CONSERVATION SECTOR

Categories of Marine Protected Area

According to regulation (Regulation of the Minister of Maritime Affairs and Fisheries of the Republic of Indonesia, 2020) Regarding Management of Conservation Areas Article 6-10 explains that marine conservation areas are defined in 3 categories, namely parks; wildlife preserve; and maritime conservation areas.

1. Park Category

The determination of water conservation areas with the category of parks is conducted to protect, preserve, and utilize biodiversity and/or fish resources. Its function is to maintain and improve the quality of biodiversity.

2. Wildlife Preserve Category

The determination of water conservation areas with the category of asylum is conducted for the purpose of protecting, preserving, and utilizing biodiversity and/or fish resources. Its function is to maintain and improve the quality of fish resources.

3. Marine Protected Area Category

The determination of marine conservation areas with the category of maritime conservation areas is carried out for the purpose of protecting, preserving, and utilizing Traditional Cultural Sites. Its function is to maintain and increase the value of maritime cultural heritage and traditional values or local wisdom.

Status of Authority and Conflict of Coastal Zone Management

The ratification of Law Number 23 of 2014 concerning Regional Government has a direct impact on regional autonomy in the management of coastal and small islands. Article 27 paragraph (1) of Law 23/2014 states that Provincial Regions are given the authority to manage marine resources in their territory. Because of this, Article 27 paragraph (1) of Law 23/2014 directly revokes the authority of the Regency/City in managing marine resources. The missing regency/municipal authorities include policy implementation, marine spatial planning, supervision and law enforcement, management and utilization coordination, and licensing (except fishery business permits). The conflict certainly makes coastal communities a passive object in every development program implemented by the Indonesian government (Ali et al., 2020).

According to Adrianto (2015), in its evaluation report also stated that the change to Law no. 23 of 2014 explains that the Regional Government is given the authority to manage the resources of the coastal marine area and small islands as far as 12 miles for the province, where the granting of authority to this area is interpreted as sovereignty. This gives rise to horizontal conflicts over sea plots in the community. Thus, conflicts that occur are not only between government institutions (sectoral conflicts) but are also related to regional authorities in managing marine coastal resources and small islands.

Matters related to management conflicts were discussed in a discussion that developed the identification of the root causes of the imbalance in the arrangement of coastal areas. It was said that the change in management authority from the district/city to the province was since coastal areas were considered better regulated at a wider level. Several previous conflict cases occurred because of claims between districts/cities and triggered conflicts between regions and involved the community. However, the granting of management authority to the provincial government is not accompanied by a commitment to regulate and supervise the use of coastal areas. This can be seen from the lack of commitment from the Provincial Government in allocating human and financial resources to support district/city governments and community groups to manage and organize coastal resources in a sustainable manner and accordance with applicable regulations (SDGS, 2021). One form of coastal resource management is in conservation efforts, one of which is the sinking of artificial reefs. As discussed in the previous chapter, the lack of cooperation between stakeholders, from the central government to local governments as well as to the community, and the unclear person in charge of the work program, caused this activity to not run optimally as it should. Whereas coral reef sinking activities need to be monitored regularly because restoring an area takes a long time.

Zoning Of Conservation Area

According to (Regulation of the Minister of Maritime Affairs and Fisheries of the Republic of Indonesia, 2020) Article 11, the zoning of the Conservation Area as intended includes:

a. Core zone, intended for absolute protection of conservation targets and must exist in every conservation area. The core zone has criteria in the form of protection of genetic resources; natural/important habitat for fish resources which includes spawning areas; care area; and foraging areas; the transfer of marine biota to fishing areas; restoration of biota and habitat; protection of biodiversity that is vulnerable to change; and/or cultural heritage protection.
b. Limited use zone; useful for fisheries and tourism activities in a sustainable manner. This zone must exist in a Conservation Area with the category of park and maritime conservation area. The limited use zone must have social, economic and cultural functions for the local community around the Conservation Area.
c. Other zones according to regional designation, are zones outside the core zone and limited use zones and do not have to exist in Conservation Areas with the categories of parks, reserves, and maritime conservation areas. Other zones according to the designated area as referred to may be in the form of rehabilitation zones; marine building and installation zones; port/mooring zone; ship traffic zone; religious zones/cultural sites; and/or zone according to regional characteristics.

Damas Beach Area Status

The East Java Coastal and Small Islands Zoning Plan or RZWP-3-K is a plan that determines the direction of resource use for each planning unit for the entire area in East Java accompanied by the determination of the structure and spatial pattern in the planning area which contains activities that may and may not be carried out and activities that can only be carried out after obtaining permits in coastal areas and small islands. This planning is regulated in East Java Provincial Regulation Number 1 of 2018 concerning Zoning Plans for Coastal Areas and Small Islands of East Java Province for 2018-2038.

RZWP-3-K aims to realize integrated, safe, efficient, and sustainable management of coastal resources and small islands in the province to improve the welfare of the people of East Java with participatory principles. The scope of regulation of the RZWP-3-K covers landward, covering the administrative boundaries of the sub-districts in the coastal areas; and towards the sea as far as 12 nautical miles measured from the coastline at the time of the highest tide towards the high seas and/or towards the archipelagic waters. The policies and strategies in the RZWP-3-K include the development of public use areas; conservation area; strategic area; and sea lane.

When viewed from the contents of the RZWP-3-K, the Damas beach area and its surroundings are not mentioned in the conservation management area because this beach is more intended for tourism. However, when reviewed, Damas Beach is an unspoiled beach and has not been managed properly by the government, and research and publications related to this location are still minimal. Therefore, there is still the possibility that Damas beach meets the requirements for establishing a conservation area, of course, it must go through several stages. The first stage of determining a candidate for a conservation area is that there is a proposal for this location initiative from the Government, Regional Government, related communities, and academics, therefore it is back again that stakeholders are particularly important in determining the status of an area. Determining the status of the area is an important first step and is related to the selection of artificial reef sinking sites. If later Damas beach is planned to become a conservation area, the suitable zone is a limited use zone, because in the zone the activities that are allowed include marine nature tourism, research, and education. The sinking of artificial reefs is expected to improve the ecosystem and bring in a variety of biota where is included in the aesthetic function that can attract the attention of tourist.

In determining the conservation area, it is necessary to carry out several stages according to the (BPSPL Padang, 2021) which refers To Regulation Of The Minister Of Marine And Fisheries 2020, article 16, which includes: (1) Proposed Initiatives for Conservation Areas; (2) Identification and Inventory; (3) Technical Consulting; (4) Public Consulting; (5) Backup includes; the purpose of establishing a Conservation Area; Area of Conservation Area; Conservation targets; Category; Provisions for the use of Conservation Areas; Zoning map; (6) Determination that consists of proposal and evaluation; (7) Boundary Setting includes Publication of Conservation Areas in Indonesian marine maps, Socialization, and Installation of boundary markers.

THE ROLE OF HUMAN RESOURCES

The main factor for the success of artificial reefs is human resources. The location of artificial reefs should be based on community conditions. Artificial reef monitoring activities by involving the community will produce maximum results by avoiding conflict. The existence of floating net cages causes a decrease in water quality and the presence of domestic waste in the bottom of the water. For maximum reef performance, it is important to pay attention to placement sites with economic or social impacts arising from the use of artificial reefs (Sirait et al., 2021).

Counseling and education to the public regarding the techniques of decreasing and monitoring artificial reefs is needed. The community around Damas beach understands the importance of conserving coral reef ecosystems for the environment and increasing community welfare through artificial reef sinking, but the realization is not optimal due to the lack of education related to correct artificial reef sinking techniques. According to Lindberg & Seaman (2011), stated stakeholders including recreation providers and related businesses, governments, tourism and economic development stakeholders, recreational and commercial scuba divers, fishermen and, academia, research and monitoring groups, artificial reef module producers, marine contractors, environmental organizations, and the media. The submergence of artificial reefs in Damas Beach is not maximal due to the lack of support and relations between related stakeholders.

POKMASWAS (Monitoring Community Group)

Decree of the Minister of Marine Affairs and Fisheries Number: Decree. 58/Minister/2001 explains the procedures for implementing a community monitoring system in the management and utilization of marine and fishery resources. Within the scope of its activities, the establishment of the SISWASMAS Network includes: (1) Community supervisory groups (POKMASWAS) are supervisors at the field level consisting of elements of community leaders, religious leaders, traditional leaders, NGOs, fishermen, fish farmers, and other maritime communities; (2) POKMASWAS is formed at the initiative of the community, facilitated by elements of the local government, and coordinated by a member of the community who functions as an intermediary between the community and the government/officers; (3) Crew fishermen as well as small fishermen and other maritime communities; (4) The management of POKMASWAS is elected by the community and registered as members.

POKMASWAS Empowerment and Capacity Building for Supervisory Groups include: (1) local traditions or culture which are environmentally friendly behavior; (2) Advancing POKMASWAS, one of which is through socialization; (3) Provision of assistance for facilities and infrastructure for supervision selectively by the government and adapted to local conditions; (4) The government is obliged to facilitate the empowerment of POKMASWAS through coaching, guidance, and training to increase the capacity of POKMASWAS.

Several POKMASWAS are known to exist in Trenggalek, among others, named Kejung Samudera, which regulates the scope of Karangandu Village, Watulimo District including Damas Beach. This POKMASWAS was designated as the only mangrove management institution in 2008 by the Department of Marine Affairs and Fisheries of Trenggalek Regency. This Pokmaswas specifically regulates mangrove areas (Susilo et al., 2015). As for Rembeng Raya, namely POKMASWAS which is engaged in the preservation of coastal areas including planting coastal vegetation, monitoring and transplanting coral reefs. POKMASWAS Rembeng Raya has received assistance with artificial coral reefs, Fish Apartments, and Floating Houses. This POKMASWAS pioneered and managed the Mutiara beach tourism area which is now crowded with tourists. Next is POKMASWAS Pantai Taman Kili pelvis, which is engaged in the conservation of turtles of the Lekang, Hawksbill, Blimbing, and green turtle species. The achievements that have been achieved are 2nd place at the Provincial level (Baraknews, 2020). There are so many Pokmaswas in the Trenggalek Regency that are under the guidance of the Trenggalek Regency Fisheries and Marine Service. However, there is no information regarding Pokmaswas that are active in conservation activities at Damas Beach, this can be used as input in the future so that conservation activities at Damas Beach can be conducted properly.

POKDARWIS (Tourism Awareness Group)

According to Rahim (2012), the Tourism Awareness Group (Pokdarwis) is domiciled in the village/district around the tourism destination. The establishment of the Tourism Awareness Group (Pokdarwis) aims;

(1) Improving the position and role of the community as important actors in tourism development, and being able to establish partners with other stakeholders in improving the quality of tourism development in the regions; (2) Building a positive perception of the community as the host through the realization of the values of Sapta Pesona for the growth and development of tourism in the region and its benefits for regional development and community welfare; (3) Introducing, preserving and utilizing the potential of existing tourist attractions in each region.

Trenggalek Regency is one of the regencies in East Java that has various tourism activities and many tours. Trenggalek has a beach length of about 96 km with a sea area of approximately 71,117 ha and a total of 57 islands. Trenggalek Regency, precisely in Watulimo District, is one of the areas in the province of East Java that has the potential for marine tourism which can be promised by the waves and the beauty of the scenery in the bay area which is the main attraction for tourists. One of the great potentials in Trenggalek is Damas beach which is famous for being unspoiled (Marli, 2019).

According to Rahayu et al. (2019), with proportional random sampling interview shows that the participation of the tourism-aware group community in tourism development in Trenggalek Regency as a whole is very high in decision-making and planning, implementation of activities, monitoring, and evaluation, while the utilization of results is high. It is also stated that formal education and motivation have a significant effect on the participation of the tourism-aware group in tourism development in Trenggalek Regency.

Although in previous studies it was stated that the participation of tourism awareness groups in Trenggalek Regency was high, in fact, it was still not implemented optimally on Damas beach. As discussed in the technical guidance for the Tourism Awareness Group with the East Java Ecotourism Forum (EJEF) as resource persons to explore the shortcomings and successes of Tourism Village destinations in Trenggalek. This activity aims to find solutions in improving the management of Tourism Villages as well as in order to share information so that Trenggalek has a standard and marketable tourist destination. In the activity, it was stated that the coordination between the Trenggalek Pokdarwis and the government was lacking, even though the coordination must be in line to increase the tourism destinations in Trenggalek. This is one proof that coordination between stakeholders, in this case, Pokdarwis and the local government is very important for the progress of a program (Suparni, 2019).

REFERENCES

Adrianto, L. (2015). *Legal analysis and evaluation of the management of coastal areas and small islands*. Academic Press.

Ali, M. M., Aditya, Z. F., & Fuadi, A. B. (2020). Coastal Communities Protection of Harmonization of Integrated Coastal Management Regulations. *Journal Of Constitution*, *17*(4).

Alwi, D., Lule, A., & Taliki, R. (2020). *Alternative Fishing Area In Tanjung Dehegila Waters*. Morotai Island.

Ampou, E. E., Hutasoit, P., & Damar, A. (2019). *The Condition of "Spider" Artificial Reefs in Nusa Dua Bali, After the IMF-WBG Meeting*. Darilaut.Id. https://darilaut.id/kajian/kondisi-terumbu-buatan-laba-laba-di-nusa-dua-bali-setelah-pertemuan-imf-wbg

Andaris, A. R., Suryanto, A., & Muskananfola, M. R. (2015). Relationship Of Physical ± Chemical Factors Of Waters To Coral Reef Covers In Karimunjawa Island. *Diponegoro Journal of Maquares, 4*(2007), 29–36.

Azis, M. F. (2006). *Movement of water in the sea*. Academic Press.

Baraknews. (2020). *Achievements Achieved by POKMASWAS Assisted by the Department of Fisheries and Marine Affairs of Trenggalek Regency.* https://www.baraknews.com/berita_jawa-timur/prestasi-yang-diraih-pokmaswas-binaan-dinas-perikanan-dan-kelautan-kabupaten-trenggalek

Corvianawatie, C., & Abrar, M. (2018). Suitability of Oceanographic Conditions in Supporting Coral Reef Ecosystems on Pari Island. *National Oceanic Journal, 13*(3), 155–162. doi:10.15578/jkn.v13i3.6322

Dhiecha, D. M., Utomo, K. P., Jati, D. R., & Tanjungpura, U. (2004). *Planning of artificial reef as reef restoration of Bengkayang Regency Lemukutan as a Conservation and Nature Tourism Area of Bengkayang Sea according to the coast and sea in the Lemukutan Island area and its surroundings, as a consequence Based on a case study*. Academic Press.

Erwanto, Z., & Masluha, U. (2019). Artificial Temple Reef Conservation Technology As Control Of Coastal Abrasion On Tabuhan Island, Bangsring Village, Wongsorejo District. *Banyuwangi., 4*(1), 103–109.

Fadilah, S., & Suprihardjo, R. (2016). Development of Marine Tourism Area Watulimo District, Trenggalek Regency. *Journal of Engineering ITs, 5*(1), 50–53.

Fathiyah, N., Pin, T. G., Saraswati, R., & Kunci, K. (2017). Spatial and Temporal Patterns of Total Suspended Solid (TSS) with SPOT Imagery in the Cimandiri Estuary, West Java. Academic Press.

Fikri, M., Isdianto, A., & Luthfi, O. M. (2021a). Environmental Conditions of Waters (Oceanographic Physics) Around Artificial Reefs at Damas Beach, Trenggalek Regency, East Java. *Journal of Marine and Coastal Science, 10*(February).

Fikri, M., Isdianto, A., & Luthfi, O. M. (2021b). Physics Oseanography Around Artificial Reef On The Pantai Of Damas, Trenggalek District, East Java. *Journal of Marine and Coastal Science, 10*(February), 35. doi:10.20473/jmcs.v10i1.25605

Giyanto, M., Hadi, T. A., Budiyanto, A., Muhammad Hafizt, Salatalohy, A., & Iswari, M. Y. (2017). *Status of Coral Reefs in Indonesia 2017*. Academic Press.

Gomez, E., Edwards, A. J., & Gomez, E. D. (2007). Reef Restoration Concepts & Guides: Making wise choices among uncertainties. Terumbu Karang Indonesia (TERANGI) Foundation.

Ilyas, I. S., Astuty, S., Harahap, S. A., & Purba, N. P. (2017). The Diversity of Targeted Reef Fish in Relation to Coral Growth Forms in the Core Zone of the Anambas Archipelago Aquatic Tourism Park. *Journal of Fisheries and Marine Affairs, 8*(2), 103–111.

Jiyah. (2016). Study Of Total Suspended Solid (Tss) Distribution In Coastal Waters Of Demak Regency Using Landsat Image. *Jurnal Geodesi Undip Januari, 6*, 41–47.

Kambey, A. D., Bataragoa, N. E., & Wantasen, A. S. (2017). Study Of Placement Of Artificial Reefs From "Bamboreef" Bamboo In Malalayang Waters Two, Malalayang District, Manado City. *Jurnal Ilmiah Platax*, *5*(1), 1–9. doi:10.35800/jip.5.1.2017.14967

Lindberg, W. J., & Seaman, W. (2011). *Guidelines and Management Practices for Artificial Reef Siting, Use, Construction, and Anchoring in Southeast Florida*. Academic Press.

Maddupa, H. H., Subhanln, B., Bachtiar, R., Ismetl, M. S., Budikartinis, Y., & Bria, D. (2007). Prospects of Biorock Artificial Reefs in Increasing Fish Resources in the Thousand Islands. *Coral Reef National Conference I,* 68–75.

Makatita, J. S., & Mangimbulude, J. (2014). Study of Phosphate and Nitrate Nutrients in Water and Sediment Seagrass Padang Tujuh Seram Island North West Central Maluku. *FMIPA-UT National Seminar 2014, 23*, 54–66. http://repository.ut.ac.id/4959/1/fmipa2014_09.pdf

Marli, Y. S. (2019). *Development of Marine Tourism Potential in Trenggalek*. https://www.kompasiana.com/yulistianasholiqhahm5879/5db6afc4097f367da76a4812/pengembangan-potensi-wisata-bahari-di-trenggalek

Ministry of Environment. (2014). *Quality sea water for marine life standarts. Decree of the State Minister of the Environment No. 51 of 2004 concerning Seawater Quality Standards*. Author.

Moira, V. S., Luthf, I. O. M., & Isdianto, A. (2020). Analysis of Relationship between Chemical Oceanography Conditions and Coral Reef Ecosystems in Damas Waters, Trenggalek, East Java. *Journal of Marine and Coastal Science*, *9*(3), 113–126. doi:10.20473/jmcs.v9i3.22294

NHRCP. (n.d.). *Artificial Reefs: What works and what doesn't*. https://newheavenreefconservation.org/marine-blog/147-artificial-reefs-what-works-and-what-doesn-t

NOAA. (n.d.). *Sea Surface Salinity*. NOAA. Retrieved October 10, 2021, from https://coastwatch.noaa.gov/cw/satellite-data-products/sea-surface-salinity.1.html

Padang, B. P. S. P. L. (2021). *Conservation Area.* https://kkp.go.id/djprl/bpsplpadang/page/262-konservasi-kawasan

Patty, S. I., Arfah, H., & Abdul, M. S. (2015). Nutrients (Phosphate, Nitrate), Dissolved Oxygen and pH Relation to Fertility in Jikumerasa Waters, Buru Island. *Journal of Coastal and Tropical Oceans*, *3*(1), 43. doi:10.35800/jplt.3.1.2015.9578

Pratiwi, R. A., Widayati, W., & Astuti, P. (2014). Evaluation of Coral Reef Conservation Policy at Polewali Mandar 2012. *Journal of Politic and Government Studies*, *3*(4), 376–385.

Purnawati, B. I., & Hartati, S. T. (2007). Condition of Artificial Reefs in the waters of Pramuka Island and Semak Daun Island, Thousand Islands. *Proceedings of the VI National Fish Seminar*, 91–96.

Rahayu, S. P. P., Anantanyu, S., & Mulyanto, M. (2019). Participation of Tourism Awareness Groups in Tourism Development in Trenggalek Regency. *Proceedings of the Industrial Research Workshop and National Seminar*, *10*(1), 1348–1356. https://jurnal.polban.ac.id/ojs-3.1.2/proceeding/article/view/1421

Rahim, F. (2012). *Tourism Awareness Group Guidelines.* Director General of Tourism Destination Development, Ministry of Tourism and Creative Economy.

Regulation of the Minister of Maritime Affairs and Fisheries of the Republic of Indonesia. (2020). *The Minister of Maritime Affairs and Fisheries of the Republic of Indonesia Number 31/Permen-Kp/2020 concerning Management of Conservation Areas.* Author.

Restu, I. W. (2016). *Coral Reef Ecosystem and Its Status (Case Study of Coral Reef Condition) in Bali Province.* Udayana.

Rizka, R. F., Purnomo, P. W., & Sabdaningsih, A. (2020). Effect of total suspended solid (tss) on zooxhanthellae density on Acropora sp. In laboratory scale. *Pasir Laut, 4*(1), 95–101.

Rizqika, C. N. A., & Latifah, N. (2018). Growth rate of coral reef *Acropora formosa* on Menjangan Small Island, Karimunjawa National Park. *Journal of Maquares, 7*(4), 315–322.

Salim, D., Yuliyanto, Y., & Baharuddin, B. (2017). Physical-Chemistry Oceanographic Parameter Characteristics of Kerumputan Island, Kotabaru Regency, South Kalimantan. *Enggano Journal, 2*(2), 218–228. doi:10.31186/jenggano.2.2.218-228

Salleh, N. H. M., Aman, A. A., & Hamid, S. A. (2018). Selection of Artificial Reef Deployment Sites by Using Evidential Reasoning. *Marine Artificial Reef Research and Development: Integrating Fisheries Management Objectives, 86*(January), 251–264.

Schutter, M., Ven, R. M., van der, Janse, M., Verreth, J. A. J., & Wijffels, R. H. (2012). Light intensity, photoperiod duration, daily light flux and coral growth of Galaxea fascicularis in an aquarium setting: A matter of photons? *Journal of the Marine Biological Association of the United Kingdom, 92*(4), 703–712. doi:10.1017/S0025315411000920

SDGS. (2021). *A Gift Or A Curse In The Management Of Coastal Conflicts.* SDGS Center Universitas Hassanudin. https://sdgscenter.unhas.ac.id/anugerah-atau-kutukan-dalam-pengelolaan-konflik-wilayah-pesisir/

Siburian, R., Simatupang, L., & Bukit, M. (2017). Analysis Of The Quality Of Sea Waters On Activities In The Port Environment Of Waingapu- Alor. *East Sumba., 23*(1), 225–232.

Sirait, A. P. R., Luthfi, O. M. M., & Isdianto, A. (2021). Physical Characteristics of Artificial Reef Post Deployed on 2017 in Trenggalek Damas Beach. *Journal of Marine and Coastal Science, 10*(1), 58. doi:10.20473/jmcs.v10i1.25607

Sumarno, D., & Muryanto, T. (2014). Salinity Levels, Dissolved Oxygen, And Water Temperature In The Artificial Coral Reef Unit (Tkb) Kotok Kecil Island and Harapan Island. *Btl, 12*(2), 121–126.

Suparni. (2019). *Develop Tourism Village, Trenggalek Tourism Office Holds Training for Pokdarwis.* https://faktualnews.co/2019/03/12/kembangkan-desa-wisata-dinas-pariwisata-trenggalek-gelar-pelatihan-untuk-pokdarwis/128534/

Suryanti, Y., & Supriharyono. (2011). Effect Of Depth On Coral Morphology On Cemara Kecil Island, Karimunjawa National Park. *Journal of Fisheries Science, 7*(1), 63–69.

Susilo, E., Purwanti, P., & Lestariadi, R. A. (2015). Utilization of Mangrove Resources at Pancer Cengkrong and Damas, Prigi Beach. *Trenggalek.*, *5*(2), 19–25.

Tanamal, Y., Tuhumury, S. F., & Sangaji, M. (2019). Analysis of Suitability and Carrying Capacity of the Great Lagoon Rehabilitation Area and Slope Reef of Kipuo Lagoon, Negeri Ihamahu. Triton. *Journal of Aquatic Resource Management*, *15*(1), 21–29.

Thaeraniza, S. T., Muzaky, O., & Andik, L. (2020). *Macroinvertebrata Recruitments in Artificial Reef After Two Years of Sinking*. Academic Press.

Trialfhianty, T. I., & Suadi. (2017). The role of the community in supporting coral reef restoration in Pemuteran, Bali, Indonesia. *Journal of Coastal Conservation*, *21*(6), 873–882. doi:10.100711852-017-0553-1

Utami, T. M. R., Maslukah, L., & Yusuf, M. (2016). Distribution of Nitrate (NO3) and Phosphate (PO4) in Karangsong Waters, Indramayu Regency. *Buletin Oseanografi Marina*, *5*(1), 31. doi:10.14710/buloma.v5i1.11293

Yonar, M., Luthfi, O. M., & Isdianto, A. (2021). Dynamics of Total Suspended Solid (TSS) Around Damas Beach Coral Reef, Trenggalek. *Journal of Marine and Coastal Science*, *10*(1), 48–57. doi:10.20473/jmcs.v10i1.25606

Yusuf, M., Handoyo, G., & Wulandari, S. Y. (2012). Characteristics of Current Patterns in Relation to Water Quality Conditions and Phytoplankton Abundance in the Waters of the Karimunjawa Marine National Park. *Buletin Oseanografi Marina*, *1*(5), 63–74. doi:10.14710/buloma.v1i5.6918

KEY TERMS AND DEFINITIONS

Aesthetics: Refers to the beauty of appearance.

Biofouling: Attachment and accumulation of living organisms on the surface of a submerged substrate in the sea.

Biorock Technology: Is a method of restoring coral reefs through the process of adding minerals by flowing a low-voltage electric current of about 1.2 volts on a steel frame that has been formed and sunk on the seabed.

Hydro-Oceanography: A marine scientific field that specifically studies the properties of the movement of sea water which includes tides, ocean waves and ocean currents.

Nitrate: The main form of nitrogen in water and is the main nutrient needed for plant and algal growth.

Non-Governmental Organization (LSM): An organization founded by an individual or a group of people who voluntarily provide services to the general public without the aim of profiting from its activities.

pH: Number of hydrogen ions in seawater expressed in hydrogen activity.

POKDARWIS: The Tourism Awareness Group (Pokdarwis) is one of the components in the community that has a significant role and contribution in the development of tourism in the region.

POKMASWAS: The community groups that actively participate in monitoring the use of marine and fishery resources.

Salinity: Degree of concentration of salt dissolved in water.

Sediment: Material derived from the breakdown of older rock or material derived from the weathering process of rock and transported by water, air and ice, or material deposited by naturally occurring processes.

Sedimentation Rate: The amount (volume) of sediment uplifted per unit area per unit time.

Total Suspended Solid (TSS): It is a solid material in suspended waters in the form of organic and inorganic substances.

APPENDIX: ORGANIZATION BACKGROUND

Marine Science Study Program, Faculty of Fisheries and Marine Science Universitas Brawijaya, Malang, East Java was established based on Decree No. 2953/D/T/2008 issued by the Directorate General of Higher Education on September 4, 2008. Marine Science Study Program, Faculty of Fisheries and Marine Science Universitas Brawijaya focuses its studies on four principal areas, namely:

- **Oceanography:** understand, explain, analyze, and model oceanographic phenomena and their impacts related to the dynamics of coastal and marine ecosystems.
- **Exploration:** mapping of coastal and marine living and non-living resources using remote sensing software and geographic information systems; operate marine instruments in observing and measuring coastal and marine phenomena in oceanographic studies, exploration, conservation, and marine biotechnology.
- **Biotechnology:** explore genetic diversity and exploit the potential for bioprospecting (marine biological materials, bioremediation, biomonitoring, biomarkers) on marine organisms.
- **Conservation:** carry out monitoring and assessment both directly and indirectly based on scientific research principles regarding the condition of coastal and marine ecosystems; carry out efforts to conserve coastal and marine biological resources in tropical waters in a sustainable manner.

Section 4
Sustainable Options

Chapter 10
Building Bio-Shelters Using Industry 4.0 Data-Driven Hybrid Fabrication Technologies:
Techniques for Customisable Complex Artificial Reef Design

Kate Dunn
University of New South Wales, Australia

Matthias Hank Haeusler
University of New South Wales, Australia

Kuai Daniel Yu
University of New South Wales, Australia

Yannis Zavoleas
University of Ioannina, Greece & University of New South Wales, Australia

Gabriel Undery
University of New South Wales, Australia

Charlotte Sarah Firth
University of New South Wales, Australia

ABSTRACT

This chapter describes the innovative design and fabrication of a complex reef system for Sydney Harbour, Australia. The Bio-Shelters project is an ongoing collaborative investigation into the application of computational design and Industry 4.0 hybrid technologies to create site-specific artificial reefs. The location of the artificial reefs is Blackwattle Bay in Sydney Harbour. The Bay has been used for heavy industry for over 100 years, resulting in polluted water and severely damaged natural marine ecosystems. The chapter discusses the design approach to the Bio-Shelters, the marine species targeted, the iterative computational design, and sustainable material investigations that took place during the development. The authors then detail the hybrid fabrication processes using laser-cut steel waffle structures and shotcrete surface treatments. The chapter concludes with the current status of the ongoing project and defines the next steps using robotic fabrication.

DOI: 10.4018/978-1-6684-2344-8.ch010

INTRODUCTION

The Bio-Shelters project started in 2016 and is an ongoing investigation into the application of computational design and Industry 4.0 hybrid technologies to create customised, site-specific artificial reefs. The developed bio-reef is intended for Blackwattle Bay in Sydney Harbour where more than 50% of the shoreline has been hardened by coastal defences such as seawalls which have a severe negative impact on native ecosystems (Dafforn et al., 2015). The bay is an urbanised estuary used by heavy industry, tourism and residents. The water quality is poor and polluted with microplastics and high levels of heavy metals (Dafforn et al., 2015; Montoya, 2015).

With the development of a new Sydney Fish Market on the Bay, the local NSW government has mandated improving the water quality as an essential remit of the developers. Through a process of consultation and collaboration with marine biologists, it was determined that building artificial reefs would be an efficient way to clean the water, promote biodiversity and increase environmental consciousness to the visitors and tourists. The intended site of the artificial reef, in the following called *'Bio-Shelters'*, is under the wharf and along the water foreshore (as a landscape architecture feature) of the Fish Markets at Blackwattle Bay. This location has complex and limiting constraints including restricted natural light, and very few artificial reefs fixture points. These constraints make the site unsuitable for more simple solutions that use a repeated 2-dimensional textured surface. A more complex structure is imperative as meanwhile, to be effective and foster adoption by indigenous marine species and ecosystem engineers, the design solution must be highly customised with multiple, complex surfaces to maximise exposure to natural light. Blackwattle Bay being determined as the location for the Bio-Shelter project, the research team went on to develop three-dimensional objects that meet its characteristics. In consequence the project team aimed to design solutions that:

1. Are site-specific, by using data from marine biologists and environmental scientists about the site and by adapting to the background landscape.
2. avoid repetition of same parts and instead present local variety appropriate to the intended species, also the local sea tide levels and sun exposure.
3. extend automation through a scripted process, so that variations for different placements be produced by minor adjustments of the model and the data inputs.
4. present consistency throughout the process, so that the design methods are linked with research in materials and fabrication techniques that are compatible with the site's requirements and benefit from the team's expertise in diverse areas.

BACKGROUND

Existing types of artificial reefs predominantly either rely on simple two-dimensional objects with one textured surface, or on repetitive three-dimensional structures submerged into the water intended for various sites with special topographies and climatic features. Some of the projects address the seafloor while others have focused on populated harbour infrastructure such as seawalls.

The design and material composition of artificial reefs can be quite diverse. For example, projects such as those by Reef Design Lab (Reef Design Lab, n.d.) use concrete blocks, stable tables, pipes and more to grow coral. The Living Seawall Project was founded by the Sydney Institute of Marine Science in collaboration with the Reef Design Lab to improve the ecological sustainability of new and existing foreshore developments to ensure the future of marine life (Sydney Institute of Marine Science, n.d.). The Living Seawall Project was originally located in the Sydney Harbour Basin and is now applied globally (Ibid). One notable project from this collaboration was the Volvo living seawall which developed 2.5D tiles that had a surface pattern aimed to mimic the root structure of a mangrove tree, native to Australia (Volvo, n.d.).

Another significant project by MARS (Modular Artificial Reef System) has developed three-dimensional lattice structures that are deployed off the side of boats and assembled by divers (Reef Design Lab, n.d.). This structure has a surface geometry to encourage the natural recruitment of juvenile coral. These surfaces are 3D printed and moulded in ceramic using a slip casting process (Ibid). A 3D tiling system has also been developed by the University of Hong Kong Faculty of Architecture, and the University of Hong Kong Swire Institute of Marine Science and is an ongoing management measure for coral restoration in the Hoi Ha Wan Marine Park in Hong Kong (Lange et al., 2020). The design of these tiles is developed through robotic 3D printing of ceramic and terracotta materials to create a system that prevents sediment build-up, which is one of the major threats for coral reefs (Ibid). They used an algorithmic design process to develop biomimicry patterns integrated with spaces for securing coral fragments. In addition to these projects by others, the authors fabricated tiles, intended to test the hybrid material compounds, and deployed them into the site of the new Sydney Fish Market for analysis and evaluation in 2019 (See Figure 1).

Figure 1. Tiles developed to test hybrid material compounds using oyster shell waste © Research Team

The modular approaches described above are not customisable and lack the complexity needed at the site of the New Sydney Fish Markets. The Bio-Shelter team's response was to employ computation to achieve more complexity and respond to the challenging parameters of the site and the organisms the structure is intended for. Holistic approaches were sought, assisted by data-driven input/output processes in meeting ecological challenges. A design imperative included the vital operations (intensive properties) of the reef and the local climate and environmental constraints (extensive properties). This information was activated into a simulation model approximating physical action by which solutions were produced with reference to the initial data.

THE DEVELOPMENT OF BIO-SHELTERS

Approach

The Bio-shelters project uses a design-led research methodology whereby data is used to inform the design parameters, designs are developed, prototypes are produced, tested, and then redeveloped until they meet the aims of the project. The initial stage involved collecting data related to the biological characteristics of key habitat-forming species, then translating these into design imperatives. The habitat requirements were studied by extensive literature research, then they were codified as typological, organisational, and morphological inputs for the development of alternative schemes.

Targeted Marine Species

The species were selected from a list of common sessile (non-moving) and mobile species found in Sydney Harbour (Hutchings et al., 2013). Marine biology experts within the group focused on organisms of natural rocky shorelines at and near Blackwattle Bay. The sessile species were then assigned into two main functional groups: 1) filter-feeders and 2) algae. Filter-feeders, which included oysters, mussels, barnacles, and a polychaete (subtidal and intertidal), improve water quality (Coen et al., 2007; Grabowski et al., 2012; Armbruster, 2018) and increase biodiversity by creating habitat for other organisms (Barnes, 2003; Jackson et al., 2008). Algae, which included both intertidal and subtidal species, also serve as habitat for mobile species (Poore & Lowry, 1997; Migné et al., 2015), provide carbon sequestration benefits (Krause-Jensen & Duarte, 2016) and can mitigate environmental stressors in the intertidal zone (Bertness et al., 1999). The literature research included a variety of snails, limpets, a sea star and a fish. Mobile species found on natural rocky shores are often rare or absent on artificial structures (Chapman, 2003). The selected organisms improve the water quality along urbanised coastlines.

The research presented both qualitative and quantitative data for use in the Bio-Shelters models. Variables included, but were not limited to, general information about the species such as size, distribution or recruitment season, information about biotic interactions (competition, predation) with other species, environmental data on tolerances to temperature, salinity, water loss, sedimentation and water quality, distribution on the shore (tidal/water depth, wave actions), as well as light, substrate and other habitat requirements. However, not all information was available for every species. Although some of the variables (i.e., salinity, water quality) were set by the environmental conditions and could be manipulated with the design, these variables were used to determine which species were likely to colonise the site.

The second stage of the development of the project was to incorporate the information gathered into the computational model. For this purpose, the findings were transformed into categories or binary data for most variables (e.g., desiccation tolerance-> can survive out of water: yes/no). Then, the focus was given to variables that are paramount for the survival of the species such as different habitats and site conditions. These included water flow (for filter feeders), light and water supply (to prevent desiccation) for algae, and access to food and shelter from environmental stressors for mobile species. This could be achieved by controlling shading through form and orientation decisions, also by allowing water flow through gaps and voids into the structure for circulation and through the addition of refuges such as water-retaining pool features and small shaded areas. Shaded spots were kept to a minimum to avoid settlement of invasive species, which favour shaded areas (Miller & Etter, 2008; Dafforn et al., 2012). Species were then grouped according to their vertical distribution on the shore landscape as subtidal, low-, mid-, and/or high intertidal, and the design of each zone was matched to the species habitual needs (Chapman, 2003; Schaefer et al., 2018). Transition zones were included to ensure a smooth change from one zone to another. Surface structure was included whenever possible, by noting however that in the final design such areas would be smaller due to structural limitations related to own weight, the manufacturing process and material testing. Information on recruitment season was also suggested to determine optimal deployment of the Bio-Shelters to facilitate colonisation by target species.

To address the material composition needs for nature-compliant waterfront seawalls, following related studies (Marcus et al., 2016), a literature review and meta-analysis was used to compare recruitment of native species between natural (e.g., wood, shell, rock) and artificial (e.g., concrete, PVC) materials. Metrics included species richness, abundance in different functional groups, density and abundance of key habit-forming taxa of invasive species. The research revealed that there was no difference in total species richness, abundance or cover between natural and artificial materials, but there was significantly greater invasive species richness and cover on artificial ones. Total species abundance varied among material types and was generally higher on Perspex and lower on PVC, plexiglass, fibreglass, metal and porcelain compared to natural materials. Concrete, acrylic, brick, fibreglass, glass, plastic, PVC, rubber, metal Perspex and steel supported similar species abundances to natural materials. Additional research suggests that selection of high pH materials can help to mitigate the impacts of ocean acidification on associated species. Under scenarios of acidification and warming, sea urchins that settled on higher pH concrete grew larger, had longer spines and greater survivorship than those that settled on lower pH granite or greywacke (Mos et al. 2019). Furthermore, the cover of algal turf was greater on concrete than on granite or high-density polyethylene under scenarios of ocean acidification and increased temperature (Davis et al., 2017). Concrete also has the benefit in the Sydney region of providing settlement cues for the native rock oyster (Anderson, 1996).

To support diversity and to attract specific organisms it was critical that the design meets with the site's special characteristics. As a start, coastal habitats are under threat from rapid urbanisation. This is because, in some estuaries including Sydney Harbour, over 50% of the shoreline has been hardened by coastal defences such as seawalls being typically flat, featureless, 2D structures lacking microhabitats such as rock pools and crevices. Green or eco-friendly artificial habitats may enhance the native biodiversity and the ecological value of seawalls along heavily urbanised coastlines. Previous approaches to green engineering have typically identified habitat features (i.e., rockpools, crevices) missing from seawalls and added these using additive (i.e., retrofit) or subtractive (i.e., drilling) approaches. Such approaches assume that these habitat features function similarly on artificial structures as in natural habitats, despite the large environmental differences. However, the repeated seawalls commonly developed are generally stressful environments for many of these key-habitat forming species because they typically lack microhabitats that protect them from predation and environmental stressors such as drying at low tide.

In response, the Bio-Shelters project first describes reefs as highly sophisticated ecosystems whose natural habitats are threatened by human-related activity especially in proximity to highly urbanised areas; then, it develops 3D structures customised to its habitant's needs, thought as a critical improvement to ameliorate existing stressors and pressures and cause population restoration of the desirable species as in their natural environments. These structures would be guided by biological data describing various inhabitants' living conditions and contextual information concerning seascape, site topography, water temperature and currents. The produced datasets established an iterative process leading to site and organism-specific design outcomes. Parameters influencing the viability of marine species such as bivalves and seaweeds were translated into design structures and then prototyped.

The project's initial foundations were discussed and presented in Zavoleas and Haeusler's 2017 paper, *"Extending the modelling - Dynamic approaches applied to reef habitats at Sydney Harbour"*. This paper outlines a critical approach to computation in architecture by using multi-agent systems and dynamic simulation tools. The authors argue that such methods reinforce viewing design as a data-driven process, whereby a problem is analysed to a set of agents and their properties. The related actions assume extensive modelling techniques, recursive experimentation and testing to assist design from the early stages until completion.

Figure 2. Various marine species placed in relation to a rocky shoreline © Research Team

In reflection, Bio-Shelters employ similar architectural methods to tackle problems other than typical architectural ones. The pathway for designing and constructing artificial reefs would further comment on their shape, fabrication, materiality, and on-site placement, additionally reinforcing that extensive modelling techniques currently applicable in architecture could also respond to different scenarios about our settlements and the environment (See Figure 2).

Moreover, Bio-Shelters' aim was to integrate the design process with material exploration and fabrication techniques. That concept was analysed in the project's comprehensive overview entitled *"Designing Bio-Shelters: Improving Water Quality and Biodiversity in the Bays Precinct through Dynamic Data-Driven Approaches"* (Zavoleas et al. 2020). This paper presents the design, material exploration and some preliminary fabrication tests. Finally, Zavoleas and Davidova (2020) positioned the thinking that went into the development of Bio-Shelters in their paper *"Post-Anthropocene: The Design after the human-centred Design Age"* arguing and exemplifying possible traces of transition towards post-Anthropocene that is envisioned through non-hierarchical organisational systems. The paper takes Morton's discussion on 'hyper objectivity' further into multi-layered codesign performed in real-time and real-life across bio digital agents, including humans. The underlying concept is that although Earth might be recently experiencing drastic times and one catastrophic scenario follows the other, a natural succession often comes after most disasters. Consequently, Bio-Shelters was established within a cross-disciplinary framing that shared architectural knowledge alongside various scientific expertise to respond to a broadly acknowledged scope.

Geometry Simulation using Site-Specific and Marine Biology Data

To achieve meaningful results, the Bio-Shelter designs had to be able to liaise with the special requirements of different sites. The area west of the Harbour Bridge in which the Bays Precinct is found, has poor water quality, high levels of heavy metal pollution and microplastics, with much of the natural rocky reef habitat replaced by seawalls (Dafforn et al., 2012; Montoya, 2015). Despite environmental degrading, a variety of habitat-forming species (e.g., barnacles, oysters, seaweeds) can persist on the remnant rocky shores of the harbour, forming complex habitat structures for other animals and seaweeds, and cleaning water through filter-feeding and removal of excess nutrients and heavy metals.

Biological and material inputs along with such site-specific information form design parameters and ought to be included in the computational design model. The idea was to employ data-driven approaches supported by scripting and dynamic simulations throughout the whole process. Accordingly, the aim was rethought so that advancements in computation would be also able to handle marine datasets, by which to generate an n-number of design outcomes (seawalls, reefs) with reference to the site and water quality (Zavoleas & Haeusler, 2017). Autodesk Maya was used as the main platform to develop the script given its advanced modelling features and ability to simulate dynamic inputs through particles, also material behaviours and a good approximation of physics through forces, fields and collisions. On a different focus, it was assumed that the results better respond to environmental requirements if they incorporated sustainable materials for the fabrication process. The designs would be physically produced through digital fabrication processes, linked with analysis and the computational model through automated and scripted procedures. Information about natural rocky reefs and seawalls was a critical factor in shaping the script so that the produced schemes would be linked back to the needs of different organisms, environmental conditions, and the site. The Bays Precinct area was the first case study to test the script, with the potential to expand to other urbanised coastal sites. The inputs were organised with regards to macro-, meso- and micro- dimensions:

- **macro-** responds to site location (geospatial data): positioning, orientation, exposure to sun/wind, wave impact, topography, and material consistency of the shoreline.
- **meso-** sets the habitat requirements of barnacles, oysters, mussels, kelps, canopy-forming / branching coralline algae, forming the typological features, and microclimate.
- **micro-** refers to the material composition. For higher compatibility and efficiency, the project uses a mixture of concrete/crushed oyster shells.

Several models were designed to test the data-driven process and to advance the variety of geometry features that the outputs could carry. The earlier findings of this process were used in the later development of one large scale prototype of 80x30cm, set as a first 'proof of concept' for this project (See Figures 3 & 4).

Figure 3. First prototypes with 3D printing exploring the complexity of the geometry that would later challenge traditional fabrication processes © Research Team

Areas with different typological features correspond to varying information about sun exposure, tide, and site's material and physical topography as macro- inputs. To include them adds biodiversity since the occurring zones attract different species according to their needs as meso- constraints. Meanwhile, having designed and fabricated a prototype that inherited all the desired features concerning the complexity of form, the focus of the work moved then towards fabrication and materiality, presented in two further publications.

Composite Material Investigation

The paper "Recycled Sustainable 3D Printing Materials for Marine Environments" (Dunn et al., 2019) discussed the testing experiments of sustainable recycled materials for large scale 3D printed construction in a marine context. A pilot experiment was undertaken to assess the optimal material type, and macro- and micro-texture. It had two factors, material composition (which also influenced micro- texture) and the presence or absence of pits and mounds. Performance criteria were the corrosion of tiles and their colonisation by native species. The trial was based on the use of concrete. Concrete can be produced at scale, has had significant testing in marine conditions, does not need firing, and can be easily altered (Dunn et al., 2019). The material mixes were (See Figure 5):

1. a standard concrete mix of 2-parts cement: 1-part sand: 2-parts crushed rock aggregate.
2. a vermiculite treatment, with 2-parts cement: 1-part sand: 1-part vermiculite, 1-part crushed rock aggregate.
3. an oyster treatment, with 2-parts cement: 1-part sand:1-part sustainably sourced crushed oyster shells, 1-part crushed rock aggregate.

Figure 4. Different materials that formed part of the novel hybrid material compounds using Oyster shell waste © Research Team

All treatments had the same standard concrete base mix, with treatments 2 and 3 also including an additive. Vermiculite (hydrated laminar magnesium, aluminium, iron silicate) is a porous, lightweight material that was investigated due to its potential to reduce the weight of the concrete structures, add texture and anchor points for different species to attach to. [REMOVED INCLUDEPICTURE FIELD]

Figure 5. The tiles attached to frames prior to deployment from a jetty in Blackwattle Bay; tiles were randomly arranged on frames then and all frames were hung in the intertidal zone

Oyster shell was tested as an additive because it encourages the recruitment of live oysters (Zimme-Faust & Tamburri, 1994), which in turn provide food and habitat to other associated species. Oyster shells were sourced from the Sydney Fish Market, crushed with a Rock Crushing Machine and a Ball Mill, then sieved through a mesh sieve. Sixty 200 x 200 x 50mm tiles were produced and placed into the water. The tiles were mounted on a steel mesh frame and submerged vertically into the intertidal zone at Blackwattle Bay, Sydney to test the structural integrity of the material in marine conditions and the interaction with marine life (See Figure 6).

These thirty tiles (See Figure 1) were collected from the ocean after 6 weeks. Photos were taken to assess the percentage cover of fouling organisms on the vertical outward-facing surface and the horizontal downward-facing surface. Overall, there were more green and brown filamentous algae on the vertical outward face and more invertebrates on the horizontal downward face (including oyster recruits and the non-indigenous species *Styela plicata*, *Bugula neritina*, *Bugula flabellata*, *Schizoporella errata*, and *Diplosoma listerianum*). No obvious differences were observed among materials or complexities. The remaining tiles are monitored for a longer period to get a better sense of their performance and behaviour in the intended context.

The next stage of the material testing involved a prototype 500x500mm section at 1:1 scale of the custom computationally designed script. Priority was given to the development of a sustainable material along with the appropriate fabrication technique. Two changes were made:

1. a higher proportion of sand and oyster shells was used to increase the natural materials. The ratio was 1-part cement, 1-part sand, 2-parts crushed oyster shell, 1-part vermiculite.

2. The oyster shell particles were scaled up to 10x10mm to increase the texture of the material to facilitate anchor points for different marine species.

Cement was tested as a binder to which different materials were subtracted and added. Subtracted materials included crushed rock from traditional concrete mixes. Additive substitute materials included vermiculite as aggregate and texture variation, iron oxide as colourant for the concrete, and locally sourced waste biomaterials as crushed oyster shells. Crushed discarded oyster shells were sourced from Sydney Fish Market. Including oyster shells into the mix encourages oyster colonisation onto structures (See Figure 7). Oyster shell disposal relies on taking the shells to landfill. By using them in the build material, the project reduces landfill.

Other projects that are incorporating oyster shells in building materials and marine structures include tabby cement, a vernacular building technique from the southern US imported from Spain in the 17th Century (Sheehan and Sickels-Taves, 2002), and artificial reefs such as the U.S. Fish and Wildlife Service Project PORTS (Brenner, 2015), the shellfish restoration project in Port Phillip Bay Australia (Victorian Fisheries Authority), the artificial oyster reef in Moreton Bay in Queensland, Australia, and the San Francisco Bay Oyster Restoration Plan (George, 2005). However, none of these uses oyster shells as a 3D-printing material component. Therefore, the Bio-Shelters team sought different ways to adapt the available 3D printing technologies to the project's site location features. The Sydney fish markets are on Blackwattle Bay, a tidal area with high marine traffic, very depleted native marine species, and in a densely urbanised part of the city with considerable tourism. Due to its versatile character, the structural solutions ought to be combined with a functional means of rehabilitating the water to stay highly visible at high and low tides, and their form compatible with the intended species' habitat requirements.

Figure 6. Samples of the novel hybrid material compounds using Oyster shell waste © Research Team

Complex Fabrication Research

To achieve the complex geometries required for the site, the 3D printing fabrication process was particularly important. Other precedents use fabrication processes such as placing oyster shells in coconut fibre bags at the reef at Moreton Bay, whole oyster shells in plastic nets in the U.S. Fish and Wildlife Service Project PORTS, and whole oyster shells mixed with granite at the Port Phillip Bay project. For Bio-Shelters, several early prototypes were produced to refine on-site performance and material/design integration techniques. The script was adjusted accordingly to give appropriate structural thickness at different sizes. Additionally, the scheme began to incorporate different solutions for tiling, connecting and mounting the model onto the site. A segment was further enlarged to a 1:5 scale to test additive and subtractive methods.

To test the fabrication options, a series of physical models were developed in ceramics using a Potterbot 3D printer and then fired in a kiln. However, the scale of the project prohibits large scale kiln firing and further, it defeats its sustainable objectives. So, it was decided to combine Free Form Fabrication (FFF) and CNC routing of formwork in tooling wax. FFF/FDM printing is relatively easy to adjust for different purposes as many of the mechanisms can be easily adjusted. This process can also be scaled up by introducing large gantry frames and robotic arms for delivering materials. CNC milling is suitable to give refined resolution to roughly printed wax mould pieces then used to cast the prototype model parts of the concrete/oyster shell composite. The wax is recyclable since it is melted and reused. The research group linked with Laing O'Rourke and their FreeFab™ technology using wax 3D printing and 5-axis CNC milling to produce a complex geometry via a commercial fabrication process. The specialists confirmed that this technology was able to build the design. In the meantime, the research team fabricated a concrete mould to build the prototype internally with its own assets (See Figures 8, 9). A blue foam mould was created with 3-axis CNC milling, then the concrete mixture was poured to produce the prototype according to the initial aims.

Figure 7. Geometry 3D model of the prototype that was built with compound materials © Research Team

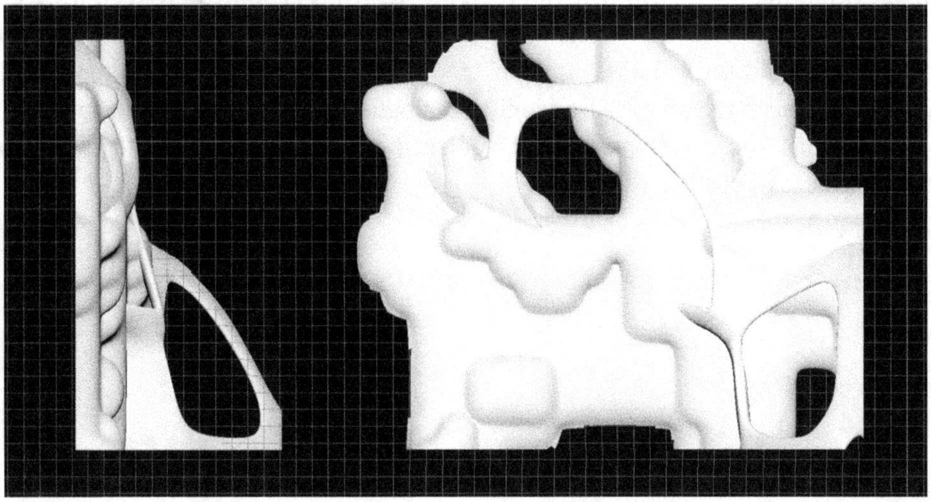

Figure 8. Close up view of the 3D model fabricated with hybrid concrete/oyster mixtures © Research Team

In parallel, the team worked together with Computational Design student Matthew Trilsbeck as part of his thesis project on first trials in Hybrid Clay 3D fabrication processes. The research paper, *Meeting in the middle: Hybrid Clay 3D fabrication processes for bio-reef structures* (Trilsbeck et al., 2019) outlined a study that seeks to improve the translation of complex geometries into physical clay artefacts through additive three-dimensional printing processes by drawing on the notion of digital craft and giving focus to human-machine interaction as a collaborative practice. For the Bio-Shelters project, one model of the first designs that were prototyped as miniatures with PLA was scaled up and built with a Z-Corp printer and ABS material (See Figure 10). The aims were then reset for physical size and compatible materials. A set of tests was conducted to print larger forms out of clay. Clay is well suited to artificial reef and seawall habitats (Rael et al., 2018), but its response to the complexity was unpredictable. Consequently, a set of hybrid digital-analogue crafting tests were made to turn complex forms into clay outputs. These attempts would be pursued in detail across data integration, materiality, and fabrication as the next challenges.

Figure 9. Prototype of 3D print that seeks to improve the translation of complex geometries into physical clay artefacts through additive three-dimensional printing processes by drawing on the notion of digital craft and giving focus to human-machine interaction as a collaborative practice © Research Team

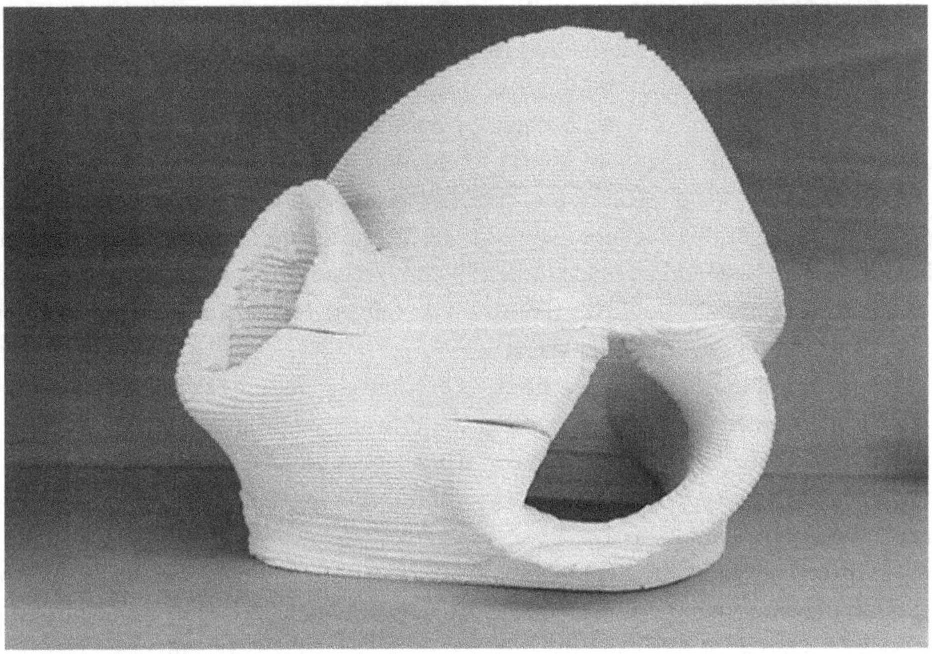

Hybrid Fabrication Technologies for Bio-Shelter 2.0

The 2021 iteration of Bio-Shelter 2.0 involved a 1-1 scale model that was 5 metres wide by 1 metre high. The model was installed at Blackwattle Bay in March 2021. Based on the existing research and experience, a more complex structure is imperative to meet the constraints of the site. The research team defines *Hybrid Fabrication Technology* as a process that combines multiple traditional and contemporary technologies to develop complex geometries and models that could not be achieved through one singular method alone. Contemporary fabrication techniques have had shortcomings and limitations in terms of transferring these complex digital models into the physical without the assistance of time-consuming and expensive processes. Further, some of the processes such as using polystyrene as a mould surface could contribute to landfill.

A series of models were made to test form, complexity, fabrication options and optimum material composition and uptake by the targeted marine species. Computational and parametric design has allowed designers to experiment and develop new forms that could have once been considered impossible to make. In building Bio-Shelters, hybrid fabrication technologies (and the use of integrated formwork systems) were experimented with as a means of producing complex artificial reef structures in cementitious materials accurately and efficiently. Testing of traditional casting techniques revealed shortcomings in the transfer of digital models into physical examples via standardised methods. Therefore, hybrid fabrication sought ways to combine both traditional and emerging fabrication technologies in the production of Bio-Shelter prototypes.

To ensure the successful testing of Bio-Shelter prototypes, a wide range of design considerations were investigated prior to fabrication. A suitably engineered reef structure is required to not only withstand the harshness of harbour environments but provide shelter to its inhabitants and not damage foreshore infrastructure. In considering the physical attributes of artificial reef panels and their subsequent impacts on biological performance, hybrid fabrication techniques sought ways to facilitate the production of habitat-forming features such as holes, hollows, and refuges wherever practicable. Such efforts provide opportunities for the settlement of a wide range of marine organisms and serve to boost ecological performance. Other important aspects of hybrid fabrication focused on composite material selection, optimisation of surface areas and textures, reef profile and orientation to natural light, and the environmental impacts of fixturing full-scale prototypes on site. These are discussed in further detail as follows:

Materials*:* Reef panels should be made of highly durable, non-toxic materials and require minimal ongoing maintenance. Materials should be suitable for prolonged use in marine environments, impermeable to water and resistant to hydrostatic pressure.

Surface Area: The total surface area of artificial reef panels (*including internal surfaces)* is important with regard to their intended function and ecological performance. Marine biology research indicates that the total area (*or areas*) of internal and external surfaces are critical to supporting biodiversity. Therefore, compound (*or double curvature*) fabrication technologies are utilised to increase internal surface areas and boost reef productivity.

Surface Texture*:* The surface texture of prototypes is modified to promote the settlement of algae and (*subsequently*) invertebrates. The presence of algae is critical to the performance of artificial reef structures as a food source for other marine organisms. Algae do not consume organic materials, instead, it feeds on the waste produced by marine organisms (and other pollutants). With this in mind, the Bio-Shelter reef panels are designed with a high level of surface complexity (*or texture*) for the timely recruitment of filter feeders (*oysters, clams, etc.*). Such efforts underpin the productive capacity of artificial reef communities as a whole.

Installation and Environmental Impacts*:* Due to changes in tides (*and tidal ranges*), timely installation of reef panels (*and subframes*) was critical. Installation of artificial reef panels should have little to no impact on existing marine infrastructure, and not pollute the harbour. In lieu of this, a subframe was installed prior to artificial reef panels so as to not impact existing settlements of marine life (*such as oyster beds*) in and around sea walls. The subframe is designed to rest on the seabed floor and be fixed to sea wall infrastructure with minimal environmental impact. This system allows for the simplified installation of large-scale compound reef panels.

Early prototyping started with the production of simple but robust, 1:1 study models. Further, early prototypes were limited in size and overall complexity compared to subsequent iterations but were fabricated using hybrid fabrication technologies wherever practicable. It was important to perform rigorous testing of composite materials to evaluate their application in the context of hybrid fabrication and apply evidence-based solutions (as required) prior to large scale production. Pilot testing sought ways to analyse waffle construction techniques, identify process improvement opportunities, and develop sustainable solutions moving forward. Optimisation of laser cutting processes was a key factor in determining the success of early prototypes and the feasibility of hybrid fabrication technologies as a whole.

Figure 10. Bio-shelter model generated from Autodesk Maya © Research Team

Due to the complicated perforations and undercuts that were generated from the Maya model and the output format as meshes, a rasterization process was employed to evaluate the geometry and simplify the overall complexity of the model for fabrication. A universal geometry processing script was developed in Grasshopper to simplify the high polygon count mesh models. The simplification of the polygon count was achieved through nurbs wireframe surface generations. The method created planes along all axes (x, y, z) at a user-specified distance and total plane count, achieving a higher resolution at a shorter distance and larger total plane count. The planes were then used to intersect any input mesh identifying the subdivided grid curves (lines) for surface generation. The curves were then simplified further, allowing the user to specify the resolution of the grid for the surface creation.

During the reconstruction of the mesh surface into a NURBS surface, the large openings in the Bio-Shelter geometry were required to remain as it was a critical element in creating an optimal habitat environment. The surface which was generated from the grid curves process did not compensate for openings as the method could only work when the grid curves were all connected. An analysis method was created to identify the openings as geometry which was then used to cut from the grid surface, resulting in a full simplification and conversion of the mesh model. This process did not only reduce the complexity of the mesh model input but also converted the mesh geometry into a NURBS geometry, a critical step that was needed for the further development of the fabrication ready model.

Double Moulds

Once the mesh geometry had been successfully rasterized and converted into a NURBS geometry, a double mould processing method was tested during the early stages of prototyping. The method in creating double moulds allowed for the shell surface to have a thickness as well as keeping the sophisticated double-curved geometry. Once the mesh geometry had been successfully rasterized and converted into a NURBS geometry, a double mould processing method was tested during the early stages of prototyping. The method, in creating double moulds, allowed for the shell surface to have a thickness as well as keeping the sophisticated double-curved geometry.

Figure 11. Section sliced from the Bio-Shelter geometry with double moulds (first perspective) © Research Team

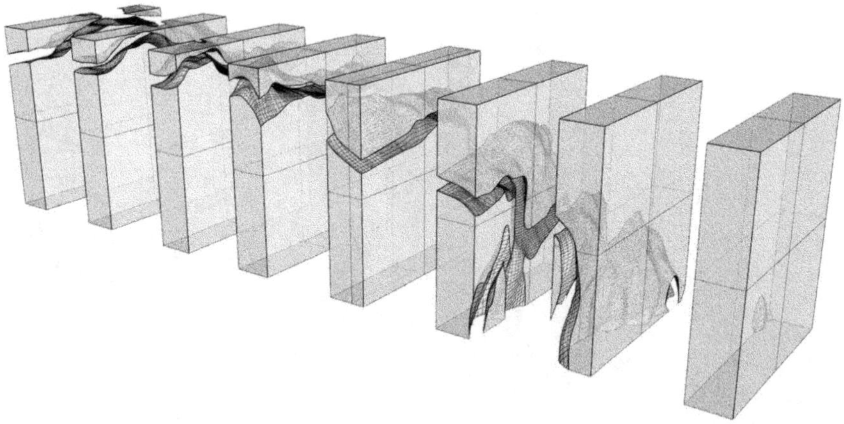

Figure 12. Section sliced from the Bio-shelter geometry with double moulds (second perspective) © Research Team

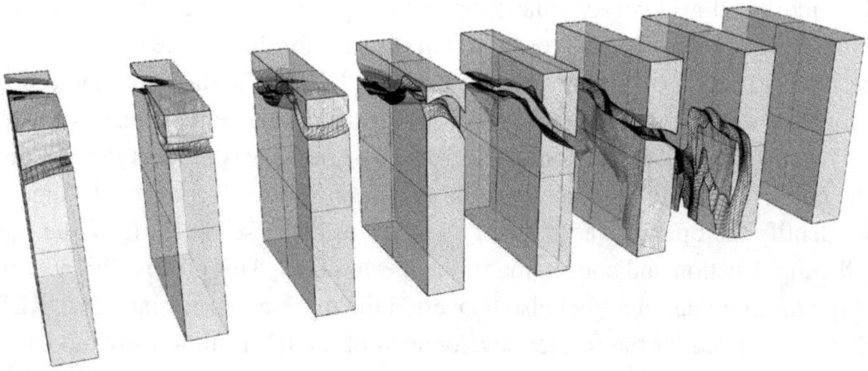

In the proposed assembly process of the double mould layers, a moulding box would be created to house and hold the moulds when pouring the material that would form the Bio-shelter sections. A critical consideration was given to the materiality of the moulds that would be housed inside these moulding boxes. Renewable moulding materials were considered as it was essential for the fabrication process to minimise waste. Foam was ultimately proposed as the material for the moulds as it allows for CNC milling and ease of assembly inside the moulding box, however, this would result in large amounts of material waste which became a key aspect in halting the double mould fabrication process.

Advanced Composite Laser Cut Fabrication

Current singular fabrication techniques, such as additive and subtractive methods had resulted in overly complex moulds and models that could not be easily produced were expensive and required considerable man-hours to create. Further, installing them in the harbour was deemed too challenging to do at any scale. Thus, the decision was made to experiment with hybrid fabrication techniques. In doing so, computational design methods utilised Grasshopper3D to slice reef panel geometries and produce a waffle layout structure or non-standard stereotomy. The individual members of waffle structures are customised to intersect and be easily joined together. Waffle construction allows for the rapid prototyping of complex double curvature surfaces and volumes, undercuts, and perforations. The resulting model is separated into individual surfaces that can be laid flat with connections or notches and laser cut onto 3.0mm mild steel sheet material. This method of fabrication is a more efficient cutting method than traditional metal cutting processes. The precision of the laser Facilitates the production of more complex parts which are easily reassembled in production environments. This base structure was welded together, and a mesh was applied to take shape and form the waffles' complex surfaces (see Figure 14).

Figure 13. Overall waffle structure separated by sections for fabrication, labelled for assembly © Research Team

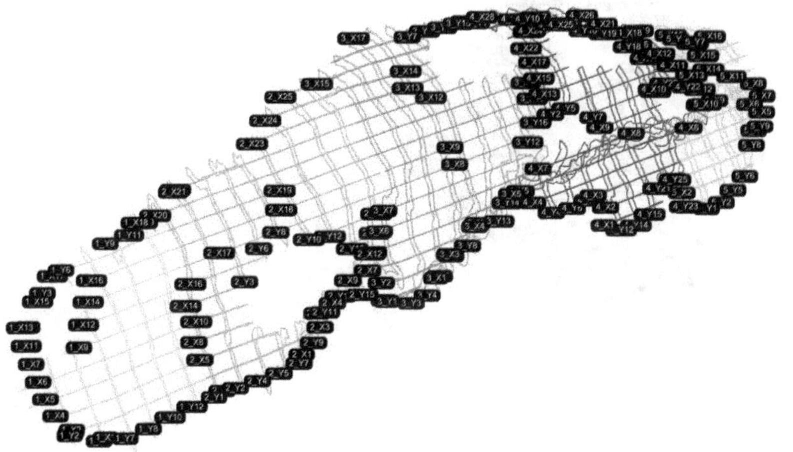

Typical construction of compound, or double curvature concrete structures is a time, labour and cost, intensive process. As demonstrated in the development of various prototypes, the project identifies laser-cut compound structures as stay-in-place formwork and presents methods of concrete placement, specifically wet-mix shotcrete, as a feasible manufacturing solution in defining reef-like characteristics. In this case, the use of hybrid technologies, as well as engineered materials, offers several advantages over more traditional methods of reinforced concrete placement. Automated industrial processes and the use of CNC laser cutting allow for rapid production of complex non-standard surfaces, while advances in concrete admixture technology mean it is possible to produce robust and flexible shotcrete mixtures customisable to digital fabrication processes. In the prototype models presented, the implementation of a stay-in-place forming system is shown to increase the structural capacity of reef panels under static and dynamic loads specific to intertidal environments and promotes ease of installation by incorporating relevant hardware and fixturing components.

Hybrid Fabrication Version 1.0

First-generation prototypes saw laser cut composite models used to investigate the process of slicing Bio-Shelter geometry from a 3D model into 2D surfaces suitable for laser cut processing. The Bio-shelter's 400 x 400mm contour surface was developed in Grasshopper3D by intersecting the model at set distances in both the X and Y axes. Contour and extrude commands were used to create the rib-like structure (see Figure 15). This produced the surfaces and the joints' location, number, and notch thickness used to assemble the structure. For version one, these waffle surfaces were positioned and joined at 150mm intervals. Each surface was assigned a "thickness" equal to the depth of the material used in the laser cutting process; this determined the notch thickness.

Figure 14. One section of waffle structure laid with each section parametrically fitted onto metal sheets for laser cutting © Research Team

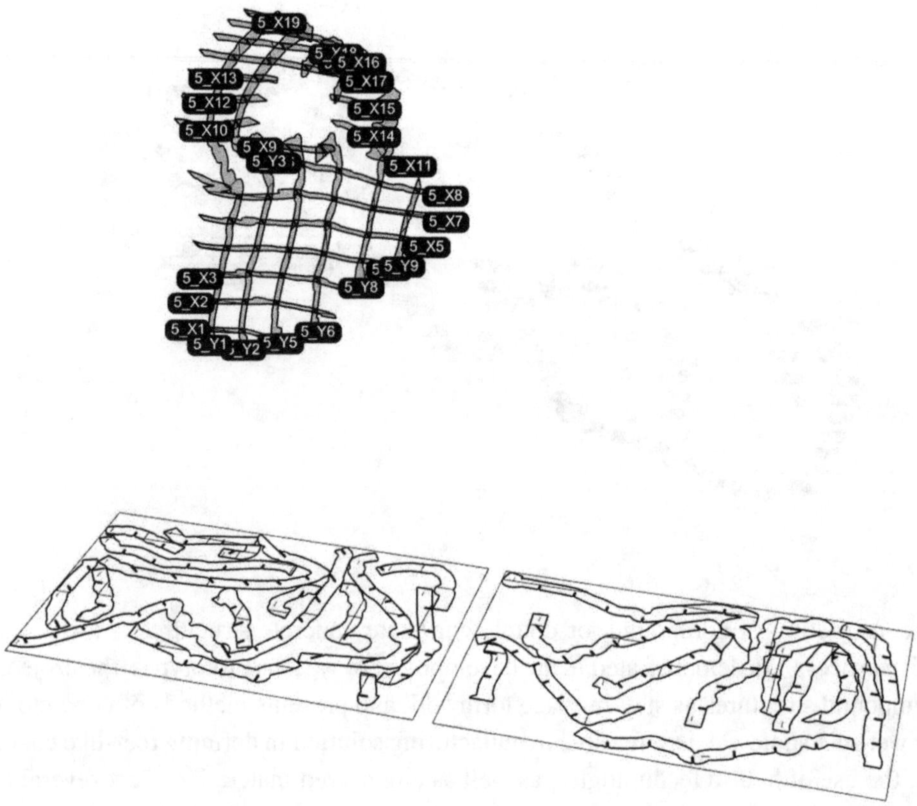

The resulting 2D lines work is extracted from the model and laid flat and nested within a material boundary on the XY plane, ready for laser cutting. Version one was laser cut out of 2.0mm basswood ply instead of mild steel to ensure the feasibility of the slicing and construction of the waffle components. Once assembled, an armature (or covering) of welded steel mesh is press-formed by bending and/or stretching to the finished basswood surfaces. Mesh is laminated, or composed of multiple layers, for additional strength and improved capture of surfaces and details. The lamination process (and application of multiple layers) allows the material to maintain its intended shape. Mesh layers are fastened together with a soft annealed steel wire. Pilot testing of applied techniques analysed the merits of the feasibility of version one prototyping, determined the process was a feasible approach, and readily applied to metal laser cutting processes (version 2.0).

Hybrid Fabrication Version 2.0

Version two was an identical 400 x 400mm waffle structure of the bio-shelter geometry that was sliced using the same Grasshopper definition. This prototype focused on development for 3.0mm Mild Steel that used a system of integrated rods that were welded in place through the waffle structure. In this version, cementitious materials and the hybrid fabrication were tested.

For version 2.0 the Kern Fibercell FC50 metal laser cutter was used to cut the interlocking waffle components in a 3.0mm Mild Steel sheet. In fibre laser cutting processes, laser light is created via a diode bank and is transmitted through a fibre optic cable. Amplified light, upon exiting the fibre cable, is collimated, or straightened and focused by a lens onto the material being cut. The concentrated energy heats the metal to above its melting point and is cut to shape.

The path of the laser is programmed using proprietary KCAM software to follow the desired geometry of the workpieces. Material settings and system parameters are adjusted within the CAM user interface. The basic setup of laser control tools requires the operator to select inputs for quick and accurate results. The correct adjustments are crucial in the metal laser cutting process as they directly affect the material removal rate and surface finish of the laser cut pieces. This process underwent rigorous testing and revisions to ensure the assembly of the structure was simple and required minimal post-processing.

When considering the surface texture and edge quality of metal laser cutting, the role of gas-assisted cutting and the selection of gas is crucial. The primary purpose of assist gas is to remove regular atmospheric air and blow away molten material. Pressurised gas also helps to cool the metal sheet, minimising deformation in heat prone areas. Version two prototypes manual grinding to finish the edges for assembly. Simple assembly of parts to create complex forms is one of the main reasons to use waffle structures. X and Y planes used to create waffle parts are sorted by laser etched part numbers, for example X2 or Y10, and assembled on a flat milled work surface prior to welding. Interlocking joints of respective parts are fastened together with a soft annealed steel wire. 1.25mm x 250mm tie wire lengths are easily formed and twisted together using end-cutting pliers. Tie wire is galvanised and offers excellent corrosion resistance in general construction and concrete placement applications.

The use of a 3D welding table allows for optimum positioning and fixing of 3.0mm mild steel components (see Figure 16). This modular system uses angle brackets, clamping pins, and spacer blocks to secure waffle components prior to welding. Individual parts are clamped during the tacking and welding process to minimise distortion (*expansion and contraction*) of weld areas and adjacent base metals. Tack welds are placed within the weld joint and spaced at 50mm intervals. The purpose of a tack weld is to hold parts of an assembly in proper alignment temporarily until the final welds are made. These smaller welds (*or tacks*) keep the workpiece from overheating and warping until the permanent weld is made. Tack welds are subsequently welded over with weld passes.

Figure 15. Laser-cut steel substructure for Bio-Shelter project © Research Team

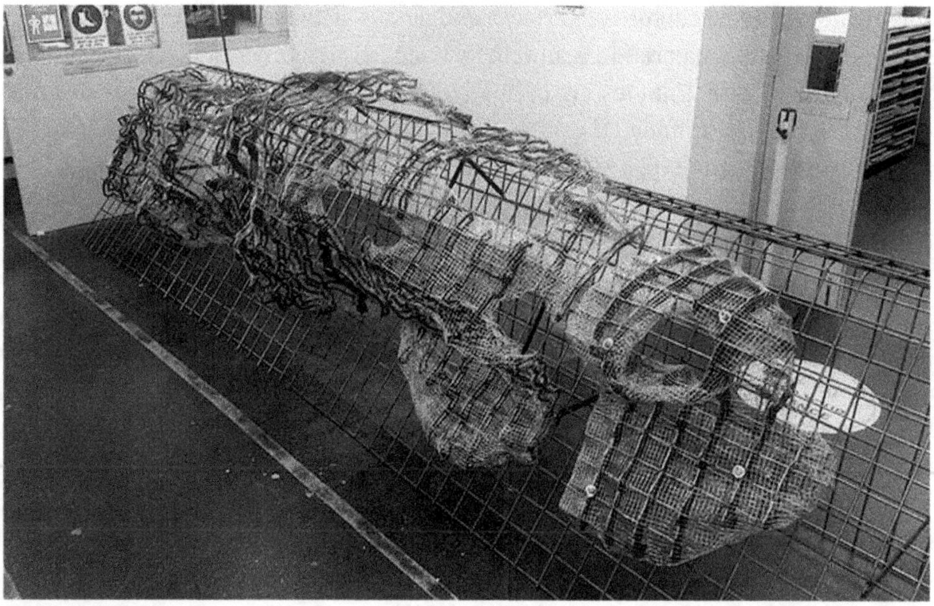

Application of Concrete

In the context of artificial reef design, cementitious materials, such as concrete, are widely used in the construction of marine infrastructure projects. In the production of Bio-Shelter prototypes, the selection of *concrete-like* materials was based on the results of earlier material testing. As discussed, concrete can be produced at scale, is easily altered, and demonstrates similar material properties to that of coral limestone. Marine biology data, and the study of artificial reef designs, have shown algae and marine invertebrates to thrive on concrete structures too. Furthermore, its workability makes it suitable for the fabrication of complex surfaces and features relative to complex data-driven design and construction projects.

In parallel, shotcrete is a process whereby concrete is sprayed at high velocity onto a surface, or in the case of Bio-Shelter, a mesh substrate. Like ferro-cement applications, whereby cement mortars are applied by hand onto steel supports, bio-shelter prototypes sought a revision of traditional methods specific to hybrid fabrication and requirements of artificial reef construction.

Shotcrete technology was ultimately utilised based on its versatility as a medium and demonstrated performance in the production of compound surfaces (see Figure 17). Application means materials can be applied directly onto waffle structures, eliminating the need for traditional formwork in the generation of composite reef panels. Therefore, the use of shotcrete significantly reduces labour and material costs, making such projects both economical and sustainable. Overall construction is reduced due to the low cost of applicable materials, and to the fact that no additional formwork or related foundations are needed.

Figure 16. Shotcrete sprayed onto steel substructure © Research Team

Findings and Observations

Through the process of prototyping using hybrid fabrication, the shotcrete waffle system performed as intended and key observations can be now made for future development and testing. The geometry developed using computational design methods allowed for complex, organic forms that mimicked the physical properties of reefs that support a range of marine life (see Figure 18). Not only do the computational design methods allow for complex forms, but allow for mass customisation, with changes being made instantaneously. A flexible, digital model would allow a direct translation from different locations and contexts, defined through marine biological data, into a hybrid fabrication process.

The design and fabrication methods outlined in this chapter allowed for a complex curved steel reinforced concrete system that will perform well in situations with heavier structural load, such as impact from waves, in comparison to traditional construction techniques and framing systems. It does not require additional vertical supports such as column posts or piers, as the waffle structure itself can be self-supportive. The interconnecting parts are prefabricated and allow for fast assembly reducing fabrication time. In addition to this, the prototyping process also provided us with an understanding of the correct techniques for laser-cutting metal, allowing for increased efficiency and automation for future projects.

Figure 17. Final project installed at Blackwattle Bay © Research Team

CONCLUSION AND FUTURE RESEARCH DIRECTIONS

The final section of this chapter concludes the hybrid fabrication techniques and the current status of the Bio-Shelter reef in Blackwattle Bay and documents the next steps for this research.

Currently it is too early to evaluate the project regarding its marine biological success. The installation has only been in the water for a few months (over winter) and no marine biological study has yet been conducted. Thus, we will concentrate on the discussion of the fabrication process. Here one can note that the project could be completed in time and within budget yet there are several fabrication issues outlined in the chapter above. While the computational design method utilised grasshopper3D to slice the reef panel geometries and produce a waffle layout structure (or non-standard stereotomy) offered a high level of precision (Fig. 16) the following step of Shotcreting (Fig. 17) did not. This lack of precision can be seen in Fig 18, where the second element of the right and its neighbour on the left do not align. This is a result of the fabrication process in which a builder, without much guidance, sprays concrete onto the waffle layout structure.

Consequently, this part of the hybrid fabrication process will be revisited in the next step. The research team proposes a dual robot arm printing process. In our proposed dual robot arm printing process one robot arm will hold the waffle layout structure (or non-standard stereotomy). As most industrial robot arms utilize six axes, this gives this robot arm the ability to position the waffle layout structure in the x, y, and z planes, as well as position itself using roll, pitch, and yaw movements - all towards optimising and assisting the 3D printing process done with the second robot arm. While several research projects apply 3D robot printing, most of them print on a static table or a table that can rotate around one axis only. We argue that this is not sufficient for our purposes. As seen in Fig 16 each element of the Bio-Shelters waffle layout structure has undercuts and openings that are hard to reach even for a robot arm with more than six axes. Further the 3D printed filament needs to be applied on all six sides of the waffle layout structure covering it completely in concrete.

Thus, as next steps the research team will develop and test a series of scripts to test this new robot fabrication process using the research team's robots. Considerations and focus will be placed here particularly on large scale fabrication processes which would allow the fabrication of larger elements, with greater precision and lower costs. The research team estimates that these next steps would allow for a mass production of Bio-Shelter reefs or seawalls and thus could contribute to regenerating on a large scale for lost marine habitat.

ACKNOWLEDGMENT

This research was supported UNSW and by a Landcom, State Government Contract [grant number RG182066]; Urban Growth NSW State Government Contract [grant number RG182616]; Infrastructure NSW, State Government Contract.

REFERENCES

Anderson, M. J. (1996). A chemical cue induces settlement of Sydney rock oysters, Saccostrea commercialis, in the laboratory and in the field. *The Biological Bulletin*, *190*(3), 350–358. doi:10.2307/1543027 PMID:29227699

Armbruster, S. (2018). *How Old Oyster Shells Are Reviving A Queensland Reef.* SBS News, Australia. www.sbs.com.au/news/how-old-oyster-shells-are-reviving-a-queensland-reef

Bertness, M. D., Leonard, G. H., Levine, J. M., Schmidt, P. R., & Ingraham, A. O. (1999). Testing the Relative Contribution of Positive and Negative Interactions in Rocky Intertidal Communities. *Ecology*, *80*(8), 2711–2726. doi:10.1890/0012-9658(1999)080[2711:TTRCOP]2.0.CO;2

Brenner, M. (2015). *Shells and Students: Building Living Reefs in Southern New Jersey.* usfwsnortheast.wordpress.com/2015/05/22/shells-and-students-in-southern-new-jersey/

Chapman, M. G. (2003). Paucity of mobile species on constructed seawalls: Effects of urbanization on biodiversity. *Marine Ecology Progress Series*, *264*, 21–29. doi:10.3354/meps264021

Coen, L. D., Brumbaugh, R. D., Bushek, D., Grizzle, R., Luckenbach, M. W., Posey, M. H., Powers, S. P., & Tolley, S. G. (2007). Ecosystem services related to oyster restoration. *Marine Ecology Progress Series*, *341*, 303–307. doi:10.3354/meps341303

Dafforn, K. A., Glasby, T. M., Airoldi, L., Rivero, N. K., Mayer-Pinto, M., & Johnston, E. L. (2015). Marine urbanization: An ecological framework for designing multifunctional artificial structures. *Frontiers in Ecology and the Environment*, *13*(2), 82–90. doi:10.1890/140050

Dafforn, K. A., Glasby, T. M., & Johnston, E. L. (2012). Comparing the invasibility of experimental "reefs" with field observations of natural reefs and artificial structures. *PLoS One*, *7*(5), e38124. doi:10.1371/journal.pone.0038124 PMID:22666459

Davidova, M., & Zavoleas, Y. (2020). Post-Anthropocene: The Design after the Human-Centered Design Age. *Re: Anthropocene: Design in the Age of Humans, 25th International Conference on Computer-Aided Architectural Design Research in Asia (CAADRIA), 2*, 203-212.

Davis, K. L., Coleman, M. A., Connell, S. D., Russell, B. D., Gillanders, B. M., & Kelaher, B. P. (2017). Ecological performance of construction materials subject to ocean climate change. *Marine Environmental Research, 131*, 177–182. doi:10.1016/j.marenvres.2017.09.011 PMID:28988853

Dunn, K., Haeusler, M. H., Zavoleas, Y., Bishop, M., & Dafforn, K. (2019). Recycled Sustainable 3D Printing Materials for Marine Environments. *Challenges: Sustainability and Cultural Heritage, 37th International Conference on Education and Research in Computer Aided Architectural Design in Europe (eCAADe), 2*, 583-592. http://papers.cumincad.org/data/works/att/ecaadesigradi2019_641.pdf

George, A. (2005). *Artificial Reefs for Oysters.* www.baynature.org/article/artificial-reefs-for-oysters/

Gibson, R. N., & Barnes, M. (2003). The use of intertidal barnacle shells. *Oceanography and Marine Biology: An Annual Review, 38*, 157–187.

Grabowski, J. H., Brumbaugh, R. D., Conrad, R. F., Keeler, A. G., Opaluch, J. J., Peterson, C. H., Piehler, M. F., Powers, S. P., & Smyth, A. R. (2012). Economic Valuation of Ecosystem Services Provided by Oyster Reefs. *Bioscience, 62*(10), 900–909. doi:10.1525/bio.2012.62.10.10

Hutchings, P., Ahyong, S., Ashcroft, M., Mcgrouther, M., & Reid, A. (2013). Sydney Harbour: Its Diverse Biodiversity. *Australian Zoologist, 36*(3), 255–320. doi:10.7882/AZ.2012.031

Jackson, A., Chapman, M., & Underwood, A. (2008). Ecological Interactions in the Provision of Habitat by Urban Development: Whelks and Engineering by Oysters on Artificial Seawalls. [x.]. *Austral Ecology, 33*(3), 307–316. doi:10.1111/j.1442-9993.2007.01818.x

Krause-Jensen, D., & Duarte, C. M. (2016). Substantial Role of Macroalgae in Marine Carbon Sequestration. *Nature Geoscience, 9*(10), 737–742. doi:10.1038/ngeo2790

Lange, C., Ratoi, L., & Co, D. L. (2020). *Reformative Coral Habitats-Rethinking Artificial Reef structures through a robotic 3D clay printing method.* Academic Press.

Marcus, A., Ikeda, M., & Jones, E. (2016). Buoyant Ecologies. In *ACADIA: Post-Human Frontiers.* Michigan Taubman College of Architecture & Urban Planning. http://acadia.org/projects/QEPKQ4

Migné, A., Golléty, C., & Davoult, D. (2015). Effect of Canopy Removal on A Rocky Shore Community Metabolism and Structure. *Marine Biology, 162*(2), 449–457. doi:10.100700227-014-2592-6

Miller, R. J., & Etter, R. J. (2008). Shading Facilitates Sessile Invertebrate Dominance in the Rocky Subtidal Gulf of Maine. *Ecology, 89*(2), 452–462. doi:10.1890/06-1099.1 PMID:18409434

Montoya, D. (2015). *Pollution in Sydney Harbour: Sewage, Toxic Chemicals and MicroPlastics.* NSW Parliamentary Research Service.

Mos, B., Dworjanyn, S. A., Mamo, L. T., & Kelaher, B. P. (2019). Building Global Change Resilience: Concrete Has the Potential to Ameliorate the Negative Effects of Climate-Driven Ocean Change on A Newly Settled Calcifying Invertebrate. *The Science of the Total Environment, 646*, 1349–1358. doi:10.1016/j.scitotenv.2018.07.379 PMID:30235620

Poore, A. G., & Lowry, J. K. (1997). New Ampithoid Amphipods from Port Jackson, New South Wales, Australia (Crustacea: Amphipoda: Ampithoidae). *Invertebrate Systematics, 11*(6), 897–941. doi:10.1071/IT95045

Rael, R., San Fratello, V., Arja, L., & Curth, S. (2018). *Coral Reef Seeding Units.* http://www.emergingobjects.com/project/coral-reef-seeding-units/

Reef Design Lab. (n.d.). *MARS.* https://www.reefdesignlab.com/mars1

Schaefer, N., Dafforn, K. A., Johnston, E. L., & Mayer-Pinto, M. (2018). Size, Depth and Position Affect the Diversity and Structure of Rock Pool Communities in An Urban Estuary. *Marine and Freshwater Research, 70*(7), 1034–1044. doi:10.1071/MF18074

Sheehan, M. S., & Sickels-Taves, L. B. (2002). Vernacular building materials and the factors conditioning their use: Tabby, a case study. *Material Culture, 34*(2), 16–28.

Sydney Institute Of Marine Science. (n.d.). *Mission. Living Seawalls.* https://www.livingseawalls.com.au/mission

Trilsbeck, M., Gardner, N., Fabbri, A., Haeusler, M. H., Zavoleas, Y., & Page, M. (2019). Meeting in the middle: Hybrid clay three-dimensional fabrication processes for bio-reef structures. *International Journal of Architectural Computing, 17*(2), 148–165. doi:10.1177/1478077119849655

Victorian Fisheries Authority. (n.d.). *Rebuilding Shellfish Reefs in Port Phillip Bay.* https://vfa.vic.gov.au/recreational-fishing/recreational-fishing-reefs/shellfish-reefs

Volvo. (n.d.). *Living Seawall - An Ocean Conservation Project.* Volvo Car Australia. https://www.volvocars.com/au/why-volvo/discover/living-seawall

Zavoleas, Y., & Haeusler, M. H. (2017). Extended Modelling: Dynamic Approaches Applied to Design Reef Habitats at Sydney Harbour. *Shock! Sharing Computational Knowledge, 35th International Conference on Education and Research in Computer Aided Architectural Design in Europe (eCAADe), 1*, 67-74. http://papers.cumincad.org/cgi-bin/works/Show?ecaade2017_146

Zavoleas, Y., Haeusler, M. H., Dunn, K., Bishop, M., Dafforn, K., Schaefer, N., Sedano, F., & Yu, K. D. (2020). Designing Bio-Shelters: Improving Water Quality and Biodiversity in the Bays Precinct through Dynamic Data-Driven Approaches. *Journal of Digital Landscape Architecture, 2020*(5), 521–532. doi:10.14627/537690053

Zimme-Faust, R. K., & Tamburri, M. N. (1994). Chemical Identity and Ecological Implications of a Waterborne, Larval Settlement Cue. *Limnology and Oceanography, 39*(5), 1075–1087. doi:10.4319/lo.1994.39.5.1075

Chapter 11 Computational Fluid Dynamics Analysis of Artificial Reefs

Prasad G.
https://orcid.org/0000-0002-5709-9182
Dayananda Sagar University, India

Sripad Kulkarni S.
Dayananda Sagar University, India

Shankar P. S.
Dayananda Sagar University, India

RakeshKumar C.
Veltech Rangarajan Dr. Sagunthala R& D Institute of Science and Technology, India

Balaji G.
Hindustan Institute of Technology and Science, India

ABSTRACT

The artificial reef is a vital submerged structure used to improve marine biological conditions and fishery resources. An artificial reef, with its different designs and structures, reshapes the local flow field by creating upwelling flow, decelerating flow, and back eddy. The shift in flow might result in nutrient exchange in the water and an increase in local primary production. Artificial reefs have been built all around the world, but there is no standardised technique for determining the effect range of an artificial reef.

INTRODUCTION

Artificial reefs (ARs) are an important and popular resource enhancement technique that can create new habitats and food resources for marine life while also breathing life into the ocean by concentrating fish (Brickhill et al., 2005; Firth et al., 2016) and increasing natural biological resource production (Bohnsack and Sutherland, 1985; Polovina and Sakai, 1989; Pickering and Whitmarsh, 1997; Seaman

DOI: 10.4018/978-1-6684-2344-8.ch011

and Jensen, 2000; Boswell et al., 2010). From 1962 to the early 1990s, the United States and Japan were the primary centers of AR research, with the primary goal of improving coastal fisheries (Lima et al., 2019). Japanese researchers have demonstrated that specially designed shallow water reefs can improve juvenile fish survival and growth (Sheehy, 1979). In 2004, 12 percent of Japan's continental shelf held 2107 m^3 of several types of ARs, particularly metal structures. Florida has many ARs and one of the most active AR programs in the United States. As a result, a growing number of countries (particularly China, Australia, Italy, Brazil, and the United Kingdom) have begun to focus on AR programs, such as a look at the future of ARs in Europe (Jensen, 2002), AR design for Korean coastal waters (Kim et al., 1994), the current state of ARs in Thailand (Kheawwongjan and Kim, 2012), and the development of concrete ARs using industrial waste in China (Huang et al., 2016). Thirty million cubic meters of artificial reefs have been installed in China's Shandong and Guangdong provinces since 2008. By 2025, relevant Chinese government departments will have deployed more than 5107 m^3 of artificial reefs in the Yellow Sea, South Sea, and East Sea. The total area of the artificial reef zones will be 1100km^2.

BACKGROUND

Millions of cubic meters of ARs are being built around the world for fishery or resource enhancement, and their deployment will be a complex and contentious social issue (Sutton, 2007). ARs' scales and purposes vary greatly (e.g., small structures by individuals in artisanal fishing communities, complex systems of heavy modules for commercial seafood production, and experimentally based structures for the restoration of the ecosystem). Le Diréach et al. (2015) investigated the effect of AR size, material, and structure on attracting fish and discovered that the mean size and density of the target species tended to increase with AR size or volume. Baine (2001) examined 249 papers on ARs and discovered that 36 papers (14% of the total) emphasized the importance of design complexity as well as the configuration, size, volume, and area of the reef. To make AR management easier, these man-made structures have been classified into scales based on module, set, group, and complexity (Grove and Sonu, 1983), as illustrated in Fig.1. The combination of architectural complexity, module design, and module layout on the bottom is regarded as the key to AR effectiveness (Sherman, 2002). AR deployment models such as the intensively stacked deployment model, flatly concentrated deployment model, and flatly distributed deployment model can be used to place ARs in specific areas of the marine environment (Yoon et al., 2016). The sizes, shapes, and volumes of ARs vary from deployment to deployment in a series of deployments.

Figure 1. Geometry and meshing of a man-made structure (module)

This study used computational fluid dynamics technology (CFD) to successfully set up an experiment to investigate the three-dimensional reconstruction of a cube-type artificial reef (CTAR) set in an underwater environment. Using the established 3D reconstruction approach, we can measure the usable volume, surface area, projected area, height, and base diameter of a CTAR set made up of 20 to 200 CTAR modules. Second, the forecast helps the government determine the best location for the ecological effects of the ARs, such as species attraction and ecosystem restoration. The fluid flow plays a prominent role in terms of better performance of an artificial reef subjected to biological and engineering aspects. In order to understand dynamic flow patterns is important thus to consult previous studies such as Rahman et al. (2020). Artificial reef performances were evaluated by calculating the performance indices of the upwelling and back eddy profile.

Biological data plays a prominent role in the development of artificial reefs. Coral reefs, occupying only about 0.2% of the ocean floor, are the most biologically diverse marine ecosystems known today (Knowlton et al. 2010). Because of the incredible number of marine species which inhabit coral reef ecosystems, having healthy reefs is crucial for the continuation of marine biodiversity. In recent years, a global increase in disease and stress throughout reef systems has led to increased mortality rates of coral. In the Indo-Pacific, coral reef cover has been declining at a rate of 1% per year for the last 20 years and a rate of 2% per year between 1997 and 2003 (Bruno and Selig 2007). A key factor in the decline of coral reefs is the growing concern of climate change, and a number of scientists are addressing the effects it will have upon coral reefs around the world (Bradbury and Seymour 2009; Hoegh-Guldberg et al. 2007). Other areas of concern include increased rates of development in coastal areas, which results in more sedimentation and nutrient runoff into coral reef ecosystems (Fabricius 2005; Rogers 1990). Overfishing and destructive fishing techniques are also having a negative impact on coral reefs (Bradbury and Seymour 2009).

MAIN FOCUS OF THE CHAPTER

The greater diversity of habitat resources in more complex reefs allows colonization to occur over a longer period of time, resulting in high species richness, density, and biomass (Charbonnel et al., 2000). According to Jordan et al. (2005), AR sets composed of several small reefs will have greater abundance than larger ones of equal volume because smaller reefs can support a higher density of fish than larger reefs. Large AR units are made up of vast, undivided empty chambers. Nonetheless, AR sets can provide a large number of relatively small, interconnected chambers as well as a large substratum area (Charbonnel et al., 2000).

Because of its impact on fish assemblage structure, the scale of an AR set, which is measured in terms of volume, height, attachment area, and base diameter, has gotten a lot of attention. The three-dimensional reconstruction method enables precise volume, surface area, and other morphometric measurements of three-dimensional objects (Cookson et al., 1993; Baillard et al., 1999; Zisserman et al., 1999; Chittaro, 2002; Williams et al., 2010). As a result, it is critical to devise a strategy for estimating the cube-type AR (CTAR) set's spatial scale via a 3D reconstruction of the set utilizing the alpha shape method.

As demonstrated in Fig.2 and Fig.3 velocity pattern CFD results, CTARs feature a simple shape, good workability, and low cost. Furthermore, because of the relatively large vacuum at each surface, CTARs are useful for fisheries and migratory fish protection (Kim et al., 2014). To create an AR set, several modules can be stacked. Small organisms can hide, forage for food, and reproduce in the gaps between adjoining reefs (Seaman and Lindberg, 2009). The stacked CTAR set can also be used in conjunction with other AR types. The AR experiment in China's Haizhou Bay, for example, coupled a CTAR set with sunken ships.

To extend the service life of the ARs, the direct influence of the water flow on the vessels' reefs is decreased; second, the complement and combination of them can occupy a broad sea space, which can enhance the effect range of ARs. Several studies have investigated the size of an AR set or other deployments. Numerous studies have been conducted on the ecological effects of ARs on their surroundings and interactions with marine species (Polovina and Sakai, 1989; Pickering and Whitmarsh, 1997; Pickering et al., 1998; Kim et al., 2008; Boswell et al., 2010; Le Diréach et al., 2015). According to these studies, the fish attracted to AR zones differ from reef to reef depending on their age, size, and structure. Because ARs are three-dimensional structures, the data on richness and biomass are based on the volume of the artificial reefs rather than the surface area.

The aim of the study pertaining to artificial reef is considered suitable for certain application, thus an application is discussed by Olivia Langhamer (2012) which is based on effect of artificial reef on offshore renewable energy conversion. In this research paper the artificial structures can create increased heterogeneity in the area important for species diversity and density hence the rapid worldwide growth of offshore renewable energy production will provide marine organisms with new hard substrate for colonization. Thus, acting as artificial reefs hence can prevent faulty structure applications also after the prevention of defective artificial reef it will have adverse effect on underwater status of the reef. In order to have control on these reefs it is important to provide additional critical habitat, which increases the environmental carrying capacity and eventually the abundance and biomass may result in complexity in the underwater status of artificial reef.

Figure 2. Velocity streamline

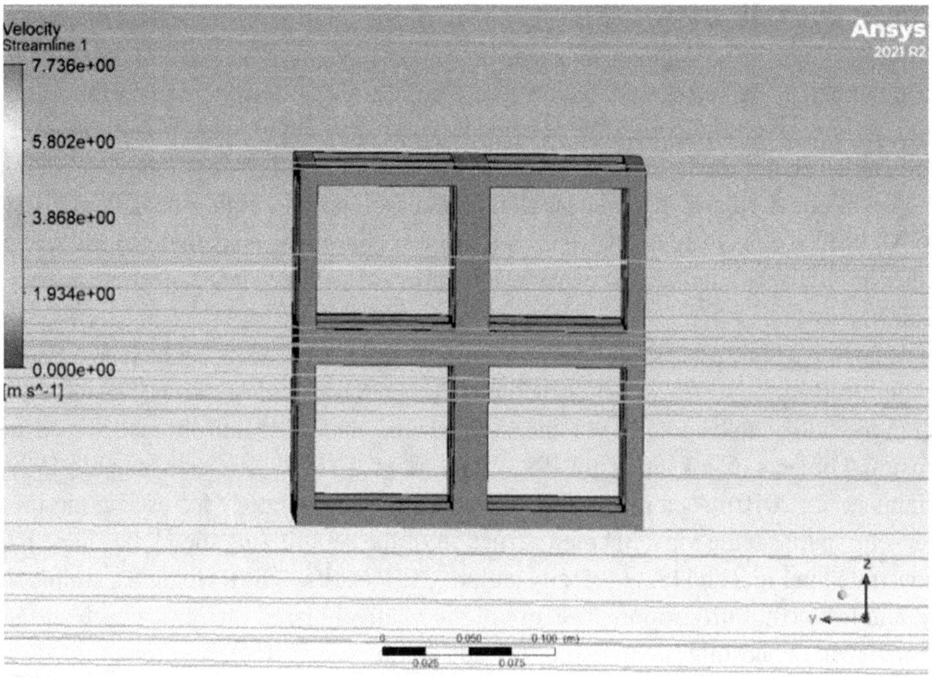

SOLUTIONS AND RECOMMENDATIONS

As a result, the volume of the AR set, particularly its irregular geometry, needs to be investigated. It is simple to calculate the volume of a single reef module. Reef sets made up of hundreds of small reef modules, on the other hand, can provide a plethora of small, interconnected chambers and substratum areas. The calculation of the reef set volume in this case is dependent not only on the shape of the reef sets, but also on the interconnected chambers. Field measurements, water tank experiments, and numerical analytical methods are the main ways to investigate the size, volume, height, and surface area of an AR set in terms of scale sizing.

Yoon et al. (2016) proposed three coefficients to predict the total loss of ARs, effective usable volume, and burial volume of the nine CTAR sets built between 1987 and 2002. However, due to the high costs of field investigations and the limitations of marine environmental conditions, the general method of surface structure surveying using multi beam scanning is difficult to employ in order to fully understand ARs and the underwater structure of the seabed. A low-cost prediction method must be developed in order to reduce measurement costs. Bruno et al. (2011) conducted tests under various turbidity conditions to demonstrate that the 3D reconstruction quality is acceptable even at high turbidity levels. Our research will set up an experiment to investigate the 3D reconstruction of a CTAR set in an underwater environment using linear structured light and binocular stereo vision technology.

Figure 3. Velocity streamline with flow pattern

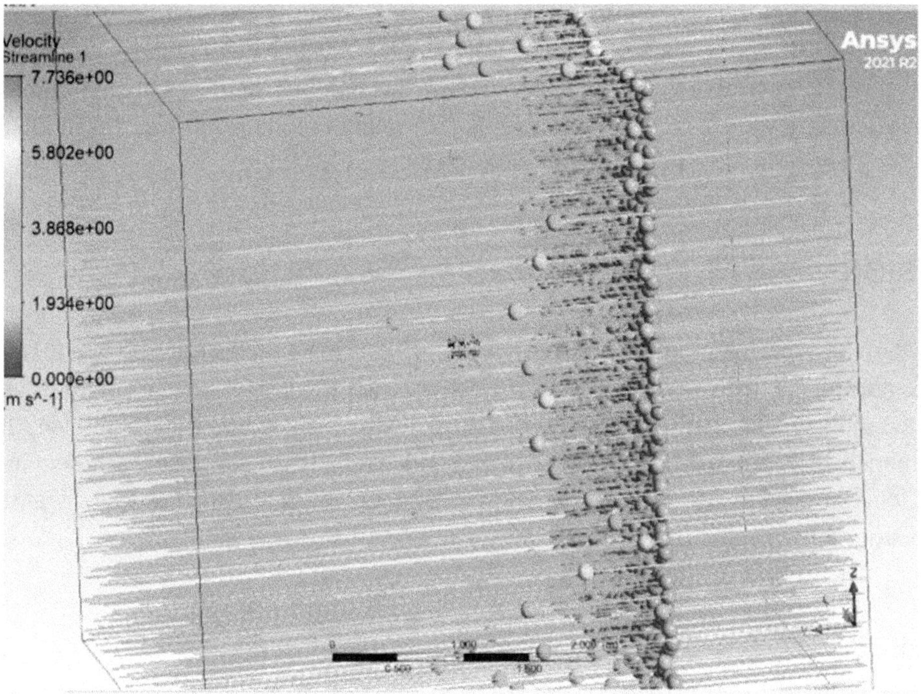

FUTURE RESEARCH DIRECTIONS

To design a surface that most closely resembles the CTAR set defined by the point clouds, the alpha shape method (Edelsbrunner et al., 1983; Edelsbrunner and Mucke, 1994) is applied. Then, using the alpha shape approach, a parameter research is undertaken based on these 3D reconstruction models to determine the scale size of a CTAR set (i.e., usable volume, surface area, projected area, height, and base diameter). Artificial reefs have been built all around the world, but there is no standardized technique for determining the effect range of an artificial reef.

The physical impact of an artificial reef on the surrounding flow is primarily reflected by three types of flow patterns: upwelling current, wake flow, and back eddy. In this study, we examine three distinct approaches based on influence length, influence area, and influence volume to determine the most effective method for indicating the impact effect of an artificial reef. The numerical approach of the renormalization group k-turbulence model is used to compute the features of the flow field around four different types of artificial reefs, and the following three types of assessment methodologies are utilized to compare the flow field. The results show that influence volume is superior to the other two in general, influence area and length can be misleading at times, especially when the artificial reef is asymmetric, influence area is a better way to show the diversity of wake flow, and influence length cannot reflect the back eddy. It is suggested that the influence volume be used as the primary measurement for measuring the impact range of an artificial reef. Meanwhile, the influence area may be a useful measure of the wake flow, and the influence length is just a comparison.

One of the purposes of this study is to explore the 3D reconstruction of a cube-type artificial reef (CTAR) set by linear structured light and binocular stereo vision technology in an underwater environment. Tang et al. (2021) in their paper developed the experimental setup composed of two cameras in a stereo vision configuration. The alpha shapes method can be used to construct a surface that most closely reflects the artificial reef set described by the points. A parameter study is conducted to assess the scales of the set (i.e., usable volume, surface area, projected area, height, and base diameter) based on 3D reconstruction. Experimental results show that the quality of 3D reconstruction in an underwater environment is acceptable for estimating the scale size of the CTAR set.

CONCLUSION

This study set up an experiment to investigate the three-dimensional reconstruction of a CTAR set in an underwater environment using CFD technology. First, we can measure the usable volume V u,s, surface area, projected area, height, and base diameter of the CTAR set made up of 20 to 200 CTAR modules using the established 3D reconstruction approach. Second, the forecast aids the government in determining the best location for the ARs to have the greatest biological impact, such as species attraction and ecosystem restoration.

REFERENCES

Baillard, C., Schmid, C., Zisserman, A., & Fitzgibbon, A. (1999). Automatic line matching and 3D reconstruction of buildings from multiple views. *ISPRS Conference on Automatic Extraction of GIS Objects from Digital Imagery*, 69-80.

Baine, M. (2001). Artificial reefs: A review of their design, application, management and performance. *Ocean and Coastal Management*, *44*(3-4), 241–259. doi:10.1016/S0964-5691(01)00048-5

Bernardini, F. (1997). *Sampling and reconstructing manifolds using alpha-shapes*. Purdue University. Report number: 97-013.

Bohnsack, J., & Sutherland, D. (1985). Artificial reef research: A review with recommendations for future priorities. *Bulletin of Marine Science*, *37*, 11–39.

Boswell, K. M., Wells, R. J. D., Cowan, J. J. H. Jr, & Wilson, C. A. (2010). Biomass, density, and size distributions of fishes associated with a large-scale artificial reef complex in the Gulf of Mexico. *Bulletin of Marine Science*, *86*(4), 879–889. doi:10.5343/bms.2010.1026

Bradbury, R. H., & Seymour, R. M. (2009). Coral reef science and the new commons. *Coral Reefs*, *28*(4), 831–837.

Brickhill, M. J., Lee, S. Y., & Connolly, R. M. (2005). Fishes associated with artificial reefs: Attributing changes to attraction or production using novel approaches. *Journal of Fish Biology*, *67*, 53–71.

Bruno, F., Bianco, G., Muzzupappa, M., Barone, S., & Rationales, A. V. (2011). Experimentation of structured light and stereo vision for underwater 3D reconstruction. *ISPRS Journal of Photogrammetry and Remote Sensing*, *66*(4), 508–518. doi:10.1016/j.isprsjprs.2011.02.009

Bruno, J. F., & Selig, E. R. (2007). Regional decline of coral cover in the Indo-Pacific: Timing, extent, and subregional comparisons. *PLoS One*, *2*(8), e711. doi:10.1371/journal.pone.0000711 PMID:17684557

Chan, Y., & Walmsley, R. P. (1997). Learning and understanding the Kruskal-Wallis one-way analysis-of-variance-by-ranks test for differences among three or more independent groups. *Physical Therapy*, *77*(12), 1755–1762. doi:10.1093/ptj/77.12.1755 PMID:9413454

Charbonnel, E., Francour, P., Harmelin, J. G., Ody, D., & Bachet, F. (2000). Effects of artificial reef design on associated fish assemblages in the Côte Bleue Marine Park (Mediterranean Sea, France). In *Artificial Reefs in European Seas* (pp. 365–377). Springer. doi:10.1007/978-94-011-4215-1_22

Chittaro, P. (2002). Species-area relationships for coral reef fish assemblages of St. Croix, US Virgin Islands. *Marine Ecology Progress Series*, *233*, 253–261. doi:10.3354/meps233253

Cookson, M. J., Davies, C. J., Entwistle, A., & Whimster, W. F. (1993). The microanatomy of the alveolar duct of the human lung imaged by confocal microscopy and visualized with computerbased 3D reconstruction. *Computerized Medical Imaging and Graphics*, *17*(3), 201–210. doi:10.1016/0895-6111(93)90044-N PMID:8402527

Edelsbrunner, H., Kirkpatrick, D., & Seidel, R. (1983). On the shape of a set of points in the plane. *IEEE Transactions on Information Theory*, *29*(4), 551–559. doi:10.1109/TIT.1983.1056714

Edelsbrunner, H., & Mucke, E. P. (1994). 3-dimensional alphashapes. *ACM Transactions on Graphics*, *13*(1), 43–72. doi:10.1145/174462.156635

Fabricius, K. E. (2005). Effects of terrestrial runoff on the ecology of corals and coral reefs: Review and synthesis. *Marine Pollution Bulletin*, *50*(2), 125–146. doi:10.1016/j.marpolbul.2004.11.028 PMID:15737355

Firth, L. B., Knights, A. M., Bridger, D., Evans, A., Mieszkowska, N., Moore, P., O'connor, N., Sheehan, E., Thompson, R., & Hawkins, S. (2016). Ocean sprawl: Challenges and opportunities for biodiversity management in a changing world. *Oceanography and Marine Biology*, *54*, 189–262. doi:10.1201/9781315368597-5

Grove, R. S., & Sonu, C. J. (1983). Review of Japanese fishing reef technology. Southern California Edison Company.

Hoegh-Guldberg, O., Mumby, P. J., Hooten, A. J., Steneck, R. S., Greenfield, P., Gomez, E., Harvell, C. D., Sale, P. F., Edwards, A. J., Caldeira, K., Knowlton, N., Eakin, C. M., Iglesias-Prieto, R., Muthiga, N., Bradbury, R. H., Dubi, A., & Hatziolos, M. E. (2007). Coral reefs under rapid climate change and ocean acidification. *Science*, *318*(5857), 1737–1742. doi:10.1126cience.1152509 PMID:18079392

Huang, X., Wang, Z., Liu, Y., Hu, W., & Ni, W. (2016). On the use of blast furnace slag and steel slag in the preparation of green artificial reef concrete. *Construction & Building Materials*, *112*, 241–246. doi:10.1016/j.conbuildmat.2016.02.088

Jensen, A. (2002). Artificial reefs of Europe: Perspective and future. *ICES Journal of Marine Science, 59*, 3–13. doi:10.1006/jmsc.2002.1298

Jordan, L. K. B., Gilliam, D. S., & Spieler, R. E. (2005). Reef fish assemblage structure affected by small-scale spacing and size variations of artificial patch reefs. *Journal of Experimental Marine Biology and Ecology, 326*(2), 170–186. doi:10.1016/j.jembe.2005.05.023

Kheawwongjan, A., & Kim, D. S. (2012). Present status and prospects of artificial reefs in Thailand. *Ocean and Coastal Management, 57*, 21–33. doi:10.1016/j.ocecoaman.2011.11.001

Kim, C. G., Lee, J. W., & Park, J. S. (1994). Artificial reef designs for Korean coastal waters. *Bulletin of Marine Science, 55*, 858–866.

Kim, D., Woo, J., Na, W. B., & Yoon, H. S. (2014). Flow and structural response characteristics of a box-type artificial reef. *Journal of Korean Society of Coastal and Ocean Engineers, 26*(3), 113–119. doi:10.9765/KSCOE.2014.26.3.113

Kim, H. S., Kim, C. G., Na, W. B., & Kim, J. K. (2008). Chemical degradation characteristics of reinforced concrete reefs in South Korea. *Ocean Engineering, 35*(8-9), 738–748. doi:10.1016/j.oceaneng.2008.02.003

Knowlton, N., Brainard, R. E., Fisher, R., Moews, M., Plaisance, L., & Caley, M. J. (2010). Coral reef biodiversity. *Life in the world's oceans: Diversity distribution and abundance*, 65-74.

Langhamer, O. (2012). Artificial reef effect in relation to offshore renewable energy conversion: State of the art. *TheScientificWorldJournal, 2012*. doi:10.1100/2012/386713 PMID:23326215

Le Diréach, L., Astruch, P., Bonhomme, D., Bonhomme, P., Fourt, M., Rouanet, E., & … . Ruitton, S. (2015). Comparative assessment of the deployment of 6 artificial reef types in Marseille Prado Bay (France) from a five-year seasonal survey of the fish fauna. *Proceedings of the RECIFS Conference on Artificial Reefs: from Materials to Ecosystems*, 27, 29.

Lima, J. S., Zalmon, I. R., & Love, M. (2019). Overview and trends of ecological and socioeconomic research on artificial reefs. *Marine Environmental Research, 145*, 81–96. doi:10.1016/j.marenvres.2019.01.010 PMID:30837123

Pickering, H., & Whitmarsh, D. (1997). Artificial reefs and fisheries exploitation a review of the 'attraction versus production' debate, the influence of design and its significance for policy. *Fisheries Research, 31*(1-2), 39–59. doi:10.1016/S0165-7836(97)00019-2

Pickering, H., Whitmarsh, D., & Jensen, A. (1998). Artificial reefs as a tool to aid rehabilitation of coastal ecosystems: Investigating the potential. *Marine Pollution Bulletin, 37*(8-12), 505–514. doi:10.1016/S0025-326X(98)00121-0

Polovina, J. J., & Sakai, I. (1989). Impacts of artificial reefs on fishery production in Shimamaki, Japan. *Bulletin of Marine Science, 44*, 997–1003.

Rahman, M. A. A., Nazri, M. N., Fitriadhy, A., Ahmad, M. F., Kasiman, E. H., Musa, M. A., ... Mohd, M. H. (2020). A Fundamental CFD Investigation of Offshore Structures for Artificial Coral Reef Development. *CFD Letters, 12*(7), 110–125. doi:10.37934/cfdl.12.7.110125

Rogers, C. S. (1990). Responses of coral reefs and reef organisms to sedimentation. *Marine ecology progress series. Oldendorf*, *62*(1), 185–202. doi:10.3354/meps062185

Seaman, J., & Jensen, A. (2000). Purposes and practices of artificial reef evaluation. In *Artificial Reef Evaluation* (pp. 1–19). CRC Press. doi:10.1201/9781420036633-6

Seaman, W., & Lindberg, W. J. (2009). Artificial reefs. In J. H. Steele (Ed.), *Encyclopedia of Ocean Sciences* (2nd ed., pp. 226–233). Academic Press. doi:10.1016/B978-012374473-9.00668-8

Sheehy, D. (1979). *Fisheries development*. Warwe Spectrum. Winter.

Sherman, R. (2002). Artificial reef design: Void space, complexity, and attractants. *ICES Journal of Marine Science*, *59*, S196–S200. doi:10.1006/jmsc.2001.1163

Sui, J., Zheng, Z., Chang, Z., & Yu, Z. (2016). Dynamics analysis on the deployment of a frame-type artificial reef. *Fishery Modernization*, *43*, 74–79.

Sutton, S. G. B., & Sally, L. (2007). Socio-economic aspects of artificial reefs: Considerations for the Great Barrier Reef Marine Park. *Ocean and Coastal Management*, *50*(10), 829–846. doi:10.1016/j.ocecoaman.2007.01.003

Tang, Y., Zhang, Z., & Wang, X. (2021). Estimation of the Scale of Artificial Reef Sets on the Basis of Underwater 3D Reconstruction. *Journal of Ocean University of China*, *20*(5), 1195–1206. doi:10.100711802-021-4751-y

Wallace, D. L. (2012). Simplified beta-approximations to the Kruskal-Wallis H test. *Journal of the American Statistical Association*, *54*(285), 225–230.

Williams, K., Rooper, C., & Towler, R. (2010). Use of stereo camera systems for assessment of rockfish abundance in untrawlable areas and for recording Pollock behavior during midwater trawls. *Fishery Bulletin–National Oceanic and Atmospheric Administration*, *108*, 352–362.

Yoon, H. S., Kim, D., & Na, W. B. (2016). Estimation of effective usable and burial volumes of artificial reefs and the prediction of cost-effective management. *Ocean and Coastal Management*, *120*, 135–147. doi:10.1016/j.ocecoaman.2015.12.007

Zisserman, A., Fitzgibbon, A., & Cross, G. (1999). VHS to VRML: 3D graphical models from video sequences. *Proceedings IEEE International Conference on Multimedia Computing and Systems*, 51-57. 10.1109/MMCS.1999.779119

Chapter 12
The Role of Oil and Gas Offshore Platform Reconversion in Creating Artificial Reefs: A Multi-Business and Socio-Economic Perspective

Roberto Vona
Federico II University of Naples, Italy

Vincenzo Basile
https://orcid.org/0000-0001-8776-0915
Federico II University of Naples, Italy

Nunzia Capobianco
https://orcid.org/0000-0001-7355-490X
Federico II University of Naples, Italy

Francesca Loia
https://orcid.org/0000-0003-1755-968X
Federico II University of Naples, Italy

ABSTRACT

In recent times, decommissioning of offshore platforms has become an even more discussed topic, for its relevant environmental, social, and economic repercussions. In particular, by carrying out economic considerations, all the divestiture possibilities applicable to an offshore platform and the relative sustainable business models (SBMs) will be analyzed in a wide framework of the circular economy and sustainable principles. In this scenario, sustainable decommissioning (SD) of offshore platforms process refers to multidimensional and interdisciplinary challenges, which requires a deep understanding of technical, legal, economic, financial, social, and environmental variables. The decommissioning of these structures is an issue that has gained a great deal of international attention and will require in the next years an open dialogue and exchange between institutions, oil and gas companies, enterprises, and the environment.

DOI: 10.4018/978-1-6684-2344-8.ch012

INTRODUCTION

In recent times, decommissioning of offshore platforms has become an even more discussed topic, for its relevant environmental, social, and economic repercussions (Basile et al., 2021a; Capobianco et al., 2021; Loia et al., 2021b). Decommissioning specifically refers to the series of processes involved in withdrawing a facility from service at the end of its life; its deconstruction and dismantling; and the removal of components for reuse, remanufacturing, recycling, storage and/or disposal. The international guidelines stated that abandoned or disused offshore installations or structures on any continental shelf are required to be removed. However, the considerable costs of decommissioning, as well as environmental considerations, have led to a gradual change in international regulations towards a more flexible approach if technical estimations allow it (Fowler et al., 2014). Indeed, experts and scholars have agreed that the partial removal options can deliver better environmental outcomes than complete removal of platforms in terms of biodiversity enhancement, provision of reef habitat, and protection from bottom trawling, aspects that instead are negatively affected by the complete removal. This awareness has led some nations to leave obsolete structures to act as artificial reefs and/or to find alternative solutions for their sustainable reuse of these assets (Margheritini, 2020). Multiuse platforms at sea (MUPS), indeed, represent an interesting solution for the development of marine infrastructures, including areas in which to start and develop various creative economic activities that are in harmony with the needs of environmental protection including sea shellfish farming, decarbonisation plants, renewable energy, tourism, and recreation (Nassar et al., 2020). Rigs-to-Reefs (RTR) refers to the conversion of decommissioned offshore oil and gas rigs into artificial reefs. In the United States, RTR is a nationwide program developed by the Bureau of Safety and Environmental Enforcement. RTR success' programs demonstrate how oil and gas offshore platforms represent an ecological opportunity for marine ecosystem enhancement or restoration. In this sense, several scholars stated that marine organisms attach themselves to the underwater portions of oil production platforms, transforming them into artificial reefs and these facilities can develop rich and various marine communities (e.g., Bell and Smith, 1999; Love et al., 2007; Macreadie et al., 2011, 2012; Todd et al., 2020). In the Gulf of Mexico, for example, platforms support populations of red snapper (*Lutjanus campechanus*) (Gallaway et al., 2009) and in Southern California rockfish (*Sebastes paucispinis*) (Love et al., 2007). In addition, in the Italian context, there is the reserve of *Paguro*, which is the wreck of a drilling platform, built by General Italian Oil Company (AGIP)[1] in Porto Corsini in 1963 for the extraction of natural gas. Recognized in 2010 as a site of community interest by the Emilia-Romagna region, it is the first reusing case of a former offshore platform as an artificial reef in the Italian context. The above has led to a revisiting of the regulations in favour of a more flexible approach, as outlined above, depending on specific technical and environmental assessments. Following these considerations, this chapter proposal inquiries about the possibility of reusing offshore extraction platforms, as artificial reefs. By carrying out economic considerations, all the divestiture possibilities applicable to an offshore platform and the relative Sustainable Business Models (SBMs) will be analysed in a wide framework of the circular economy and sustainable principles. The reconversion of these platforms in favour of multi-purpose initiatives, indeed, can have significant potential also in terms of reduction of operational costs for the offshore energy and aquaculture industry by means of concerted spatial planning and sharing of infrastructure (Abhinav et al., 2020). In addition, the increasingly urgent issues related to sustainability have triggered the need to balance all the social, organizational, and environmental concerns and have prompted future organizational efforts to define and adopt sustainable business models aimed at the well-being of the population and the improvement of environmental conditions. In this scenario, sustainable decommissioning (SD) of offshore platforms refers to multidimensional and interdisciplinary challenges, which requires a deep understanding of technical, legal, economic, financial, social, and environmental variables. The decommissioning of these structures is an issue that has gained a great deal of international attention and will require in the next years an open dialogue and exchange between institutions, oil and gas companies, enterprises, and environmental state agencies.

THE DECOMMISSIONING OF OIL AND GAS OFFSHORE PLATFORM: PROBLEM AND OPPORTUNITIES

Decommissioning can be deemed the last phase of the lifetime of an offshore platform. It essentially consists of the deactivation of an infrastructure, which often occurs because it is no longer economically viable. Globally it has been estimated that >7500 offshore oil and gas structures in the water of 53 countries will become obsolete over the next several decades, and most will require total removal under current regulations (Parente et al., 2006; Lakhal et al., 2009; Techera and Chandler, 2015). According to the market outlook by Visiongain[2] analysis, the global offshore oil & gas decommissioning market was valued at US$9,235 million in 2020 and is projected that it will reach a market value of US$17,099 million by 2031. The global market is expected to grow at a CAGR[3] of 5.8% during the forecast period 2021-2031. The offshore infrastructure includes pipelines, wellheads, platforms, artificial islands, and power cables. In some cases, state agencies may require the removal of onshore facilities used for processing and other post-extraction activities. During the 2021-30 period, more than 18,500 wellheads, 2,850 subsea trees, and 83,000 km of offshore pipelines and umbilicals currently in operation are subject to decommissioning and a large programme is underway. Europe is the largest market with 33% of spending, followed by the Asia Pacific and North America with 23% and 17% respectively. The UK is expected to be the largest spending country on decommissioning, followed by the US, Brazil, Norway, and Australia. In Italy, 138 platforms are nearing the end of their production cycle, which will thus have to be dismantled. In the Northern Adriatic Sea alone, 21 platforms will be decommissioned by 2021/2022 (MISE-UNMIG[4], 2017). More specifically, each offshore platform structure is different and therefore requires its customized approach to decommissioning. Generally, the end-of-life options for offshore infrastructure include (I) total removal, (II) *in situ* decommissioning (leaving the infrastructure in place either completely intact or with the topsides removed and legs toppled), (III) removal and relocation offshore (e.g., dive site or fisher), as well as (IV) partial removal (removing some parts of the infrastructure while leaving others *in situ*) (Ekins et al., 2005) (Fig.1).

Figure 1. The possible options for the decommissioning of oil and gas platforms and the area of possibilities to apply reconversion strategies
Source: adapted from Martins et al., 2020.

Abandoning the removal option for these structures represents the best practice from an environmental, ecological, and managerial point of view. Indeed, their other perspectives related to the decommissioning options consider environmental opportunities and challenges. They include, beyond the complete removal, reefing in *situ* (including leave in place, topping, and horizontal reefing), reefing elsewhere, and alternative uses. In this sense, a Sustainable Decommissioning (SD) of offshore platforms, thus, can bring value by contributing to improving environmental protection, creating new entrepreneurship initiatives, and generating a positive impact on collective well-being. Based on this consideration, otherwise, they are slated for total removals following the closure of their related hydrocarbon deposits.

MULTIUSE PLATFORMS AT SEA AS NOVEL MARINE INFRASTRUCTURES

Over recent years, sustainability has become a key consideration for oil and gas companies that aim to implement serious changes in their competitive strategies and business models. The Multi-Use Platforms at Sea (MUPS) summarized in Figure 2 could have huge potentialities related to the economic, technological, and environmental considerations on the Sustainable Decommissioning (SD) of offshore platforms. As highlighted previously, a gradual change in the international legislation is taking place regarding offshore platforms and a more flexible approach is emerging based on partial removal or readaptation of facilities according to case-specific circumstances. This has been caused especially because global influences related to sustainable approach and circular economy principles have pushed towards a more sustainable reconversion of offshore platforms. For instance, Baltic countries have already made great strides in making their energy systems cleaner. Growing evidence suggests the Baltic countries have enormous potential for offshore wind energy, which could have a revolutionary effect on the entire region. In addition, in the context of the North Sea, already several initiatives based on the readaptation of oil and gas platforms in photovoltaic and wind energy systems have been carried out. These initiatives are well accepted by the local population because the energy wind plants are positioned on offshore platforms, which are located far away from the coast. However, governance should apply a bottom-up decision-making process by being aware of the difficulties of local companies and by helping concretely individual actions to create sustainable and economic value and improve social wellbeing. A sustainable reconversion of offshore platforms can bring value by creating new entrepreneurship initiatives as well as to better social welfare thanks to the production of green energy. In this sense, a strategic role is covered by a public-private partnership, which should act as a driver of value co-creation (Capobianco et al., 2021; Loia et al., 2021a, b). This is in line with several European and international projects based on a partnership between local governments and private business which have actively contributed to stimulate new knowledge and common understanding on SD of offshore platforms, such as 4POWER[5] and ENTROPI[6] projects which aims to advance Key Enabling Technologies (KETs)[7] along the value chain to accelerate the deployment of multi-use offshore platforms, particularly for renewables and aquaculture. Furthermore, an interesting recently closed project of the Horizon 2020 program is the so-called "The Blue Growth Farm" (O'Reilly and O'Sullivan, 2014), started in June 2018 and coordinated by Rina Consulting S.p.A. and many other organizations and universities from Italy, Spain, France, and the UK contributing, proportionally to a total of EU$7,6 billion from EU funding. This long-term project aims to produce advanced industrial knowledge with a fully integrated and efficient offshore multipurpose floating platform, to host an automated aquaculture system, capable of producing high-quality fish. Within the framework of the MUPS concept, this project has become one of the EU's most interesting and ambitious projects by ensuring the integrated and sustainable exploitation of open sea resources (Eikeset et al., 2018). There are a different number of European projects funded by the European Commission under the Seventh Framework Programme for Research and Development, more specifically under the "Ocean of Tomorrow". In particular, the TROPOS8 project gathers 20 partners from nine countries[9], under the coordination of PLOCAN10. Linked to these projects, the H2OCEAN11 project aims at developing an innovative design for an economically and environmentally sustainable open-sea MUPS. The H2OCEAN platform will harvest wind and wave power, using part of the energy on-site for multiple applications including a multi-trophic aquaculture farm, and convert on-site the excess energy into hydrogen that can be stored and shipped to shore as a green energy carrier. This R&D project is supported by an ongoing industrial and academic partnership of five countries within the fields of renewable energy, fish farming, hydrogen generation, maritime transports, and related research disciplines. The unique feature of the H2OCEAN concept, besides the integration of different activities into a shared multi-use platform, lies in the novel approach for the transmission of offshore-generated renewable electrical energy through hydrogen (Brennan and Kolios, 2014). According to also to MERMAID12 project, the European oceans will be subject to massive development of marine infrastructure shortly. Marine structures for offshore wind farms and aquaculture must be installed at various sites and on a much larger scale than the earlier implementation of offshore structures to fulfil EU strategies based on the reduction of fossil-based energy and the sustainable use of marine resources, including (e.g.) aquaculture (Rasenberg, 2013; Christensen et al., 2015). Novel innovative design concepts should address different physical conditions to make the best use of the ocean space. Both for offshore renewable energy and aquaculture a substantial part of the costs are variable costs related to operations and maintenance of the plants. Optimising the use of ocean space for different purposes might benefit from shared resources such as staff allocation, transportation of staff and material from and to the platforms, use of forecasting systems, ships etc. (Koundouri et al., 2017). In addition, the MARIBE[13] is exploring cooperation opportunities for companies that combine different Blue Growth (BG) and Blue Economy (BE) sectors (Aguilar-Manjarrez, 2016; Williams et al., 2017; Johnson and Dalton, 2018). The project aims at unlocking the potential of multi-use platforms in the BE to support sustainable growth in the marine and maritime sectors by investigating cooperation opportunities for companies within the four key BG sectors e.g., "marine renewable energy", "aquaculture", "marine biotechnology" and "seabed mining". On the same course of action, the MUSES[14] research project which explored the real opportunities for a multi-use action plan in European seas (Przedrzymirska et al., 2018; Bocci et al., 2019; Van den Burg et al., 2020), provides an understanding of environmental, spatial, economic, and societal benefits of co-location, and highlight inappropriate regulatory, operational, environmental, health and safety, societal and legal aspects. There are 10 project partners from across Europe with a mix of consultancies, academia and government bodies that provide both wide geographical coverage and broad depth of knowledge/expertise, with strong links to stakeholder groups. In the same direction, PlaCE (Offshore Platform Conversion for Eco-Sustainable Multiples Use), under the direction of the Italian Ministry of Instruction and in collaboration with ENI S.p.A. and a few Italian Universities, aims to test cutting-edge technologies and solutions for the eco-sustainable re-use of offshore platforms located in the Adriatic Sea at the end of their productive phase. In nutshells, as reported by Capobianco et al. (2021), the reconversion of these structures can be considered as an economic stimulus from different points of view, which includes (i) marine weather centres, (ii) extraction plant of salts or minerals from seawater (e.g., the extraction of magnesium from seawater), (iii) construction of a regasification station (the liquid gas would arrive with ships, treated on the platform and sent ashore with the same sealine that transported the gas extracted in the previous life of the plant), (iv) wind turbines, solar panels, and tidal energy, (v) popular point of interest for tourism initiatives and sports activities (e.g., scuba diving, water-skiing, parasailing), (vi) creation of an artificial reef suitable for marine repopulation, which would also have a function of attraction, and (vii) development of sustainable aquaculture business.

Figure 2. Integrated multipurpose platform system for offshore oil and gas decommissioning
Source: adapted from Dol (2020).

RIGS-TO-REEFS POLICY AND ARTIFICIAL HABITAT

Artificial reefs have been used for centuries, but coordinated construction and design only started in the 17th century in Japan. Rigs-to-reefs (RTR) is the process of converting obsolete offshore oil and gas structures into artificial reefs (Kaiser and Pulsipher, 2005). The first examples of RTR occurred in the 1980s when platforms were removed from production in Louisiana and transported to Florida where they were repurposed as artificial reefs (Kaiser, 2006; Jørgensen, 2009). Since then, other states and countries have started rigs-to-reefs programmes, in which reefing of offshore oil and gas structures is a well organised and documented alternative to full removal to shore. Table 2 in the appendix describes the background of RTR programmes and initiatives worldwide. In 1984, the American Congress signed the National Fishing Enhancement Act, which recognised how beneficial could be by encouraging states to turn decommissioned oil and gas rigs into artificial reefs. Such biotic reefs have been created from oil rigs in the United States, Brunei, and Malaysia. In the United States, where the practice started and is most common, RTR is a nationwide program developed by the former Minerals Management Service, now Bureau of Safety and Environmental Enforcement, of the U.S. Department of the Interior. The program has involved fishers, the oil industry, and government regulators in the Gulf of Mexico, where offshore platforms develop a diverse reef community. In September 2012, 420 former oil platforms, about 10 per cent of decommissioned platforms, have been converted to permanent artificial reefs. Opposition in California has prevented an RTR program on the West Coast of the US. Similarly, environmental opposition has prevented the implementation of RTR in the North Sea. By April 2018, approximately 532 offshore platforms have been re-purposed as artificial reefs in the Gulf of Mexico, mostly in Louisiana and Texas (Ajemian et al., 2015; Bureau of Safety and Environmental Enforcement, 2018). This represents just over 11% of the total number of platforms decommissioned in the Gulf of Mexico (Bull and Love, 2019).

As showed in table 1, the impact of manufactured constructions offshore has proven to influence the surrounding marine life (Love et al., 2007) and to enhance marine ecological conditions: for this reason, it has become a popular tool adopted by governments and private parties for different purposes (Nautilus Consultants, 2003; OSPAR, 2009).

Table 1. Different aims of the RTR approach

Nature conservation, habitat restoration and compensation of habitat loss.	Artificial reefs like other habitats support the development of some species, but not of all. They can increase ecological connectivity and can have important biogeographical consequences. Depending on the type of species, this can be positive, neutral, or negative. Sessile species cover the hard substrate, varying with location and depth, so for these species artificial reefs are supportive. The reef can develop into a small, well-functioning ecosystem where various species reside and spawn. The bigger pelagic fish species are primarily attracted for food, shelter, and do not seem to spawn at the artificial reefs (Pickering and Whitmarsh, 1997).
Fisheries and mariculture.	Artificial reefs provide habitat for crustacean fisheries (e.g., lobster), for aquacultures of molluscs, algae, sponges, worms, etc. and pelagic fish stocks. Reefs are associated with high catch rates of economically important fish species (Grossman et al., 1997). Increases in fish catch around artificial reefs by 5% to 4000% have been documented (Santos and Monteiro, 1998). From a fish experiment at the Ekofisk platform in the North Sea, the researchers concluded that platforms enhance fishing operations, because aggregations of large fishes such as cod and saithe are significant (Soldal et al., 2002).
Sea defence and wave reduction device	Artificial reefs can support the creation of shelter areas for shipping and other offshore activities, and the expansion of the barrier structures near-shore.
Recreational activities	Artificial reefs can support sport fishing or diving. For instance, in the Gulf of Mexico, many people approach the platforms to spend their weekends diving and fishing around the structures.

Source: Authors' elaboration

Based on these considerations, the beneficial effects of artificial reefs creation are several: i) provide hard substrate habitat; ii) former a barrier structure and locally excludes trawl fisheries; iii) fosters environmental and sea resilience; iv) become like a magnet that creates life fauna and flora. Artificial reefs can function as stepping-stones for both sessile and mobile species (Langhamer, 2007), by providing the necessary habitat for hard-substrate species that would otherwise be absent. It also provides food, shelter, and nesting sites for fish. As a result, fish stocks may increase. If a species population is habitat limited (low availability of suitable habitat) or recruitment-limited (bounded by survival, dispersion, and settlement of larvae) due to lack of hard substrates, the addition of artificial reefs can result in a sustained increase in biomass (Macreadie, 2011; Bohnsack, 1989). The regional impact depends on i) interconnectivity of populations, ii) the total platform reef area compared to total reef areas and iii) the effect on specific key species that favour the conditions of platforms (Holbrook et al., 2000). Simultaneously, artificial reefs can form shelter areas, excluding trawl fishing where the material is located. The presence of oil and gas structures also function as fisheries exclusion zones, in the first place by creating a network of steel beams and offering shelter space for marine organisms, and in the second place by its safety zone with a radius of 500 meters around the structure (Schroeder & Love, 2004). Lastly, for the environment, repurposing obsolete structures saves fuel emissions that otherwise would be expended transporting and disposing of the structure. It also enriches the marine life in the area. For oil and gas companies, repurposing obsolete structures saves them the costs of removing, transporting, and disposing of them onshore. In addition, the artificial reefs attract marine life that enhances fisheries and contributes to the economy by attracting recreational and commercial fishing and diving. For divers and recreational and commercial fishers, artificial reefs create a rich diversity of marine life. For marine species, the artificial reefs provide habitat, shelter, food, and other necessary elements for biodiversity and a productive ocean. Interesting for the future potentiality of artificial reefs creation is the work of the Blue Latitudes Foundation[15]. The organization is currently testing the use of remotely operated vehicles (ROVs) to explore depths beyond the limitations of recreational divers to research and better understand the value of offshore oil and gas structures as reefs. However, it is important to note that just because RTR has been successful in a certain area (e.g., the Gulf of Mexico), it does not mean it would automatically be an ecologically beneficial exercise in the North Sea, California, or Australia. Every ecosystem is different and needs to be evaluated as such; creating a reef, simply because there is a platform that needs to be decommissioned, is indeed little more than waste disposal (Macdonald, 1994; Salcido, 2005). A major obstacle in the path of RTR legislation is the relative lack of ecological research on offshore structures. For example, despite the presence of over 40 offshore oil and gas installations on the continental shelf of north-west Australia, there has been a limited number of published studies on the ecology of the structures in this region (e.g., Fowler and Booth, 2012; Pradella et al., 2014; McLean et al., 2017; Bond et al., 2018). Macreadie et al. (2012) concluded that environmental research must be part of the development of the RTR policy, pointing to the case of California, where an RTR bill was vetoed in 2001 based on a lack of evidence that reefed platforms produce net environmental benefits. Macreadie et al. (2012) argue that the subsequent successful passing of an RTR bill in 2010 was due mainly to the studies of American Scholars (Schroeder and Love, 2002, 2004).

A SUSTAINABLE PERSPECTIVE ON THE ARTIFICIAL REEF: THE ITALIAN CASE OF *PAGURO* GAS PLATFORM'S WRECK

The Paguro platform was built by the General Italian Oil Company between 1962 and 1963 in Porto Corsini in Italy. It was created as a mobile platform for the perforation of wells for the extraction of methane gas in the Adriatic Sea. In 1965 it was placed on the new site known as PC7 (Porto Corsini 7), a dozen miles from the mouth of the "Fiumi Uniti" river. On September 28th, 1965, a drill reached 2900 meters of depth, a sudden eruption of gas caused the good walls to sink. The platform collapsed, consumed by flames fed by the gas leakage, and the following day it sank into the sea._Over time, the site has become an Artificial Reef, an oasis of life with wildlife (Figure 3). The platform is now twelve miles from the port of Marina di Ravenna at a depth of twenty-five meters. In July 1995, it was declared a Biological Protection Zone (BPZ) by the Ministry of Agriculture, Food and Forestry (MiPAAF). This recognition, together with the constitution of the Association *Paguro* in Ravenna in 1996, has contributed to the safeguard of marine life in the protected area. The association was created near the area to preserve, protect, and enhance the wreck, as well as to regulate sports, educational and scientific diving. By Commission Decision 2012/14/EU, the *Paguro* gas platform wreck was also designated as a Site of Community Importance (SCI code: IT 4070026) and a management plan outlining further conservation constraints was drawn up. In 2019, it was declared a Special Area of Conservation for biodiversity. It has become a valuable habitat and is now a European protected site under the Habitats Directive Natura 2000 Ecological Network. Between 1990-1991 and 1999-2000, jackets dismantled from about twenty other platforms were disposed of in the area, creating an artificial reef system, which is now a destination for intense diving activity. The Management Plan of the new SCI includes specific objectives of conservation and protection of the habitat and species, as well as social awareness and involvement of the local population. The related measures are therefore linked to strict control of access to the site now dedicated specifically for recreational, educational, and scientific purposes, including continuous monitoring activities. Any kind of fishery is prohibited. The non-governmental organization (NGO) "Associazione Paguro" oversees authorizing diving and managing recreational activities. The NGO acts to promote the cultural and natural heritage of the site, liaises with several diving schools, and organizes educational events (e.g., the dive days). It is also connected to the National Museum of Underwater Activities in Marina di Ravenna, where historical, cultural, and educational activities take place. Public institutions, research institutes and associations are responsible for the correct implementation of the Management Plan and are involved in the environmental conservation of the site. Diving associations, diving clubs and small and medium enterprises such as diving centres, boat rentals, etc., are involved in the process of enhancing the value of the *Paguro* wreck as well as aiming to offer a valuable touristic proposal. Educational events, such as diving days to the site, or underwater watching projects, for monitoring the marine environment, have been organized to support environmental education (Loia et al., 2021a). In addition, the *Paguro* platform hosted several training courses for becoming diving guides. These activities were fundamental for training those who were interested in working within the *Paguro* wreck context as an artificial reef. The relational aspect recovers a significant role. In several online platforms (such as Emilia Romagna region web portals, tourism websites, scuba online portals, Italian marine biology website), a section on the *Paguro* platform has been developed on one hand to promote the project and strengthen social participation; on the other hand, to gain attention and create interactions with local interests in working in on behalf of the *Paguro* ecosystem. In addition, the Paguro Association website was created to regulate sports, education, and scientific diving (Loia et al., 2021). Therefore, the *Paguro* wreck is now representing a valuable combination of environmental protection and underwater cultural heritage, Consequently, it protects the marine environment, increases underwater activities and tourism offer, develop new job profiles, and improve local environmental education. In this way, a converted offshore platform improves local tourism, the economic profitability in the area and local employment initiatives. On a national and international level, the reuse of the *Paguro* platform is a valuable case of promoting blue growth and sustainable development, aimed at increasing the protection of marine ecosystems (Loia et al., 2021a).

Figure 3. Paguro Platform: a reef off the coasts of Ravenna
Source: Ravenna Tourism website (2021)

ARTIFICIAL REEF RECONVERSION FOSTERED BY A LIFE CYCLE APPROACH

A sustainable project of reconversion offshore platforms can generate positive effects by safeguarding marine flora and fauna, reducing CO_2 (Jansen et al., 2020), understanding and engaging citizens and the local community, and supporting social impacts associated with conversion project employment. Additionally, reconversion projects can bring with them opportunities in terms of both employment and regeneration of raw materials with the potential to be converted and reused for other purposes. In the last few years, there are many successful decommission projects to support the circular economy, throughout the oil and gas industry. High-value components such as valves and turbines finding new homes, to entire topsides or Floating Production Storage and Offloading Units finding new life being repurposed in new projects (e.g., Sillimanite gas field in the UK and the Dutch continental, 2019). As well as opportunities to re-use offshore infrastructure for oil and gas, there are also opportunities to re-purpose infrastructure for use in the industry such as carbon capture, utilisation and storage, hydrogen production and offshore wind. The re-focusing of skills and expertise present in the oil industry will be essential for success in energy transition projects. The potential benefit emerging from the reconversion of end-of-life offshore platforms as artificial reefs stem from the evaluation of the costs considering the entire asset life cycle. More specifically, not only economic costs but also environmental and social costs resulting from a lacking procedure in sustainable decommissioning of the plant. In a reconversion project adoption of Sustainable and Circular Business Models (SBCM), based on value creation in the ecosystem perspective (Haeuber and Franklin, 1996; Browman et al. 2004) shed light on the possibility to reconfigure the value chain activities and reuse these structures to generate an opportunity from an economic, ecological, and social perspective. The novel scenario for the MUPS can represent an innovative solution to meet the growing pushes towards sustainability and circular economy associated with marine resources. Life Cycle Assessment (LCA) aims at determining the environmental effects of a product/function of a product based on a "from cradle to grave" view. LCA can be used to make a "strengths and weaknesses" analysis, product improvement and product comparison. It may contribute to remedies in the design stage and provide environmental and economic benefits. LCA developed in the stages that include: (i) identifying and quantifying the environmental loads involved (energy and raw materials used, emissions, wastes), (ii) assessing and evaluating potential environmental impacts of the loads, and (iii) assessing the opportunities available to bring about environmental improvements (Klöpffer, 2003; Basile et al., 2021b). This stage continues to the end of the study because LCA is an iterative process. LCA will be used in the assessment of the MUPS as a comparison tool between single-use and multi-use to evaluate the feasibility of MUPS using environmental impacts. Therefore, MUPS represents a great opportunity for re-using or reconverting the end life structure, restricting the impacts on marine (artificial reef creation) ecosystems as well as offering socio-economic benefits. By re-using the MUPS, some oil and gas platforms can be reconverted; a part of their structure could be useful for activities related to a particular industry such as touristic-recreational, aquaculture (e.g., fish farming and shellfish farming), renewable energy, and hydrogen storage. The shifting towards sustainability requires that firms should find innovative ways to be profitable and at the same time expand the traditional frontiers of business to include the environmental and social dimensions, in other words, take account of *"Life Cycle Thinking"*. Life Cycle Management (LCM) aims to minimize the environmental and socio-economic burdens associated with product or product portfolio throughout its entire life cycle and value chain. The studies on life cycle cost (Fuller, 2010; Lam et al., 2018) integrate the Asset Integrity Management contribution (Chandima Ratnayake et al., 2012; Kusumawardhani et al., 2016) into a holistic framework of Sustainable Life Cycle Management as illustrated in Figure 4 below (Basile et al., 2021b). Generally, the LCM can be used to define, analyse, and manage product-related information and activities towards continuous improvement along the product life cycle (Remmen et al., 2007; Basile et al. 2021b). Artificial reefs could be a great opportunity considering the LCM of MUPS in terms of sustainable decommissioning and reusing of industrial materials avoiding some dismantling and ecological costs or negative externality for the oil and gas companies.

Figure 4. Sustainable life cycle management framework
Source: Basile et al., 2021b.

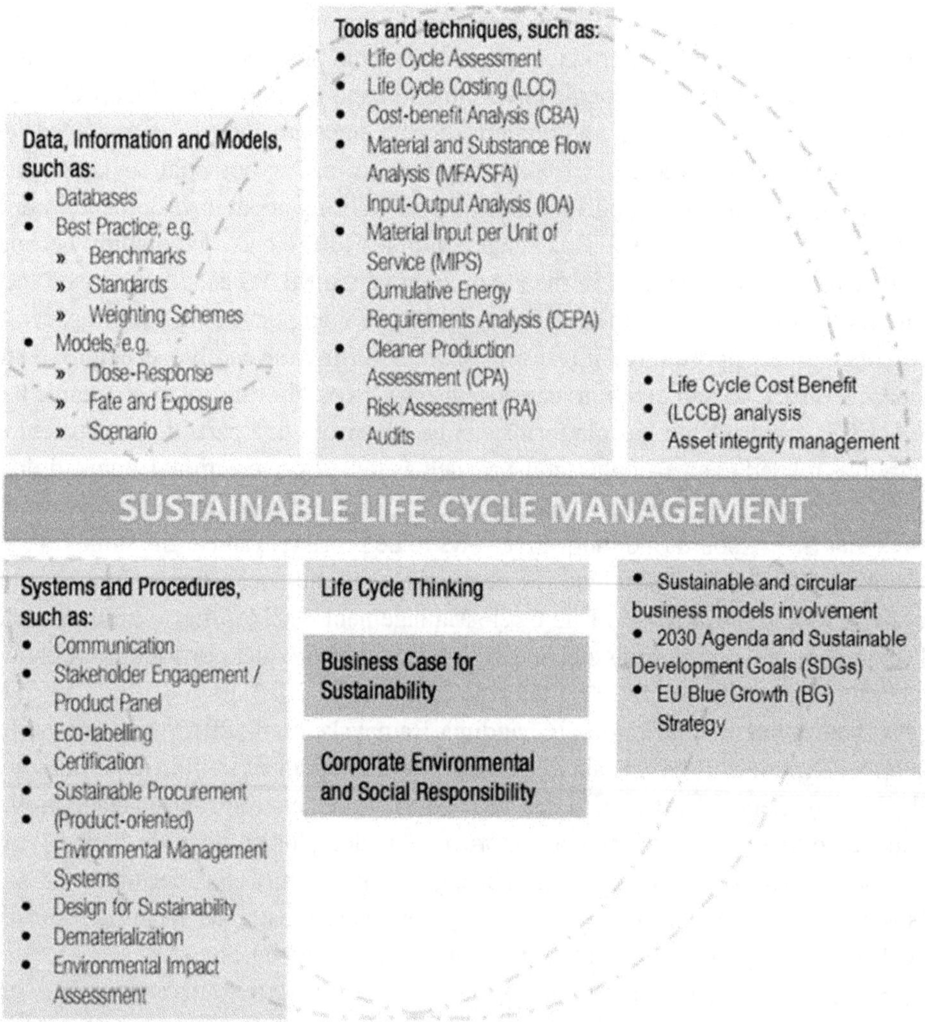

Towards Sustainable and Circular Business Models in the Decommissioning of Offshore Platforms

MUPS concepts integrate different maritime economic activities within the same space. In line with the EU's Blue Growth Strategy, this new type of business model provides a series of potential advantages: efficient use of marine space, sharing of risks and costs, sharing resources, reduced environmental impacts, and enhanced socio-economic benefits. Delivering this vision will require tools that identify viable multi-use combinations allowing for the optimal use of sea space (Legorburu et al., 2018). To create economic and environmental (?) value, the artificial reefs are linked to MUPS and marine aquaculture. All the ideas will contribute to making a profit besides reducing costs from the decommissioning process. Deep-sea fish farming is a prospective business for coastal countries providing half of all seafood consumed in the world (Dol, 2018). Since the farm will be located offshore in deep waters, it is required to be operated with minimal human interaction and will be able to internally generate the required amount of energy to remain self-sustainable. The energy requirement for the activities on the platform can be generated by wind, tidal currents, waves, vortex-induced ocean current and solar energy. To be cost-effective, the design of this offshore mariculture farm should be integrated with renewable energy sources. This approach can contribute to the creation of public awareness on environmental issues (Loia et al., 2021b): it can stimulate the knowledge on environmental educational issues associated with deep-sea fish farming as well as offshore renewable energy application to marine life sustainability. Figure 5 below shows how the combined vision of economic, social, and environmental might support the reconversion of the offshore platforms to the artificial reefs. The canvas model opens the door for the identification of additional factors that might influence the development of this new business model; for example, differences in energy markets, climate policies, or sustainable development patterns must be highlighted. The main contribution of the SCBMC for the oil and gas offshore platform is that it links the logic of how an organization creates, offers, and delivers value to its wider stakeholders with the minimisation of environmental and social costs. The iteration of sustainability and circular economy in the proposed model in conjunction with a more detailed cost-benefit analysis (as a decision management tool) will help strengthen the model. At this early stage, it can be seen as a good way to communicate a business model to stakeholders, including funders, the public sector, and the media. Indeed, the SCBMC is close to the logic of Triple-Layered Business Model Canvas (TLBMC) a tool for exploring sustainability-oriented business model innovation. According to Joyce and Paquin (2016), the TLMBC adds two layers to the original Business Model Canvas: an environmental layer based on a life cycle perspective and a social layer based on a stakeholder perspective. Taken together, the three-view level of the SCBMC makes it clearer how an organization generates multiple types of economic, environmental, and social value. The starting point of the SCBMC is economic sustainability, i.e., the revenue streams that result from the activities triggered by the decommissioning industry. However, in the current scenario also environmental sustainability is fundamental. In this sense, carbon capture and utilization and storage (CCUS) represent a huge opportunity and an important emission reduction technology that can be applied across the energy industry and provide benefits to society. According to several authors (Basile et al., 2021a; Capobianco et al., 2021) leaving these structures offshore or dismantling them without effective strategic planning that takes the long view represent a huge risk to the environment and the community. In this sense, after reclaiming the platforms and making them usable and safe, it is essential to define strategically how to manage them in the long term. Therefore, activities such as integrated multitrophic aquaculture (IMTA) and artificial reefs which promote underwater tourism represent sustainable and entrepreneurial opportunities to readapt these platforms. Basile and Vona (2021) and Capobianco et al. (2021) highlight those offshore platforms could become Sea Farms, also taking advantage of the market potential arising from the increased demand for fish products. In a hypothesis of reconversion and alternative use of offshore platforms, the development of IMTA technology, which is a production technology capable of reducing pollution while increasing productivity and profits by converting waste streams into new products, is strategic because of its environmental and economic impact and can allow production improvement, supply system diversification, and cost reduction.

Figure 5. From the Business Ecosystem level to the Sustainability impact: a framework for Sustainable and Circular Business Model in the oil and gas offshore platform decommissioning and conversion
Source: Basile and Vona, 2021.

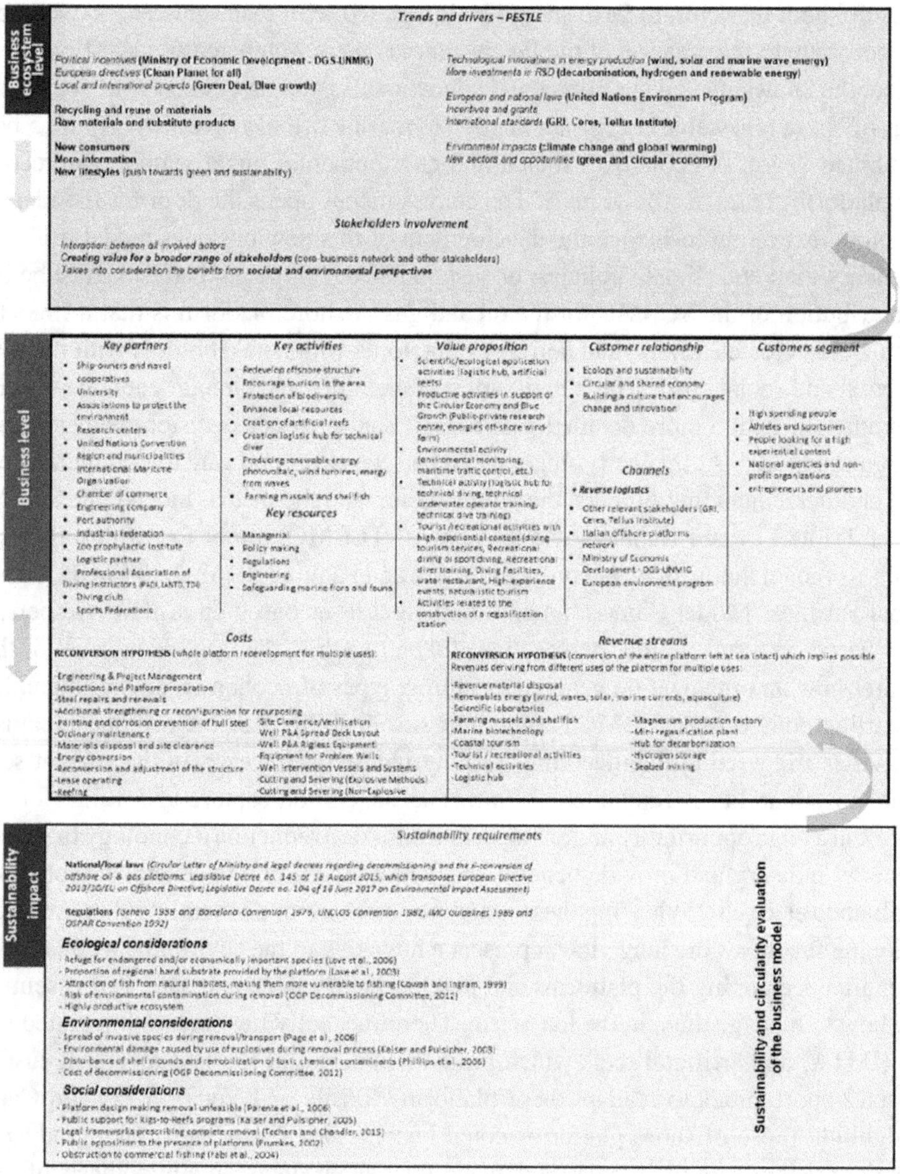

CONCLUSION AND FUTURE STREAMS OF RESEARCH

The prospective of the offshore decommissioning structures have changed over the years, as the oil and gas industry is becoming more integrated with the increasing energy, demand of global society. In the last 20 years, growing attention has been given to projections, environmental impact assessments and public awareness on the sustainable exploitation of natural resources. Upon the 2030 Agenda which stated that *"The sustainable development is defined as development that meets the needs of the present without compromising the ability of future generations to meet their own needs"* - a sustainable approach has become a fundamental issue in the global market strongly characterized by innovation, ecological concerns, energy transition, and internationalization. Therefore, governments and institutions involved in the decommissioning process are re-thinking their strategy through actions related to energy transition and the reconversion of offshore platforms based on safety principles and sustainable development. The complexity of managing a platform end of life production, the high costs involved as well as the regulatory structure pushes to implement the CE principles during the decommissioning process by hypothesizing to reuse of certain assets, alternative use of the platform or reconversion in Multi-Use Platforms at Sea (MUPS (Nassar et al., 2020) that would have a positive impact on various marine activities such as tourism, aquaculture, transport, oil and gas, and leisure. From an SD perspective, a reconversion strategy represents an opportunity considering MUPS and artificial reefs as an interesting solution for the creation of marine protected areas in which to start and develop sustainable economic activities that are in harmony with CE. In this way, it could be possible to protect the environment including initiatives based on renewable energy, shellfish farming, *"decarbonization"* plants, tourism, and recreation. In this scenario, a coordinated dialogue between offshore operators, stakeholders and scientists engaged in fisheries, aquaculture, shipping, tourism, renewable energy, and other activities stimulate cooperative exchange for the sustainable use of the marine environment. This represents a key driver condition for developing a sustainable blue economy internationally. The SD projects focus on issues of knowledge building, applied research, technological development, and innovation in line with the Horizon 2020 Programme for Research and Innovation. Although derogations from complete removal are allowed in some circumstances, most infrastructure (oil and gas jackets and wind turbines) will have to be completely removed from the seas under current policies and regulations. Cooperation between the offshore energy industry and academic researchers is necessary to fill knowledge gaps and could involve the use of industry vessels for scientific investigations, re-analysis of historical industry data, basic scientific training of industry personnel, the pursuit of joint research funding opportunities, and pilot decommissioning projects. Some of these approaches have already been successfully implemented worldwide. Environmentally optimal decommissioning will be achieved if stakeholders can pool their resources, knowledge, and experience to further scientific understanding of these novel marine ecosystems.

ACKNOWLEDGMENT

This research is part of a larger European offshore platform conversion project called PON-PlaCE (Offshore Platform Conversion for Eco-Sustainable Multiples Use - cod. ARS01_00891).

REFERENCES

Abhinav, K. A., Collu, M., Benjamins, S., Cai, H., Hughes, A., Jiang, B., Jude, S., Leithead, W., Lin, C., Liu, H., Recalde-Camacho, L., Serpetti, N., Sun, K., Wilson, B., Yue, H., & Zhou, B. Z. (2020). Offshore multi-purpose platforms for a Blue Growth: A technological, environmental and socio-economic review. *The Science of the Total Environment*, *734*, 138256. doi:10.1016/j.scitotenv.2020.138256 PMID:32470664

Ajemian, M. J., Wetz, J. J., Shipley-Lozano, B., Shively, J. D., & Stunz, G. W. (2015). An analysis of artificial reef fish community structure along the northwestern Gulf of Mexico shelf: Potential impacts of "Rigs-to-Reefs" programs. *PLoS One, 10*(5), e0126354. doi:10.1371/journal.pone.0126354 PMID:25954943

Basile, V., Capobianco, N., Loia, F., & Vona, R. (2021b). Sustainable decommissioning of offshore platforms: a proposal of life-cycle cost-benefit analysis in Italian oil and gas industry. In *Proceedings IFKAD 2021 Conference, "Managing Knowledge in Uncertain Times"*. University of Rome Tre.

Basile, V., Capobianco, N., & Vona, R. (2021a). The usefulness of sustainable business models: Analysis from oil and gas industry. *Corporate Social Responsibility and Environmental Management, 28*(6), 1801–1821. doi:10.1002/csr.2153

Basile, V., & Vona, R. (2021). Sustainable and Circular Business Model for Oil & Gas Offshore Platform Decommissioning. *International Journal of Business and Management, 16*(10), 10. doi:10.5539/ijbm.v16n10p1

Bell, N., & Smith, J. (1999). Coral growing on North Sea oil rigs. *Nature, 402*(6762), 601–601. doi:10.1038/45127 PMID:10604464

Bocci, M., Sangiuliano, S. J., Sarretta, A., Ansong, J. O., Buchanan, B., Kafas, A., Caña-Varona, M., Onyango, V., Papaioannou, E., Ramieri, E., Schultz-Zehden, A., Schupp, M. F., Vassilopoulou, V., & Vergílio, M. (2019). Multi-use of the sea: A wide array of opportunities from site-specific cases across Europe. *PLoS One, 14*(4), e0215010. doi:10.1371/journal.pone.0215010 PMID:30973895

Bohnsack, J. A. (1989). Are high densities of fishes at artificial reefs the result of habitat limitation or behavioural preference? *Bulletin of Marine Science, 44*(2), 631–645.

Bond, T., Langlois, T. J., Partridge, J. C., Birt, M. J., Malseed, B. E., Smith, L., & McLean, D. L. (2018). Diel shifts and habitat associations of fish assemblages on a subsea pipeline. *Fisheries Research, 206*, 220–234. doi:10.1016/j.fishres.2018.05.011

Brennan, F., & Kolios, A. (2014, January). Structural integrity considerations for the H2Ocean multi-modal wind-wave platform. In *European Wind Energy Association Conference and Exhibition 2014, EWEA 2014*. European Wind Energy Association.

Browman, H. I., Stergiou, K. I., Cury, P. M., Hilborn, R., Jennings, S., Lotze, H. K., & Mace, P. M. (2004). Perspectives on ecosystem-based approaches to the management of marine resources. *Marine Ecology Progress Series, 274*, 269–303. doi:10.3354/meps274269

Bull, A. S., & Love, M. S. (2019). Worldwide oil and gas platform decommissioning: A review of practices and reefing options. *Ocean and Coastal Management, 168*, 274–306. doi:10.1016/j.ocecoaman.2018.10.024

Bureau of Safety and Environmental Enforcement. (2018). *Rigs to Reefs*. Available at: https://www.bsee.gov/what-we-do/environmental-focuses/rigsto-reefs

Capobianco, N., Basile, V., Loia, F., & Vona, R. (2021). Toward a sustainable decommissioning of offshore platforms in the oil and gas industry: A pestle analysis. *Sustainability, 13*(11), 6266. doi:10.3390u13116266

Chandima Ratnayake, R. M., & Markeset, T. (2012). Asset integrity management for sustainable industrial operations: Measuring the performance. *International Journal of Sustainable Engineering*, *5*(2), 145–158. doi:10.1080/19397038.2011.581391

Christensen, E. D., Stuiver, M., Guanche, R., Møhlenberg, F., Schouten, J. J., Pedersen, O. S., He, W., Zanuttigh, B., & Koundouri, P. (2015). *Go offshore-Combining food and energy production*. The Technical University of Denmark. Department of Mechanical Engineering.

Dokken, Q. R. (2012). *Ecological impact of platform structures as artificial reefs, Paper in development, used with permission of Q*. Dokken, Gulf of Mexico Foundation.

Dol, S. S. (2018). Design of Self-sustaining Mariculture Farm on Offshore Rig. *10th Int'l Conference on Advances in Engineering, Science, Technology & Sustainable Development (ESTSD-18)*, 25-27.

Fowler, A. M., & Booth, D. J. (2012). Evidence of sustained populations of small reef fish on artificial structures. Does depth affect production on artificial reefs? *Journal of Fish Biology*, *80*(3), 613–629. doi:10.1111/j.1095-8649.2011.03201.x PMID:22380556

Fowler, A. M., Macreadie, P. I., Jones, D. O. B., & Booth, D. J. (2014). A multi-criteria decision approach to decommissioning of offshore oil and gas infrastructure. *Ocean and Coastal Management*, *87*, 20–29. doi:10.1016/j.ocecoaman.2013.10.019

Fuller, S. (2010). *Life-cycle cost analysis (LCCA)*. National Institute of Building Sciences.

Gallaway, B. J., Szedlmayer, S. T., & Gazey, W. J. (2009). A life history review for red snapper in the Gulf of Mexico with an evaluation of the importance of offshore petroleum platforms and other artificial reefs. *Reviews in Fisheries Science*, *17*(1), 48–67. doi:10.1080/10641260802160717

Grossman, G. D., Jones, G. P., & Seaman, W. J. Jr. (1997). Do artificial reefs increase regional fish production? A review of existing data. *Fisheries (Bethesda, Md.)*, *22*(4), 17–23. doi:10.1577/1548-8446(1997)022<0017:DARIRF>2.0.CO;2

Haeuber, R., & Franklin, J. (1996). Perspectives on ecosystem management. *Ecological Applications*, *6*(3), 692–693. doi:10.2307/2269461

Holbrook, S. J., Forrester, G. E., & Schmitt, R. J. (2000). Spatial patterns in abundance of a damselfish reflect the availability of suitable habitat. *Oecologia*, *122*(1), 109–120. doi:10.1007/PL00008826 PMID:28307947

IMSA Amsterdam. (2013). *Worldwide rigs-to-reefs experiences: considerations and motivation for the Living North Sea*. Report Feasibility Update Living North Sea Initiative. https://ecoeffective.biz/wp-content/uploads/2016/02/LNSF250_Worldwide-rigs-to-reefs-experiences_LiNSI_D02.pdf

Jansen, M., Staffell, I., Kitzing, L., Quoilin, S., Wiggelinkhuizen, E., Bulder, B., Riepin, I., & Müsgens, F. (2020). Offshore wind competitiveness in mature markets without subsidy. *Nature Energy*, *5*(8), 614–622. doi:10.103841560-020-0661-2

Jørgensen, D. (2012). OSPAR's exclusion of rigs-to-reefs in the North Sea. *Ocean and Coastal Management*, *58*, 57–61. doi:10.1016/j.ocecoaman.2011.12.012

Joyce, A., & Paquin, R. L. (2016). The triple-layered business model canvas: A tool to design more sustainable business models. *Journal of Cleaner Production, 135*, 1474–1486. doi:10.1016/j.jclepro.2016.06.067

Kaiser, M. J. (2006). The Louisiana artificial reef program. *Marine Policy, 30*(6), 605-623. doi:10.1016/j.marpol.2005.04.005

Kaiser, M. J., & Pulsipher, A. G. (2005). Rigs-to-reef programs in the Gulf of Mexico. *Ocean Development and International Law*, *36*(2), 119–134. doi:10.1080/00908320590943990

Klöpffer, W. (2003). Lifecycle based methods for sustainable product development. *The International Journal of Life Cycle Assessment*, *8*(3), 157–159. doi:10.1007/BF02978462

Koundouri, P., Giannouli, A., Airoldi, L., Bas, B., Broszeit, S., Elginoz, N., Giannakis, E., Zagonari, F., Krontira, Y., Moussoulides, A., Tsani, S., Troianos, D., Xepapadeas, P., Xepapadeas, A., & Zanuttigh, B. (2017). Socio-economic Analysis of a Selected Multi-use Offshore Site in the Mediterranean Sea. In *The Ocean of Tomorrow: Investment Assessment of Multi-Use Offshore Platforms: Methodology and Applications* (pp. 85–101). Springer International Publishing. doi:10.1007/978-3-319-55772-4_6

Kusumawardhani, M., Kumar, R., & Tore, M. (2016). Asset integrity management: Offshore installations challenges. *Journal of Quality in Maintenance Engineering*, *22*(3), 238–251. doi:10.1108/JQME-06-2015-0023

Lakhal, S. Y., Khan, M. I., & Islam, M. R. (2009). An "Olympic" framework for a green decommissioning of an offshore oil platform. *Ocean and Coastal Management*, *52*(2), 113–123. doi:10.1016/j.ocecoaman.2008.10.007

Lam, C. M., Iris, K. M., Medel, F., Tsang, D. C., Hsu, S. C., & Poon, C. S. (2018). Life-cycle cost-benefit analysis on sustainable food waste management: The case of Hong Kong International Airport. *Journal of Cleaner Production*, *187*, 751–762. doi:10.1016/j.jclepro.2018.03.160

Langhamer, O. (2012). Artificial reef effect in relation to offshore renewable energy conversion: State of the art. *TheScientificWorldJournal*, *2012*(386713), 8. doi:10.1100/2012/386713 PMID:23326215

Legorburu, I., Johnson, K. R., & Kerr, S. A. (2018). Multi-use maritime platforms-North Sea oil and offshore wind: Opportunity and risk. *Ocean and Coastal Management*, *160*, 75–85. doi:10.1016/j.ocecoaman.2018.03.044

Loia, F., Basile, V., Capobianco, N., & Vona, R. (2021a). Value Co-Creation Practices in the Decommissioning of Offshore Platforms: A Case Study Approach. *Journal of Creating Value*, *7*(2), 238–251. doi:10.1177/23949643211032040

Loia, F., Capobianco, N., & Vona, R. (2021b). Towards a resilient perspective for the future of offshore platforms. Insights from a data-driven approach, *Transforming Government: People. Process and Policy*, *15*(3). Advance online publication. doi:10.1108/TG-04-2021-0067

Love, M. S., Brothers, E., Schroeder, D. M., & Lenarz, W. H. (2007). Ecological performance of young-of-the-year blue rockfish (*Sebastes mystinus*) associated with oil platforms and natural reefs in California as measured by daily growth rates. *Bulletin of Marine Science*, *80*(1), 147–157.

MacDonald, J. M. (1994). Artificial reef debate: Habitat enhancement or waste disposal? *Ocean Development and International Law*, *25*(1), 87–118. doi:10.1080/00908329409546027

Macreadie, P. I., Allen, K., Kelaher, B. P., Ralph, P. J., & Skilbeck, C. G. (2012). Paleoreconstruction of estuarine sediments reveal human-induced weakening of coastal carbon sinks. *Global Change Biology*, *18*(3), 891–901. doi:10.1111/j.1365-2486.2011.02582.x

Macreadie, P. I., Fowler, A. M., & Booth, D. J. (2011). Rigs-to-reefs: Will the deep-sea benefit from artificial habitat? *Frontiers in Ecology and the Environment*, *9*(8), 455–461. doi:10.1890/100112

Margheritini, L., Colaleo, G., Contestabile, P., Bjørgård, T. L., Simonsen, M. E., Lanfredi, C., Dell'Anno, A., & Vicinanza, D. (2020). Development of an eco-sustainable solution for the second life of decommissioned oil and gas platforms: The mineral accretion technology. *Sustainability*, *12*(9), 3742. doi:10.3390u12093742

Martins, I. D., Moraes, F. F., Távora, G., Soares, H. L. F., Infante, C. E., Arruda, E. F., Bahiense, L., Caprace, J., & Lourenço, M. I. (2020). A review of the multicriteria decision analysis applied to oil and gas decommissioning problems. *Ocean and Coastal Management*, *184*, 105000. doi:10.1016/j.ocecoaman.2019.105000

McLean, D. L., Partridge, J. C., Bond, T., Birt, M. J., Bornt, K. R., & Langlois, T. J. (2017). Using industry ROV videos to assess fish associations with subsea pipelines. *Continental Shelf Research*, *141*, 76–97. doi:10.1016/j.csr.2017.05.006

Nassar, W. M., Anaya-Lara, O., Ahmed, K. H., Campos-Gaona, D., & Elgenedy, M. (2020). Assessment of multi-use offshore platforms: Structure classification and design challenges. *Sustainability*, *12*(5), 1860. doi:10.3390u12051860

Nautilus Consultants. (2003). *Artificial Reefs, Scotland, Benefit costs and risks*. Highlands & Islands Enterprise. http://www.nautilus-consultants.co.uk/sites/default/files/Nautilus%20artificial%20reefs%20report_0.pdf

OSPAR Commission. (2009). *Assessment of construction or placement of artificial reefs*. OSPAR, Biodiversity Series. https://www.ospar.org/documents?v=7143

Parente, V., Ferreira, D., dos Santos, E. M., & Luczynski, E. (2006). Offshore decommissioning issues: Deductibility and transferability. *Energy Policy*, *34*(15), 1992–2001. doi:10.1016/j.enpol.2005.02.008

Pickering, H., & Whitmarsh, D. (1997). Artificial reefs and fisheries exploitation: A review of the 'attraction versus production'debate, the influence of design and its significance for policy. *Fisheries Research*, *31*(1-2), 39–59. doi:10.1016/S0165-7836(97)00019-2

Pradella, N., Fowler, A. M., Booth, D. J., & Macreadie, P. I. (2014). Fish assemblages associated with oil industry structures on the continental shelf of north-western Australia. *Journal of Fish Biology*, *84*(1), 247–255. doi:10.1111/jfb.12274 PMID:24344929

Przedrzymirska, J., & Zaucha, J. (2018). *Multi-use concept in European Sea Basins*. MUSES Project. https://muses-project.com/wp-content/uploads/sites/70/2018/06/D2-6-Final-Report.pdf

Remmen, A., Astrup-Jensen, A., & Frydendal, J. (2007). Life cycle management: A business guide to sustainability. *UNEP and Danish Standards, 51.* https://www.lifecycleinitiative.org/wp-content/uploads/2013/09/UNEP_Background_document_LCM_2006_Febr.pdf

Rosenberg, L. (2013). Sea-spike detection in high grazing angle X-band sea-clutter. *IEEE Transactions on Geoscience and Remote Sensing, 51*(8), 4556–4562. doi:10.1109/TGRS.2013.2239112

Salcido, R. E. (2005). Enduring optimism: Examining the rig-to-reef bargain. *Ecology Law Quarterly, 32*, 863. doi:10.15779/Z38HZ7Q

Santos, M. N., & Monteiro, C. C. (1998). Comparison of the catch and fishing yield from an artificial reef system and neighbouring areas off Faro (Algarve, south Portugal). *Fisheries Research, 39*(1), 55–65. doi:10.1016/S0165-7836(98)00169-6

Schroeder, D. M., & Love, M. S. (2002). Recreational fishing and marine fish populations in California. *California Cooperative Oceanic Fisheries Investigations Report, 43*, 182-190. https://lovelab.msi.ucsb.edu/Schroeder_Love2002.pdf

Schroeder, D. M., & Love, M. S. (2004). Ecological and political issues surrounding decommissioning of offshore oil facilities in the Southern California Bight. *Ocean and Coastal Management, 47*(1-2), 21–48. doi:10.1016/j.ocecoaman.2004.03.002

Soldal, A. V., Svellingen, I., Jørgensen, T., & Løkkeborg, S. (2002). Rigs-to-reefs in the North Sea: Hydroacoustic quantification of fish in the vicinity of a "semi-cold" platform. *ICES Journal of Marine Science, 59*(suppl), S281–S287. doi:10.1006/jmsc.2002.1279

Sommer, B., Fowler, A. M., Macreadie, P. I., Palandro, D. A., Aziz, A. C., & Booth, D. J. (2019). Decommissioning of offshore oil and gas structures–Environmental opportunities and challenges. *The Science of the Total Environment, 658*, 973–981. doi:10.1016/j.scitotenv.2018.12.193 PMID:30583191

Techera, E. J., & Chandler, J. (2015). Offshore installations, decommissioning and artificial reefs: Do current legal frameworks best serve the marine environment? *Marine Policy, 59*, 53–60. doi:10.1016/j.marpol.2015.04.021

Todd, V. L., Williamson, L. D., Cox, S. E., Todd, I. B., & Macreadie, P. I. (2020). Characterizing the first wave of fish and invertebrate colonization on a new offshore petroleum platform. *ICES Journal of Marine Science, 77*(3), 1127–1136. doi:10.1093/icesjms/fsz077

Van den Burg, S. W. K., Schupp, M. F., Depellegrin, D., Barbanti, A., & Kerr, S. (2020). Development of multi-use platforms at sea: Barriers to realising Blue Growth. *Ocean Engineering, 217*, 107983. doi:10.1016/j.oceaneng.2020.107983

ENDNOTES

1 General Italian Oil Company (AGIP) is Italian automotive gasoline, diesel, LPG, lubricants, fuel oil, and bitumen retailer established in 1926. It has been a subsidiary of the multinational petroleum company National Hydrocarbons Board (ENI) since 2003.

2 Based in London, UK, Vision gain is an independent business intelligence provider helping managers and firms to stay informed of key changes within their industry.

3 Compound annual growth rate (CAGR) is a business and investing specific term for the geometric progression ratio that provides a constant rate of return over the period.

4 The Ministry of Economic Development (MiSE) is a department of the Italian government. He deals with industrial policy, trade, and communications. UNMIG is the national mining office for hydrocarbons and geo-resources.

5 The Policies and Public-Private Partnerships for Offshore Wind Energy (4POWER) project promoted the exchange of knowledge between different European regions and aimed to create a common knowledge base on current and future regional challenges in offshore wind energy development. The project sought to provide an improved framework for the development of offshore wind energy in the involved regions by means of policy recommendations on EU legislation.

6 The Enabling Technologies & Roadmaps for Offshore Platform Innovation ENTROPI project analyzed the capabilities enabling multi-use offshore platforms to highlight where targeted innovation could achieve a significant cost reduction and define investment propositions to make those innovations happen. For more information: https://clustercollaboration.eu/content/entropi

7 Key Enabling Technologies (KETs) a group of six technologies: micro and nanoelectronics, nanotechnology, industrial biotechnology, advanced materials, photonics, and advanced manufacturing technologies increase industrial innovation to address societal challenges and create advanced and sustainable economies.

8 The Modular Multi-Use Deep Water Offshore Platform Harnessing and Servicing Mediterranean, Subtropical and Tropical Marine and Maritime Resources (TROPOS) project aimed at developing a floating modular multi-use platform system for use in deep waters, with an initial geographic focus on the Mediterranean, Tropical and Sub-Tropical regions, but designed to be flexible enough so as not to be limited in geographic scope.

9 Spain, United Kingdom, Germany, Portugal, France, Norway, Denmark, Greece, and Taiwan.

10 The Oceanic Platform of the Canary Islands (PLOCAN) is a multipurpose technical-scientific service infrastructure that provides support for research, technological development and innovation in the marine and maritime sectors, available to public and private users. PLOCAN offers both onshore and offshore experimental facilities and laboratories, operational throughout the whole year thanks to the Canary Islands excellent climatic conditions. PLOCAN also brings broad experience in large national and EU marine/maritime projects. For more information: www.plocan.eu/en/description/

11 Development of a wind-wave power open-sea platform equipped for hydrogen generation with support for multiple users of energy.

12 MERMAID is one of three EU-FP7 funded projects selected for funding in response to Ocean 2011 on multi-use offshore platforms (FP7-OCEAN.2011-1 "Multi-use offshore platforms"). This project shall run from 2012 till 2016 and have a cost of 7,4 million euro. The European Union has granted a financial contribution of 5,5 million euro. The 28 partner institutes forming MERMAID are Universities (11), Research institutes (8), Industries (5) and Small and Medium Enterprises (4 SME's), from many regions in EU. The group represents a broad range of expertise in hydraulics, wind engineering, aquaculture, renewable energy, marine environment, project management as well as socio-economics. For more information: www.vliz.be/projects/mermaidproject/

13 The Marine Investment for the Blue Economy (MARIBE) project aims to unlock the potential of multi-use of space in the offshore economy (also referred to as Blue Economy). This forms

part of the long-term Blue Growth (BG) strategy to support sustainable growth in the marine and maritime sectors as a whole; something which is at the heart of the Integrated Maritime Policy, the EU Innovation Union, and the Europe 2020 strategy for smart, sustainable growth. For more information: www.maribe.eu/links

14 The Multi-Use in European Seas (MUSES) project is a Horizon 2020 funded project that is exploring the opportunities for Multi-Use in European Seas across five EU sea basins (Baltic Sea, North Sea, Mediterranean Sea, Black Sea and Eastern Atlantic). For more information: www.muses-project.com/

15 Blue Latitudes unites science, communication, and design to solve complex ecological challenges associated with re-purposing offshore oil and gas platforms as reefs. Provides the necessary information, market assessments, and biological assessments to allow purchasers to make an informed decision on the Rigs-to-Reefs decommissioning option, and whether or not it is feasible within a selected area.

APPENDIX: SITOGRAPHY

www.aquabt.com/horizon-2020-closed-projects
www.bluegrowth-place.eu
www.bsee.gov/what-we-do/environmental-compliance/environmental-programs/rigs-to-reefs
www.cordis.europa.eu/project/id/288192
www.cordis.europa.eu/project/id/288710
www.cordis.europa.eu/project/id/652629/reporting/it
www.cordis.europa.eu/project/rcn/102016/factsheet/en
www.maribe.eu/links/
www.muses-project.eu
www.ogauthority.co.uk/decommissioning
www.plocan.eu/es
www.rig2reefexploration.org
www.turismo.ra.it/
www.unibo.it/it/ricerca/progetti-e-iniziative/progetti-pon-pnr/place
www.unmig.mise.gov.it/index.php/it
www.unric.org/it/agenda-2030
www.vliz.be/projects/mermaidproject/

Table 2. Rigs-to-reefs worldwide programmes and initiatives

Locations	*Description and status*
Gulf of Mexico	Since the late 80s and 90s, the states of Louisiana, Alabama, Texas and Mississippi have had artificial reef plans that allow rigs to be towed to designated reefing areas. Of the more than 4000 platforms, about 2000 structures have already been decommissioned: 10% of these platforms have been reefed and the others have been removed. The initiative Save the Blue is promoting to change the current policy of the Gulf of Mexico rigs-to-reefs programmes to a format in which platforms are always left in place unless they appear to be contaminated. They claim the ecological and recreational value of the platforms is so high that they should be protected instead of removing them.
California	Off the California coast, 27 platforms will need to be decommissioned in the coming years. Since 2010 the artificial reef plan and administrative and legal frameworks are in place, but no rigs-to-reefs have been applied so far.
Brunei Darussalam	Two major rigs to reef programmes also took place in the waters of Malaysia and Brunei. In 1985, Brunei Darussalam launched an Artificial Reef Programme to enhance local fish productivity. Artificial reefs were established near Berakas, far away from commercial shipping lanes. In 1988, BSP placed two rigs to build a Rig Reef, and in 1994, five extra jackets formed a second Rig Reef. These structures were in water depths ranging from 16m to 60m. The artificial reef functions as a breeding ground for fish and supply to surrounding coastal waters. The reef was also made accessible as a diving hotspot to recreational divers (Yus et al., 1999; Twomey, 2012).
Malaysia	In 1975, the Baram 8 structure of Shell Malaysia was damaged in a storm and collapsed on the seabed. In 2004, the structure was sunk near the natural Sarawak Siwa reef to become Malaysia's first artificial reef. Malaysia does not have a clear regulatory framework on decommissioning and reefing. In this specific case, many stakeholders worked together to get permission to leave the rig in place, including local and federal governments and the Sarawak Tourist Board. Production Sharing Contracts (PSC) now does the area selection for rig-to-reef and cost/benefit analyses are conducted on a case-by-case basis (Twomey, 2012; Boothby, 2010).

Continued on following page

Table 2. Continued

Locations	*Description and status*
Australia	In Australia, about 60 oil platforms are to be decommissioned, but reefing options are not allowed yet. Government, oil, and gas industry stakeholders are considering the potential for rigs-to-reefs but are not rushing the process. Scientists of the University of Technology Sydney (UTS) urge the stakeholders to start with scientific studies to test the actual impact of rigs-to-reefs. They have preliminary data that suggest that platforms are important sites of fish production and propose leaving part of them in place or towing them to reef areas for ecological purposes. They even assess rigs-to-reefs at locations of more than 500 meters deep for deep-sea habitat creation, meaning a relocation of oil and gas structures to deeper waters (Macreadie et al., 2011; Macreadie et al., 2012).
Italy	The history of the Paguro, now given a new life, however, is based on a tragic event dating back to the 1960s. In 1963, Agip [General Italian Oil Company] decided to build a platform for the extraction of methane off the coast of Ravenna, 11 miles from the shore of Porto Corsini. After two years of regular operation, in September 1965, the Paguro drill happened upon a very high-pressure methane deposit, which quickly caused a fire and the sinking of the platform. Since then, the wreckage of the Paguro has gradually become the ideal habitat for many specimens of marine flora and fauna, so much so that it earned the title of a site of community interest, attributed by the Emilia-Romagna region in 2010. In 1995, the Paguro Association was founded in Ravenna to monitor the biological conservation area and regulate diving. Upon closer inspection, the Paguro Platform is not only a sunken wreck but also a real underwater citadel composed of various environments and arranged on various levels. The area of the site covers as much as 15,000 square metres, which includes the framework, accommodation buildings and other components of the platform, along with some disused structures from ENI-Agip's offshore platforms that were moved here between 1999 and 2000. There are two places to dive, and the depth ranges from 10 metres at the highest point of the wreckage to 33 metres below the surface at the crater generated by the 1965 explosion. A scuba diving paradise that is well suited to experienced divers, given the technical difficulties of diving, such as the visibility and size of the site. The richness of the life forms that populate the wreck, however, more than makes up for the difficulties involved. The Paguro Platform is an artificial reef, with an environment similar to that of submerged reefs, with extraordinary biological diversity that makes it unique in the region.

Source: adapted starting from Report Feasibility Update Living North Sea Initiative (2103).

Compilation of References

Abdullah, L., & Najib, L. (2016). Sustainable energy planning decision using the intuitionistic fuzzy analytic hierarchy process: Choosing energy technology in Malaysia. *International Journal of Sustainable Energy*, *35*(4), 360–377. doi:10.1080/14786451.2014.907292

Abecasis, D., Bentes, L., Lino, P. G., Santos, M. N., & Erzini, K. (2013). Residency, movements and habitat use of adult white seabream (Diplodus sargus) between natural and artificial reefs. *Estuarine, Coastal and Shelf Science*, *118*, 80–85. doi:10.1016/j.ecss.2012.12.014

Abelson, A. (2006). Artificial reefs vs coral transplantation as restoration tools for mitigating coral reef deterioration: Benefits, concerns, and proposed guidelines. *Bulletin of Marine Science*, *78*(1), 151–159.

Abhinav, K. A., Collu, M., Benjamins, S., Cai, H., Hughes, A., Jiang, B., Jude, S., Leithead, W., Lin, C., Liu, H., Recalde-Camacho, L., Serpetti, N., Sun, K., Wilson, B., Yue, H., & Zhou, B. Z. (2020). Offshore multi-purpose platforms for a Blue Growth: A technological, environmental and socio-economic review. *The Science of the Total Environment*, *734*, 138256. doi:10.1016/j.scitotenv.2020.138256 PMID:32470664

Acarlı, D., Semih, K., & Kocabaş, S. (2020). TCSG-132 Gemi Batığı Yapay Resifinin (Gökçeada, Kuzey Ege Denizi) Biyoçeşitliliği. Acta Aquatica Turcica, 16(3), 313–329. doi:10.22392/actaquatr.677175 doi:10.22392/actaquatr.677175

Achituv, Y., & Dubinsky, Z. (1990). Evolution and zoogeography of coral reefs. In Ecoystems of the World, Volume 25 — Coral Reefs. Elsevier Science.

Adams, C., Lindberg, B., & Stevely, J. (2004). *The Economic Benefits Associated with Florida's Artificial Reef.* Available from: https://edis.ifas.ufl.edu/fe649

Adams, C., Lindberg, B., & Stevely, J. (2006). The economic benefits associated with Florida's artificial reefs. *EDIS*, *2006*(14).

Adrianto, L. (2015). *Legal analysis and evaluation of the management of coastal areas and small islands*. Academic Press.

Ajemian, M. J., Wetz, J. J., Shipley-Lozano, B., Shively, J. D., & Stunz, G. W. (2015). An analysis of artificial reef fish community structure along the northwestern Gulf of Mexico shelf: Potential impacts of "rigs-to-reefs" programs. *PLoS One*, *10*(5), 1–22. doi:10.1371/journal.pone.0126354 PMID:25954943

Albayrak, T., Caber, M., & Cater, C. (2021). Mass tourism underwater: A segmentation approach to motivations of scuba diving holiday tourists. *Tourism Geographies*, *23*(5-6), 5–6, 985–1000. doi:10.1080/14616688.2019.1696884

Alcala, G. (1986). Lázaro Cárdenas, Michoacán: Living conditions and perspective of the fishermen. Michoacán Studies. The College of Michoacán, Mexico.

Ali, M. M., Aditya, Z. F., & Fuadi, A. B. (2020). Coastal Communities Protection of Harmonization of Integrated Coastal Management Regulations. *Journal Of Constitution, 17*(4).

Almeida, M., Raposo, A. C., Guerra, M. T., Gaudêncio, M., Maranhão, P., Rodrigues, N. V., & Leandro, S. (2016). Macrobenthic Assemblages in an Artificial Reef in Central-West Portugal: Preliminary Results. *Front. Mar. Sci. Conference Abstract: XIX Iberian Symposium on Marine Biology Studies.*

Alwi, D., Lule, A., & Taliki, R. (2020). *Alternative Fishing Area In Tanjung Dehegila Waters*. Morotai Island.

Ampou, E. E., Hutasoit, P., & Damar, A. (2019). *The Condition of "Spider" Artificial Reefs in Nusa Dua Bali, After the IMF-WBG Meeting*. Darilaut.Id. https://darilaut.id/kajian/kondisi-terumbu-buatan-laba-laba-di-nusa-dua-bali-setelah-pertemuan-imf-wbg

Ananda, J., & Herath, G. (2003). The use of Analytic Hierarchy Process to incorporate stakeholder preferences into regional forest planning. *Forest Policy and Economics, 5*(1), 13–26. doi:10.1016/S1389-9341(02)00043-6

Andaris, A. R., Suryanto, A., & Muskananfola, M. R. (2015). Relationship Of Physical ± Chemical Factors Of Waters To Coral Reef Covers In Karimunjawa Island. *Diponegoro Journal of Maquares, 4*(2007), 29–36.

Anderson, M. J. (1996). A chemical cue induces settlement of Sydney rock oysters, Saccostrea commercialis, in the laboratory and in the field. *The Biological Bulletin, 190*(3), 350–358. doi:10.2307/1543027 PMID:29227699

Andriesse, E. (2018). Persistent fishing amidst depletion, environmental and socio-economic vulnerability in Iloilo Province, the Philippines. *Ocean and Coastal Management, 157*(January), 130–137. doi:10.1016/j.ocecoaman.2018.02.004

Anielski, M. (2001). Measuring the sustainability of nations: The genuine progress indicator system of sustainable wellbeing accounts. In *The Fourth Biennial Conference of the Canadian Society for Ecological Economics: Ecological Sustainability of the Global Market Place*. Pembina Institute for Appropriate Development.

Armbruster, S. (2018). *How Old Oyster Shells Are Reviving A Queensland Reef.* SBS News, Australia. www.sbs.com.au/news/how-old-oyster-shells-are-reviving-a-queensland-reef

Asafu-Adjaye, J., & Tapsuwan, S. (2008). A contingent valuation study of scuba diving benefits: Case study in Mu Ko Similan Marine National Park. *Tourism Management, 29*(6), 1122–1130. doi:10.1016/j.tourman.2008.02.005

Astier, M.; Maass, M. & Etchevers, J. (2002). Derivation of soil quality indicators in the context of sustainable agriculture. *Agriscience, 36*(5).

Azis, M. F. (2006). *Movement of water in the sea*. Academic Press.

Badii, M. H. (2008). La huella ecológica y sustentabilidad (Ecological footprint and sustainability). *Daena: International Journal of Good Conscience, 3*(1), 672-678.

Bailey, K. D. (2007). *Methods of Social Research* (4th ed.). Free Press.

Baillard, C., Schmid, C., Zisserman, A., & Fitzgibbon, A. (1999). Automatic line matching and 3D reconstruction of buildings from multiple views. *ISPRS Conference on Automatic Extraction of GIS Objects from Digital Imagery*, 69-80.

Baine, M. (2001). Artificial reefs: A review of their design, application, management and performance. *Ocean and Coastal Management, 44*(3-4), 241–259. doi:10.1016/S0964-5691(01)00048-5

Baine, M., & Side, J. (2003). Habitat modification and manipulation as a management tool. *Reviews in Fish Biology and Fisheries, 13*(2), 187–199. doi:10.1023/B:RFBF.0000019480.95010.67

Baptista, J., Martinho, F., Martins, R., Carneiro, M., Azevedo, M., Vieira, A. R., Gomes, P., & Pardal, M. A. (2019). Water temperature gradient shapes the structure and composition of nearshore marine fish communities in southern Europe. *Journal of Sea Research, 154*, 101807. doi:10.1016/j.seares.2019.101807

Baraknews. (2020). *Achievements Achieved by POKMASWAS Assisted by the Department of Fisheries and Marine Affairs of Trenggalek Regency.* https://www.baraknews.com/berita_jawa-timur/prestasi-yang-diraih-pokmaswas-binaan-dinas-perikanan-dan-kelautan-kabupaten-trenggalek

Barber, J. S., Chosid, D. M., Glenn, R. P., & Whitmore, K. A. (2009). A systematic model for artificial reef site selection. *New Zealand Journal of Marine and Freshwater Research, 43*(1), 283–297. doi:10.1080/00288330909510001

Barbosa-Filho, M. L. V., Seminara, C. I., Tavares, D. C., Siciliano, S., Hauser-Davis, R. A., & da Silva Mourão, J. (2020). Artisanal fisher perceptions on ghost nets in a tropical South Atlantic marine biodiversity hotspot: Challenges to traditional fishing culture and implications for conservation strategies: Ghost net impacts: Artisanal fisher perceptions and challenges to. *Ocean and Coastal Management, 192*(March), 105189. Advance online publication. doi:10.1016/j.ocecoaman.2020.105189

Barnabe, Charbonnel, Marinaro, Ody, & Francour. (2000). Artificial reefs in France: analysis, assessments and prospects. In Artificial Reefs in European Seas. Kluwer Academics Publishers.

Basile, V., Capobianco, N., Loia, F., & Vona, R. (2021b). Sustainable decommissioning of offshore platforms: a proposal of life-cycle cost-benefit analysis in Italian oil and gas industry. In *Proceedings IFKAD 2021 Conference, "Managing Knowledge in Uncertain Times"*. University of Rome Tre.

Basile, V., Capobianco, N., & Vona, R. (2021a). The usefulness of sustainable business models: Analysis from oil and gas industry. *Corporate Social Responsibility and Environmental Management, 28*(6), 1801–1821. doi:10.1002/csr.2153

Basile, V., & Vona, R. (2021). Sustainable and Circular Business Model for Oil & Gas Offshore Platform Decommissioning. *International Journal of Business and Management, 16*(10), 10. doi:10.5539/ijbm.v16n10p1

Bateman, I., & Willis, K. G. (2002). *Valuing Environmental Preferences: Theory and Practice of Contingent Valuation Method in the US, EU, and Developing Countries*. Oxford University Press.

Becker, A., Taylor, M. D., Folpp, H., & Lowry, M. B. (2018). Managing the development of artificial reef systems: The need for quantitative goals. *Fish and Fisheries, 19*(4), 740–752. doi:10.1111/faf.12288

Begossi, A., May, P.H., Lopes, P.F., Oliveira, L.E.C., Vinha, V., & Silvano, R.A.M. (2011). Compensation for environmental services from artisanal fisheries in SE Brazil: policy and technical strategies. *Ecol. Econ., 71*, 25–32. .2011.09.008 doi:10.1016/j.ecolecon

Begossi, A. (2014). Ecological, cultural, and economic approaches to managing artisanal fisheries. *Environment, Development and Sustainability, 16*(1), 5–34. doi:10.100710668-013-9471-z

Begossi, A., Salivonchyk, S., Lopes, P. F. M., & Silvano, R. A. M. (2016). Fishers' knowledge on the coast of Brazil. *Journal of Ethnobiology and Ethnomedicine, 12*(1), 20. doi:10.118613002-016-0091-1 PMID:27245047

Bell, F. W., Bonn, M. A., & Leeworthy, V. R. (1998). *Economic Impact and Importance of Artificial Reefs in Northwest Florida.* NOAA Paper Contract Number MR235 Office of Fisheries Management and Assistance Service, Florida Department of Environmental Administration.

Bell, N., & Smith, J. (1999). Coral growing on North Sea oil rigs. *Nature, 402*(6762), 601–601. doi:10.1038/45127 PMID:10604464

Bernardini, F. (1997). *Sampling and reconstructing manifolds using alpha-shapes*. Purdue University. Report number: 97-013.

Bertness, M. D., Leonard, G. H., Levine, J. M., Schmidt, P. R., & Ingraham, A. O. (1999). Testing the Relative Contribution of Positive and Negative Interactions in Rocky Intertidal Communities. *Ecology*, *80*(8), 2711–2726. doi:10.1890/0012-9658(1999)080[2711:TTRCOP]2.0.CO;2

Bicudo, P., Custódio, M., Cardoso, N., Fortes, C., Neves, D. G., Mendes, L., ... Miguel, L. (2008, July). Viability Study of an Artificial Surf Reef in S. Pedro do Estoril, Cascais, Portugal. In *The Eighteenth International Offshore and Polar Engineering Conference*. OnePetro.

Bideci, C., & Cater, C. (2019). Dive and Novelty Seeking in Experimental Artificial Reefs. *Journal On Tourism & Sustainability*, *3*(1), 22–31.

Bideci, C., & Cater, C. (2019). In search of underwater atmosphere: A new diving world on artificial reefs. In *Atmospheric turn in culture and tourism: Place, design and process impacts on customer behaviour, marketing and branding*. Emerald Publishing Limited. doi:10.1108/S1871-317320190000016023

Boaventura, D., Moura, A., Leitao, F., Carvalho, S., Curdia, J., Pereira, P., & Monteiro, C. C. (2006). Macrobenthic colonisation of artificial reefs on the southern coast of Portugal (Ancão, Algarve). In *Marine Biodiversity* (pp. 335–343). Springer. doi:10.1007/1-4020-4697-9_29

Bocci, M., Sangiuliano, S. J., Sarretta, A., Ansong, J. O., Buchanan, B., Kafas, A., Caña-Varona, M., Onyango, V., Papaioannou, E., Ramieri, E., Schultz-Zehden, A., Schupp, M. F., Vassilopoulou, V., & Vergílio, M. (2019). Multi-use of the sea: A wide array of opportunities from site-specific cases across Europe. *PLoS One*, *14*(4), e0215010. doi:10.1371/journal.pone.0215010 PMID:30973895

Bohnsack, J. A., Johnson, D.L., & Ambrose, R.E. (1991). *Ecology of artificial reef habitats and fishes*. Academic Press.

Bohnsack, J. A. (1989). Are high densities of fishes at artificial reefs the result of habitat limitation or behavioral preference? *Bulletin of Marine Science*, (44), 631–645.

Bohnsack, J. A. (1989). Are High Densities of Fishes at Artificial Reefs the Result of Habitat Limitation or Behavioral Preference? *Bulletin of Marine Science*, *44*(2), 631–645.

Bohnsack, J. A. (1989). Are high densities of fishes at artificial reefs the result of habitat limitation or behavioural preference? *Bulletin of Marine Science*, *44*(2), 631–645.

Bohnsack, J. A. (1991). Habitat structure and the design of artificial reefs. In *Habitat Structure* (pp. 412–426). Springer. doi:10.1007/978-94-011-3076-9_20

Bohnsack, J. A., & Sutherland, D. L. (1985). Artificial reef research: A review with recommendations for future priorities. *Bulletin of Marine Science*, *37*(1), 11–39.

Bohnsack, J. A., & Sutherland, D. L. (1985). Artificial Reef Research: A review with Recommendations for Future Priorities. *Bulletin of Marine Science*, *37*(1), 11–39.

Bombace, G. (1989). Artificial reefs in the Mediterranean Sea. *Bulletin of Marine Science*, *44*, 1023–1032.

Bombace, G., Fabi, G., & Fiorentini, L. (2000). Artificial reefs in the Adriatic Sea. In A. C. Jensen, K. J. Collins, & A. P. M. Lockwood (Eds.), *Artificial Reefs in European Seas* (pp. 31–63). Kluwer Academic Publications. doi:10.1007/978-94-011-4215-1_3

Bombace, G., Fabi, G., Fiorentini, L., & Speranza, S. (1994). Analysis of the efficacy of artificial reefs located in five different areas of the Adriatic Sea. *Bulletin of Marine Science*, *55*, 559–580.

Bond, T., Langlois, T. J., Partridge, J. C., Birt, M. J., Malseed, B. E., Smith, L., & McLean, D. L. (2018). Diel shifts and habitat associations of fish assemblages on a subsea pipeline. *Fisheries Research*, *206*, 220–234. doi:10.1016/j.fishres.2018.05.011

Boonstra, W. J., Björkvik, E., Joosse, S., & Hanh, T. T. H. (2020). From Anthrome to Refugium? A Short History of Small-Scale Fisheries in the Anthropocene. Encyclopedia of the World's Biomes, 5, 180–187. doi:10.1016/B978-0-12-409548-9.11931-1

Bortone, S. A., Cody, R. P., Turpin, R. K. & Bundrick, C. M. (1998). The impact of artificial-reef fish assemblages on their potential forage area. *Italian Journal of Zoology*, 265-267.

Bortone, S. A., Brandini, P. P., Fabi, S., & Otake, S. (2011). *Artificial reefs in fishery management*. CRC Press. doi:10.1201/b10910

Bortone, S. A., & Otake, S. (Eds.). (2020). *Modern Fisheries Engineering: Realizing a Healthy and Sustainable Marine Ecosystem*. CRC Press. doi:10.1201/9780429328039

Bortone, S. A., Samoilys, M. A., & Francour, P. (2000). Fish and Macroinvertebrate Evaluation Methods. In W. Seaman Jr., (Ed.), *Artificial Reef Evaluation: With Application to Natural Marine Habitats* (pp. 127–164). CRC Press. doi:10.1201/9781420036633.ch5

Boswell, K. M., Wells, R. J. D., Cowan, J. J. H. Jr, & Wilson, C. A. (2010). Biomass, density, and size distributions of fishes associated with a large-scale artificial reef complex in the Gulf of Mexico. *Bulletin of Marine Science*, *86*(4), 879–889. doi:10.5343/bms.2010.1026

Bradbury, R. H., & Seymour, R. M. (2009). Coral reef science and the new commons. *Coral Reefs*, *28*(4), 831–837.

Branden, K. L., Pollard, D. A., & Reimers, H. A. (1994). A review of recent artificial reef developments in Australia. *Bulletin of Marine Science*, *55*(2–3), 982–994.

Brennan, F., & Kolios, A. (2014, January). Structural integrity considerations for the H2Ocean multi-modal wind-wave platform. In *European Wind Energy Association Conference and Exhibition 2014, EWEA 2014*. European Wind Energy Association.

Brenner, M. (2015). *Shells and Students: Building Living Reefs in Southern New Jersey*. usfwsnortheast.wordpress.com/2015/05/22/shells-and-students-in-southern-new-jersey/

Breton, Y., & Lopez Estrada, E. (1989). Social sciences and fisheries development. Models and methods applied to the case of Mexico. National Institute of Anthropology and History-Secretariat of Fisheries, Mexico.

Brickhill, M. J., Lee, S. Y., & Connolly, R. M. (2005). Fishes associated with artificial reefs: Attributing changes to attraction or production using novel approaches. *Journal of Fish Biology*, *67*(SUPPL.), 53–71. doi:10.1111/j.0022-1112.2005.00915.x

Brochier, T., Auger, P., Thiam, N., Sow, M., Diouf, S., Sloterdijk, H., & Brehmer, P. (2015). Implementation of artificial habitats: Inside or outside the marine protected areas? Insights from a mathematical approach. *Ecological Modelling*, *297*, 98–106. doi:10.1016/j.ecolmodel.2014.10.034

Brock, R. E. (1994). Beyond fisheries enhancement: Artificial reefs and ecotourism. *Bulletin of Marine Science*, *55*(2–3), 1181–1188.

Brotto, D. S., Krohling, W., & Zalmon, I. R. (2006). Fish community modeling agents on an artificial reef on the northern coast of Rio de Janeiro - Brazil. *Brazilian Journal of Oceanography*, *54*(39), 205–212. doi:10.1590/S1679-87592006000300004

Browman, H. I., Stergiou, K. I., Cury, P. M., Hilborn, R., Jennings, S., Lotze, H. K., & Mace, P. M. (2004). Perspectives on ecosystem-based approaches to the management of marine resources. *Marine Ecology Progress Series*, *274*, 269–303. doi:10.3354/meps274269

Brundltand, R. (2016). *Sustainability and Development*. Retrieved on June 16, 2016, from https://sustentabilidadydesarrollo.com/2011/06/25/82/

Bruno, F., Bianco, G., Muzzupappa, M., Barone, S., & Rationales, A. V. (2011). Experimentation of structured light and stereo vision for underwater 3D reconstruction. *ISPRS Journal of Photogrammetry and Remote Sensing*, *66*(4), 508–518. doi:10.1016/j.isprsjprs.2011.02.009

Bruno, J. F., & Selig, E. R. (2007). Regional decline of coral cover in the Indo-Pacific: Timing, extent, and subregional comparisons. *PLoS One*, *2*(8), e711. doi:10.1371/journal.pone.0000711 PMID:17684557

Bucaram, S. J., Hearn, A., Trujillo, A. M., Rentería, W., Bustamante, R. H., Morán, G., Reck, G., & García, J. L. (2018). Assessing fishing effects inside and outside an MPA: The impact of the Galapagos Marine Reserve on the Industrial pelagic tuna fisheries during the first decade of operation. *Marine Policy, 87*(June), 212–225. doi:10.1016/j.marpol.2017.10.002

Bull, A. S., & Love, M. S. (2019). Worldwide oil and gas platform decommissioning: A review of practices and reefing options. *Ocean and Coastal Management, 168*(December), 274–306. doi:10.1016/j.ocecoaman.2018.10.024

Bureau of Safety and Environmental Enforcement. (2018). *Rigs to Reefs*. Available at: https://www.bsee.gov/what-we-do/environmental-focuses/rigsto-reefs

Burt, J., Bartholomew, A., & Firth, L. (2021). Policy and management considerations for artificial reefs in the Arabian Gulf. *Sheikh Saud bin Saqr Al Qasimi Foundation for Policy Research. Policy Paper*, *51*, 1–31.

Cancela da Fonseca, L., Boaventura, D., Ré, P., Pereira, P., & Santos, M. N. (2007). Caracterização das Comunidades de Macroinvertebrados Bentónicos no Sistema Recifal do Alvor (Costa Sul do Algarve). *13º Congresso do Algarve*, 403-413.

Capobianco, N., Basile, V., Loia, F., & Vona, R. (2021). Toward a sustainable decommissioning of offshore platforms in the oil and gas industry: A pestle analysis. *Sustainability*, *13*(11), 6266. doi:10.3390u13116266

Cardenas, E. R. B. (2011). Artificial Reefs in Artisanal Fisheries. In S. A. Bortone, F. P. Brandini, G. Fabi, & S. Otake (Eds.), *Artificial Reefs in Fisheries Management* (pp. 209–220). CRC Press Taylor & Francis Group.

Carral Couce, L. M., Guerreiro, M. J. R., Fraguela Formoso, J. A., Alvarez-Feal, J. C., & Vizoso, A. F. (2017, October). Green artificial reef PROARR: Repopulation of coastal ecosystems and waste recycler of the maritime industries. In *Pan-American Conference of Naval Engineering* (pp. 363–373). Springer.

Carr, M. H., & Hixon, M. A. (1997). Artificial reefs: The importance of comparisons with natural reefs. *Fisheries (Bethesda, Md.)*, *22*(4), 28–33. doi:10.1577/1548-8446(1997)022<0028:ARTIOC>2.0.CO;2

Cater, C. (2007). Perceptions of and interactions with marine environments: Diving attractions from great whites to pygmy seahorses. In *New frontiers in marine tourism* (pp. 71–86). Routledge.

CBI. (2019). *Dive tourism from Europe. Centre for the Promotion of Imports from Developing Countries*. Retrieved from https://www.cbi.eu/market-information/tourism/dive-tourism/europe/

Ceyhan, T., Akyol, O., & Ünal, V. (2006). Edremit Körfezi (Ege Denizi) Kıyı Balıkçılığı Üzerine Bir Araştırma [An investigation on the artisanal fisheries in Edremit Bay (Aegean Sea)]. *Ege University Journal of Fisheries & Aquatic Science*, *23* (1/3), 373-375.

Chamber of Shipping of Turkey. (2020). https://www.denizticaretodasi.org.tr/media/SharedDocuments/sektorraporu/MaritimeSectorReport2021.pdf

Chandima Ratnayake, R. M., & Markeset, T. (2012). Asset integrity management for sustainable industrial operations: Measuring the performance. *International Journal of Sustainable Engineering*, *5*(2), 145–158. doi:10.1080/19397038.2011.581391

Chan, Y., & Walmsley, R. P. (1997). Learning and understanding the Kruskal-Wallis one-way analysis-of-variance-by-ranks test for differences among three or more independent groups. *Physical Therapy*, *77*(12), 1755–1762. doi:10.1093/ptj/77.12.1755 PMID:9413454

Chapman, M. G. (2003). Paucity of mobile species on constructed seawalls: Effects of urbanization on biodiversity. *Marine Ecology Progress Series*, *264*, 21–29. doi:10.3354/meps264021

Charbonnel, E., Francour, P., Harmelin, J. G., Ody, D., & Bachet, F. (2000). Effects of artificial reef design on associated fish assemblages in the Côte Bleue Marine Park (Mediterranean Sea, France). In *Artificial Reefs in European Seas* (pp. 365–377). Springer. doi:10.1007/978-94-011-4215-1_22

Charles, A. (2001). *Sustainable Fishery Systems*. Wiley Press.

Chittaro, P. (2002). Species-area relationships for coral reef fish assemblages of St. Croix, US Virgin Islands. *Marine Ecology Progress Series*, *233*, 253–261. doi:10.3354/meps233253

Choi, C. G., Takeuchi, Y., Terawaki, T., Serisawa, Y., Ohno, M., & Sohn, C. H. (2002). Ecology of seaweed beds on two types of artificial reef. *Journal of Applied Phycology*, *14*(5), 343–349. doi:10.1023/A:1022126007684

Chou, L. M. (1997). Artificial reefs of southeast Asia–do they enhance or degrade the marine environment? *Environmental Monitoring and Assessment*, *44*(1/3), 45–52. doi:10.1023/A:1005759818050

Christensen, E. D., Stuiver, M., Guanche, R., Møhlenberg, F., Schouten, J. J., Pedersen, O. S., He, W., Zanuttigh, B., & Koundouri, P. (2015). *Go offshore-Combining food and energy production*. The Technical University of Denmark. Department of Mechanical Engineering.

Cirik, Ş., & Neşer, M. (1999, July). *Ülkemiz deniz teknolojisinde yeni bir uygulama alanı: yapay barınaklar* [A new field of application in marine technology in our country: artificial reefs]. *National Congress on Shipbuilding and Marine Technology*, İstanbul, Turkey.

Cisneros-Montemayor, A. M., Zetina-Rejón, M. J., Espinosa-Romero, M. J., Cisneros-Mata, M. A., Singh, G. G., & Fernández-Rivera Melo, F. J. (2020). Evaluating ecosystem impacts of data-limited artisanal fisheries through ecosystem modelling and traditional fisher knowledge. *Ocean and Coastal Management*, *195*(February), 105291. Advance online publication. doi:10.1016/j.ocecoaman.2020.105291

Claudet, J., & Pelletier, D. (2004). Marine protected areas and artificial reefs: A review of the interactions between management and scientific studies. *Aquatic Living Resources*, *17*(2), 129–138. doi:10.1051/alr:2004017

Cocklin, C., Craw, M., & McAuley, I. (1998). Marine reserves in New Zealand: Use rights, public attitudes, and social impacts. *Coastal Management*, *26*(3), 213–231. doi:10.1080/08920759809362353

Coen, L. D., Brumbaugh, R. D., Bushek, D., Grizzle, R., Luckenbach, M. W., Posey, M. H., Powers, S. P., & Tolley, S. G. (2007). Ecosystem services related to oyster restoration. *Marine Ecology Progress Series, 341*, 303–307. doi:10.3354/meps341303

Collart, D. & E. Charbonnel, (1998*). Impact des re ´cifs artificiels de Marseillan et d'Agde sur le milieu marin et la pe ^che professionnelle-Bilan du suivi 1996/1997.* Contrat Conseil Re ´gional Languedoc-Roussillon et Conseil Ge ´ne ´ral de l'He ´rault. CEGEL et GIS Posidonie, France.

Cookson, M. J., Davies, C. J., Entwistle, A., & Whimster, W. F. (1993). The microanatomy of the alveolar duct of the human lung imaged by confocal microscopy and visualized with computerbased 3D reconstruction. *Computerized Medical Imaging and Graphics, 17*(3), 201–210. doi:10.1016/0895-6111(93)90044-N PMID:8402527

Coolen, J. W., Bittner, O., Driessen, F. M., van Dongen, U., Siahaya, M. S., de Groot, W., Mavraki, N., Bolam, S. G., & van der Weide, B. (2020). Ecological implications of removing a concrete gas platform in the North Sea. *Journal of Sea Research, 166*, 101968. doi:10.1016/j.seares.2020.101968

Côrtes, L. H. de O., Zappes, C. A., & Di Beneditto, A. P. M. (2014). Ethnoecology, gathering techniques and traditional management of the crab Ucides cordatus Linnaeus, 1763 in a mangrove forest in south-eastern Brazil. *Ocean and Coastal Management, 93*, 129–138. doi:10.1016/j.ocecoaman.2014.03.021

Corvianawatie, C., & Abrar, M. (2018). Suitability of Oceanographic Conditions in Supporting Coral Reef Ecosystems on Pari Island. *National Oceanic Journal, 13*(3), 155–162. doi:10.15578/jkn.v13i3.6322

Cousteau, J. Y., & Dumas, F. (1953). The silent world. Academic Press.

CREOCEAN. (2003). *Impact des re ´cifs artificiels sur le milieu marin et la pe ^che professionnelle dans le golfe d'AiguesMortes*. CREOCEAN.

Crouch, M., & McKenzie, H. (2006). The logic of small samples in interview based qualitative research. *Social Sciences Information. Information Sur les Sciences Sociales, 45*(4), 483–499. doi:10.1177/0539018406069584

Dafforn, K. A., Glasby, T. M., Airoldi, L., Rivero, N. K., Mayer-Pinto, M., & Johnston, E. L. (2015). Marine urbanization: An ecological framework for designing multifunctional artificial structures. *Frontiers in Ecology and the Environment, 13*(2), 82–90. doi:10.1890/140050

Dafforn, K. A., Glasby, T. M., & Johnston, E. L. (2012). Comparing the invasibility of experimental "reefs" with field observations of natural reefs and artificial structures. *PLoS One, 7*(5), e38124. doi:10.1371/journal.pone.0038124 PMID:22666459

Daly, H. E., & Cobb, J. B. (1993). For the common good: Reorienting the economy towards community, the environment, and a sustainable future. Economic Culture Fund.

Davidova, M., & Zavoleas, Y. (2020). Post-Anthropocene: The Design after the Human-Centered Design Age. *Re: Anthropocene: Design in the Age of Humans, 25th International Conference on Computer-Aided Architectural Design Research in Asia (CAADRIA), 2*, 203-212.

Davidson-Hunt, I. J., & Berkes, F. (2003). *Nature and Society through the Lens of Resilience: toward a Human-in-Ecosystem Perspective. Navigating Social-ecological Systems: Building Resilience for Complexity and Change.* Cambridge University Press.

Davis, D., & Tisdell, C. (1996). Economic Management of Recreational Scuba Diving and the Environment. *Journal of Environmental Management*, (February), 229–248.

Davis, D., & Tisdell, C. (1995). Recreational scuba-diving and carrying capacity in marine protected areas. *Ocean and Coastal Management*, *26*(1), 19–40. doi:10.1016/0964-5691(95)00004-L

Davis, K. L., Coleman, M. A., Connell, S. D., Russell, B. D., Gillanders, B. M., & Kelaher, B. P. (2017). Ecological performance of construction materials subject to ocean climate change. *Marine Environmental Research*, *131*, 177–182. doi:10.1016/j.marenvres.2017.09.011 PMID:28988853

De Groot, J., & Bush, S. R. (2010). The potential for dive tourism led entrepreneurial marine protected areas in Curacao. *Marine Policy*, *34*(5), 1051–1059. doi:10.1016/j.marpol.2010.03.004

Dearden, P., Bennett, M., & Rollins, R. (2007). Perceptions of diving impacts and implications for reef conservation. *Coastal Management*, *35*(2-3), 305–317. doi:10.1080/08920750601169584

Delgadillo-Garz6n & Camilo. (2009). Impact of two artificial reefs on the diurnal artisanal fishing of the Gulf of Morrosquillo, Caribbean of Colombia. *Revista de Biología Tropical*, *57*(4), 993–1007. PMID:20073330

Dhiecha, D. M., Utomo, K. P., Jati, D. R., & Tanjungpura, U. (2004). *Planning of artificial reef as reef restoration of Bengkayang Regency Lemukutan as a Conservation and Nature Tourism Area of Bengkayang Sea according to the coast and sea in the Lemukutan Island area and its surroundings, as a consequence Based on a case study*. Academic Press.

Dias, D. A. C. M. (2011). *Estudo da valorização de cinzas de biomassa na produção de materiais para a construção de recifes artificiais* (Doctoral dissertation). Faculdade de Ciências e Tecnologia.

Dimmock, K. (2007). Scuba diving, snorkeling, and free diving. In *Water-based tourism, sport, leisure, and recreation experiences* (pp. 146–166). Routledge. doi:10.1016/B978-0-7506-6181-2.50013-0

Dimmock, K., & Cummins, T. (2013). History of scuba diving tourism. In G. Musa & K. Dimmock (Eds.), *Scuba Diving Tourism* (pp. 14–28). Routledge.

Dimmock, K., & Musa, G. (2015). Scuba diving tourism system: A framework for collaborative management and sustainability. *Marine Policy*, *54*, 52–58. doi:10.1016/j.marpol.2014.12.008

Dimopoulos, D., Queiros, D., & van Zyl, C. (2019). Sinking deeper: The most significant risks impacting the dive tourism industry in the East African Marine Ecoregion. *Ocean and Coastal Management*, *181*, 104897. doi:10.1016/j.ocecoaman.2019.104897

Ditton, R. B., & Baker, T. L. (1999). Demographics, Attitudes, Management Preferences, and Economic Impacts of Sport Divers using Artificial Reefs in Offshore Texas Waters. Report prepared for the Texas Parks and Wildlife Department through a research contract with Texas A&M University.

Ditton, R. B., Thailing, C. E., Riechers, R., & Osburn, H. (2001). The Economic impacts of sport divers using Artificial Reefs in Texas Offshore Waters. *Proceedings of the Annual Gulf and Caribbean Fisheries Institute*, *54*, 349-360.

Ditton, R. B., Osburn, H. R., Baker, T. L., & Thailing, C. E. (2002). Demographics, attitudes, and reef management preferences of sport divers in offshore Texas waters. *ICES Journal of Marine Science*, *59*, 186–191. doi:10.1006/jmsc.2002.1188

Do Carmo, J. S. A., Neves, M. G., & Ten Voorde, M. (2011). Designing a multifunctional artificial reef: Studies on the influence of parameters with most influence in the vertical plane. *Journal of Coastal Conservation*, *15*(1), 99–112. doi:10.100711852-010-0124-1

DOF. (2017). Rules of operation for the Program for the Promotion of Fisheries and Aquaculture Productivity. Official Gazette of the Federation.

Dokken, Q. R. (2012). *Ecological impact of platform structures as artificial reefs, Paper in development, used with permission of Q.* Dokken, Gulf of Mexico Foundation.

Dol, S. S. (2018). Design of Self-sustaining Mariculture Farm on Offshore Rig. *10th Int'l Conference on Advances in Engineering, Science, Technology & Sustainable Development (ESTSD-18)*, 25-27.

Dowling, R. K., & Nichol, J. (2001). The HMAS Swan artificial dive reef. *Annals of Tourism Research*, *28*(1), 229–232. doi:10.1016/S0160-7383(00)00019-0

DR (2008). Decreto Regulamentar n.º 9/2008 de 18 de março. Ministério da Agricultura, do Desenvolvimento Rural e das Pescas. Diário da República, 1ª série, N.º 55 de 18 de março de 2008.

Drew, J. A. (2005). Use of traditional ecological knowledge in marine conservation. *Conservation Biology*, *19*(4), 1286–1293. doi:10.1111/j.1523-1739.2005.00158.x

Dunn, K., Haeusler, M. H., Zavoleas, Y., Bishop, M., & Dafforn, K. (2019). Recycled Sustainable 3D Printing Materials for Marine Environments. *Challenges: Sustainability and Cultural Heritage, 37th International Conference on Education and Research in Computer Aided Architectural Design in Europe (eCAADe), 2*, 583-592. http://papers.cumincad.org/data/works/att/ecaadesigradi2019_641.pdf

Dunn, K., Haeusler, M. H., Zavoleas, Y., Bishop, M., Dafforn, K., Sedano, F., Yu, D., & Schaefer, N. (2019). Recycled Sustainable 3D Printing Materials for Marine Environments. In Blucher Design Proceedings. Editora Blucher.

Düzbastılar, F. O., & Tokaç, A. (2003). Determination of effects of artificial reef size on local scouring phenomena resulting from wave action *Ege. Journal of Fisheries and Aquatic Science*, *20*(3-4), 373–381.

Edelsbrunner, H., Kirkpatrick, D., & Seidel, R. (1983). On the shape of a set of points in the plane. *IEEE Transactions on Information Theory*, *29*(4), 551–559. doi:10.1109/TIT.1983.1056714

Edelsbrunner, H., & Mucke, E. P. (1994). 3-dimensional alphashapes. *ACM Transactions on Graphics*, *13*(1), 43–72. doi:10.1145/174462.156635

Edney, J., & Spennemann, D. H. (2015). Can artificial reef wrecks reduce diver impacts on shipwrecks? The management dimension. *Journal of Maritime Archaeology*, *10*(2), 141–157. doi:10.100711457-015-9140-5

Encarnação, J., & Calado, G. (2018). Effects of recreational diving on early colonization stages of an artificial reef in North-East Atlantic. *Journal of Coastal Conservation*, *22*(6), 1209–1216. doi:10.100711852-018-0630-0

Epstein, N., Bak, R. P. M., & Rinkevich, B. (1999). Implementation of a small-scale "no-use zone" policy in a reef ecosystem: Eilat's reef-lagoon six years later. *Coral Reefs*, *18*(4), 327–332. doi:10.1007003380050207

Epstein, N., Bak, R. P. M., & Rinkevich, B. (2003). Applying forest restoration principles to coral reef rehabilitation. *Aquatic Conservation*, *13*(5), 387–395. doi:10.1002/aqc.558

Erwanto, Z., & Masluha, U. (2019). Artificial Temple Reef Conservation Technology As Control Of Coastal Abrasion On Tabuhan Island, Bangsring Village, Wongsorejo District. *Banyuwangi.*, *4*(1), 103–109.

Erzini, K., Monteiro, C. C., Ribeiro, J., Santos, M. N., Gaspar, M., Monteiro, P., & Borges, T. C. (1997). An experimental study of gill net and trammel net\'ghost fishing\'off the Algarve (southern Portugal). *Marine Ecology Progress Series*, *158*, 257–265. doi:10.3354/meps158257

Espinosa, T. P. (2018). *Snowball Sampling*. Department of probability and statistics. UNAM. http://www.dpye.iimas.unam.mx/patricia/muestreo/datos/trabajos%20alumnos/Proyectofinal_Bola%20de%20Nieve.pdf

Evans, A. J., Garrod, B., Firth, L. B., Hawkins, S. J., Morris-Webb, E. S., Goudge, H., & Moore, P. J. (2017). Stakeholder priorities for multi-functional coastal defence developments and steps to effective implementation. *Marine Policy, 75*(October), 143–155. doi:10.1016/j.marpol.2016.10.006

Fabi, G., Scarcella, G., Spagnolo, A., Bortone, S. A., Charbonel, E., Goutayer, J. J., Haddad, N., Lök, A., & Trommelen, M. (2015). Practical Guidelines for the Use of Artificial Reefs in the Mediterranean and the Black Sea. FAO-GFCM, Studies and Reviews, No. 96.

Fabi, G., Spagnolo, A., Bellan-Santini, D., Charbonnel, E., Çiçek, B. A., García, J. J. G., ... Santos, M. N. D. (2011). Overview on artificial reefs in Europe. *Brazilian Journal of Oceanography, 59*(spe1), 155-166.

Fabi, G., Spagnolo, A., Bellan-Santini, D., Charbonnel, E., Çiçek, B. A., García, J. J. G., Jensen, A. C., Kallianiotis, A., & dos Santos, M. N. (2011). Overview on artificial reefs in Europe. *Brazilian Journal of Oceanography, 59*(SPEC. ISSUE 1), 155–166. doi:10.1590/S1679-87592011000500017

Fabi, G., Scarcella, G., Spagnolo, A., Bortone, S. A., Charbonnel, E., Goutayer, J. J., ... Trommelen, M. (2015). Practical guidelines for the use of artificial reefs in the Mediterranean and the Black Sea. *General Fisheries Commission for the Mediterranean. Studies and Reviews (General Fisheries Commission for the Mediterranean)*, (96), I.

Fabi, G., & Spagnolo, A. (2011). *Artificial reefs in the management of Mediterranean Sea fisheries*. CRC Press.

Fabricius, K. E. (2005). Effects of terrestrial runoff on the ecology of corals and coral reefs: Review and synthesis. *Marine Pollution Bulletin, 50*(2), 125–146. doi:10.1016/j.marpolbul.2004.11.028 PMID:15737355

Fadilah, S., & Suprihardjo, R. (2016). Development of Marine Tourism Area Watulimo District, Trenggalek Regency. *Journal of Engineering ITs, 5*(1), 50–53.

Falcão, M., Santos, M. N., Drago, T., Serpa, D., & Monteiro, C. (2009). Effect of artificial reefs (southern Portugal) on sediment–water transport of nutrients: Importance of the hydrodynamic regime. *Estuarine, Coastal and Shelf Science, 83*(4), 451–459. doi:10.1016/j.ecss.2009.04.028

Falcão, M., Santos, M. N., Vicente, M., & Monteiro, C. C. (2007). Biogeochemical processes and nutrient cycling within an artificial reef off Southern Portugal. *Marine Environmental Research, 63*(5), 429–444. doi:10.1016/j.marenvres.2006.12.001 PMID:17239434

FAO (2018). *The State of World Fisheries and Aquaculture 2018. Meeting the Sustainable Development Goals*. License: CC BY-NC-SA 3.0 IGO.

FAO. (2004). *Application of development strategies focused on people within the scope of FAO*. FAO.

FAO. (2016). The State of World Fisheries and Aquaculture 2016. Contributing to food security and nutrition for all. Rome, Italy: FAO.

Fathiyah, N., Pin, T. G., Saraswati, R., & Kunci, K. (2017). Spatial and Temporal Patterns of Total Suspended Solid (TSS) with SPOT Imagery in the Cimandiri Estuary, West Java. Academic Press.

Fernandes, L. P., Keunecke, K. A., & Di Beneditto, A. P. M. (2014). Produção e socioeconomia da pesca do camaração sete-barbas no norte do estado do Rio De Janeiro [Production and socioeconomic of the seven-barb shrimp fishery in the north state of Rio de Janeiro]. *Boletim do Instituto de Pesca, 40*(4), 541–555.

Fikri, M., Isdianto, A., & Luthfi, O. M. (2021a). Environmental Conditions of Waters (Oceanographic Physics) Around Artificial Reefs at Damas Beach, Trenggalek Regency, East Java. *Journal of Marine and Coastal Science, 10*(February).

Fikri, M., Isdianto, A., & Luthfi, O. M. (2021b). Physics Oseanography Around Artificial Reef On The Pantai Of Damas, Trenggalek District, East Java. *Journal of Marine and Coastal Science*, *10*(February), 35. doi:10.20473/jmcs.v10i1.25605

FIPERJ. (2015). Relatório da Fundação Instituto de Pesca do Estado do Rio De Janeiro. Secretaria de Desenvolvimento Regional, Abastecimento e Pesca.

Firth, L. B., Knights, A. M., Bridger, D., Evans, A. J., Mieszkowska, N., Moore, P. J., O'Connor, N. E., Sheehan, E. V, Thompson, R. C., & Hawkins, S. J. (2016). Ocean Sprawl: Challenges and Opportunities for Biodiversity Management in a Changing World. *Oceanography and Marine Biology: An Annual Review, 54*(September), 193–269.

Firth, L. B., Knights, A. M., Bridger, D., Evans, A., Mieszkowska, N., Moore, P., O'connor, N., Sheehan, E., Thompson, R., & Hawkins, S. (2016). Ocean sprawl: Challenges and opportunities for biodiversity management in a changing world. *Oceanography and Marine Biology*, *54*, 189–262. doi:10.1201/9781315368597-5

Firth, L. B., Thompson, R. C., Bohn, K., Abbiati, M., Airoldi, L., Bouma, T. J., Bozzeda, F., Ceccherelli, V. U., Colangelo, M. A., Evans, A., Ferrario, F., Hanley, M. E., Hinz, H., Hoggart, S. P. G., Jackson, J. E., Moore, P., Morgan, E. H., Perkol-Finkel, S., Skov, M. W., ... Hawkins, S. J. (2014). Between a rock and a hard place: Environmental and engineering considerations when designing coastal defence structures. *Coastal Engineering*, *87*, 122–135. doi:10.1016/j.coastaleng.2013.10.015

Fletcher, S., Bateman, P., & Emery, A. (2011). The governance of the Boscombe artificial surf reef, UK. *Land Use Policy*, *28*(2), 395–401. doi:10.1016/j.landusepol.2010.08.002

Folke, C., Berkes, F., & Colding, J. (1998). Ecological practices and social mechanisms for building resilience and sustainability. In F. Berkes & C. Folke (Eds.), *Linking Social and Ecological Systems: Management Practices and Social Mechanisms for Building Resilience.* Cambridge University Press.

Fowler, A. M., & Booth, D. J. (2012). Evidence of sustained populations of small reef fish on artificial structures. Does depth affect production on artificial reefs? *Journal of Fish Biology*, *80*(3), 613–629. doi:10.1111/j.1095-8649.2011.03201.x PMID:22380556

Fowler, A. M., Macreadie, P. I., Jones, D. O. B., & Booth, D. J. (2014). A multi-criteria decision approach to decommissioning of offshore oil and gas infrastructure. *Ocean and Coastal Management*, *87*, 20–29. doi:10.1016/j.ocecoaman.2013.10.019

Francour, P., Harmelin, J.-G., Pollard, D., & Sartoretto, S. (2001). Mediterranean region: Siting, usage, zonation and management. *Aquatic Conservation : Marine and Freshwater Ecocsystems*, *11*(3), 155–188. doi:10.1002/aqc.442

Frank, M. (2018). *Artificial reefs: marine and freshwater applications.* CRC Press.

Frid, O., & Belmaker, J. (2019). Catch dynamics of set net fisheries in Israel. *Fisheries Research, 213*(March), 1–11. doi:10.1016/j.fishres.2018.12.021

Frikha, A., & Moalla, H. (2015). Analytic hierarchy process for multi-sensor data fusion based on belief function theory. *European Journal of Operational Research*, *241*(1), 133–147. doi:10.1016/j.ejor.2014.08.024

Froese, R., & Pauly, D. (2021). *FishBase.* www.fishbase.org

Fuller, S. (2010). *Life-cycle cost analysis (LCCA).* National Institute of Building Sciences.

Gallaway, B. J., Szedlmayer, S. T., & Gazey, W. J. (2009). A life history review for red snapper in the Gulf of Mexico with an evaluation of the importance of offshore petroleum platforms and other artificial reefs. *Reviews in Fisheries Science*, *17*(1), 48–67. doi:10.1080/10641260802160717

Gallopín, G. C. (2001). Science for the 21st Century: From Social Contract to the Scientific Core. *Int. Journal Social Science*, *168*, 219–229.

Garrod, B. (2007). Market segments and tourist typologies for diving tourism. In *New Frontiers in Marine Tourism* (pp. 53–70). Routledge. doi:10.4324/9780080551081-13

Garrod, B., & Gössling, S. (2008). Introduction. In B. Garrod & S. Gössling (Eds.), *New Frontiers in Marine Tourism: Diving experiences, Sustainability, Management* (pp. 8–29). Elsevier Press. doi:10.1016/B978-0-08-045357-6.50004-8

George, A. (2005). *Artificial Reefs for Oysters*. www.baynature.org/article/artificial-reefs-for-oysters/

Gerhardinger, L. C., Godoy, E. A. S., & Jones, P. J. S. (2009). Local ecological knowledge and the management of marine protected areas in Brazil. *Ocean and Coastal Management*, *52*(3–4), 154–165. doi:10.1016/j.ocecoaman.2008.12.007

Gibson, R. N., & Barnes, M. (2003). The use of intertidal barnacle shells. *Oceanography and Marine Biology: An Annual Review*, *38*, 157–187.

Giddy, J. K., & Rogerson, C. M. (2018). Tracking SCUBA diving adventure tourism in South Africa. *EuroEconomica*, *37*(2), 47–62.

Giyanto, M., Hadi, T. A., Budiyanto, A., Muhammad Hafizt, Salatalohy, A., & Iswari, M. Y. (2017). *Status of Coral Reefs in Indonesia 2017*. Academic Press.

Gomes-Pereira, J. N., Tempera, F., Ribeiro, P. A., & Porteiro, F. M. (2012). Notes on fauna associated with an opportunistic artificial reef near cold-water corals. *ARQUIPÉLAGO. Life and Marine Sciences*.

Gomez, E., Edwards, A. J., & Gomez, E. D. (2007). Reef Restoration Concepts & Guides: Making wise choices among uncertainties. Terumbu Karang Indonesia (TERANGI) Foundation.

Gonzalez, L., & Maria, P. (1996). Reflections on coastal fishing. Effort and Capture. Technology and overexploitation of living marine resources, El Colegio de México, México.

Gordon, W. R. Jr, & Ditton, R. B. (1986). A user-resource planning framework for offshore recreational artificial reefs. *Coastal Management*, *13*(3-4), 369–395.

Grabowski, J. H., Brumbaugh, R. D., Conrad, R. F., Keeler, A. G., Opaluch, J. J., Peterson, C. H., Piehler, M. F., Powers, S. P., & Smyth, A. R. (2012). Economic Valuation of Ecosystem Services Provided by Oyster Reefs. *Bioscience*, *62*(10), 900–909. doi:10.1525/bio.2012.62.10.10

Grace, R. A. (2001). The factors and processes that influence artificial reef longevity. *Marine Technology Society Journal*, *35*(2), 3–13. doi:10.4031/002533201788001910

Green, E. P., & Shirley, F. (1999). The global trade in coral. WCMC, World Conservation Press.

Grossman, G. D., Jones, G. P., & Seaman, W. J. Jr. (1997). Do artificial reefs increase regional fish production? A review of existing data. *Fisheries (Bethesda, Md.)*, *22*(4), 17–23. doi:10.1577/1548-8446(1997)022<0017:DARIRF>2.0.CO;2

Grove, R. S., & Sonu, C. J. (1983). Review of Japanese fishing reef technology. Southern California Edison Company.

Guillaume, S., & Charnomordic, B. (2011). Learning interpretable fuzzy inference systems with FisPro. *Information Sciences*, *181*(20), 4409–4427. doi:10.1016/j.ins.2011.03.025

Haab, T. C. (2002). *Valuing Environmental and Natural Resources, Economics of Non-Market Valuation*. Edward Elgar Publishing Limited. doi:10.4337/9781843765431

Haddock-Fraser, J., & Hampton, M. P. (2012). Multistakeholder values on the sustainability of dive tourism: Case studies of Sipadan and Perhentian Islands, Malaysia. *Tourism Analysis, 17*(1), 27–41. doi:10.3727/108354212X13330406124016

Haeuber, R., & Franklin, J. (1996). Perspectives on ecosystem management. *Ecological Applications, 6*(3), 692–693. doi:10.2307/2269461

Hanemann, W. M. (1994). Valuing the Environment Through Contingent Valuation. *The Journal of Economic Perspectives, 8*(4), 19–43. doi:10.1257/jep.8.4.19

Hanni, E., & Mathews, H. H. (1977). *Benefit-Cost Study of Pinellas County Artificial Reefs. FLSGP-T-77-005, Florida Sea Grant College Program.* University of Florida.

Harmelin-Vivien, M., Cottalorda, J. M., Dominici, J. M., Harmelin, J. G., Le Diréach, L., & Ruitton, S. (2015). Effects of reserve protection level on the vulnerable fish species Sciaena umbra and implications for fishing management and policy. *Global Ecology and Conservation, 3*, 279–287. doi:10.1016/j.gecco.2014.12.005

Harriott, V. J., Davis, D., & Banks, S. A. (1997). Recreational diving and its impact in marine protected areas in eastern Australia. *Ambio, 26*(3), 173–179.

Hawkins, J. P., & Roberts, C. M. (1992). Effects of recreational SCUBA diving on fore-reef slope communities of coral reefs. *Biological Conservation, 62*(3), 171–178. doi:10.1016/0006-3207(92)91045-T

Hernandez Sampieri, R., Fernández Collado, C., & Baptista Lucio, P. (2014). Investigation methodology. McGraw Hill.

Herrera, C. M. (2011). *Formula for calculating a sample of finite populations.* Retrieved from: https://investigacionpediahr.files.wordpress.com/2011/01/formula-para-cc3a1lculo-de-la-programa-poblaciones-finitas-var-categorica.pdf

Hiett, R., & Milon, J. W. 2002. Economic Impact of Recreational fishing and Diving Associated with Offshore Oil and Gas Structures in the Gulf of Mexico. Minerals Management Service Document MMS Study.

Hiscock, K., Sharrock, S., Highfield, J., & Snelling, D. (2010). Colonization of an artificial reef in south-west England—ex-HMS 'Scylla'. *Journal of the Marine Biological Association of the United Kingdom, 90*(1), 69–94. doi:10.1017/S0025315409991457

Hoegh-Guldberg, O., Mumby, P. J., Hooten, A. J., Steneck, R. S., Greenfield, P., Gomez, E., Harvell, C. D., Sale, P. F., Edwards, A. J., Caldeira, K., Knowlton, N., Eakin, C. M., Iglesias-Prieto, R., Muthiga, N., Bradbury, R. H., Dubi, A., & Hatziolos, M. E. (2007). Coral reefs under rapid climate change and ocean acidification. *Science, 318*(5857), 1737–1742. doi:10.1126cience.1152509 PMID:18079392

Holbrook, S. J., Forrester, G. E., & Schmitt, R. J. (2000). Spatial patterns in abundance of a damselfish reflect the availability of suitable habitat. *Oecologia, 122*(1), 109–120. doi:10.1007/PL00008826 PMID:28307947

Holland, M. M., Becker, A., Smith, J. A., Everett, J. D., & Suthers, I. M. (2021). Characterizing the three-dimensional distribution of schooling reef fish with a portable multibeam echosounder. *Limnology and Oceanography, Methods, 19*(5), 340–355. doi:10.1002/lom3.10427

Ho, W., & Ma, X. (2018). The state-of-the-art integrations and applications of the analytic hierarchy process. *European Journal of Operational Research, 267*(2), 399–414. doi:10.1016/j.ejor.2017.09.007

Hsui, C. Y., & Wang, C. C. (2013). Synergy between fractal dimension and lacunarity index in design of artificial habitat for alternative SCUBA diving site. *Ecological Engineering, 53*, 6–14. doi:10.1016/j.ecoleng.2013.01.014

Huang, I. B., Keisler, J., & Linkov, I. (2011). Multi-criteria decision analysis in environmental sciences: Ten years of applications and trends. *The Science of the Total Environment*, *409*(19), 3578–3594. doi:10.1016/j.scitotenv.2011.06.022 PMID:21764422

Huang, X., Wang, Z., Liu, Y., Hu, W., & Ni, W. (2016). On the use of blast furnace slag and steel slag in the preparation of green artificial reef concrete. *Construction & Building Materials*, *112*, 241–246. doi:10.1016/j.conbuildmat.2016.02.088

Huddart, D. (2019). Recreational SCUBA diving and snorkelling. In *Outdoor Recreation* (pp. 361–394). Palgrave Macmillan. doi:10.1007/978-3-319-97758-4_14

Hunt, J. C. (1995). Divers' accounts of normal risk. *Symbolic Interaction*, *18*(4), 439–462. doi:10.1525i.1995.18.4.439

Hushak, L. J., Kelch, D. O., & Glenn, S. J. (1999). The Economic Value of the Lorain County, Ohio, Artificial Reef. Ohio Sea Grant College Program. *American Fisheries Society Symposium*, *22*, 348–362.

Hutchings, P., Ahyong, S., Ashcroft, M., Mcgrouther, M., & Reid, A. (2013). Sydney Harbour: Its Diverse Biodiversity. *Australian Zoologist*, *36*(3), 255–320. doi:10.7882/AZ.2012.031

Hylkema, A., Hakkaart, Q. C. A., Reid, C. B., Osinga, R., Murk, A. J., & Debrot, A. O. (2021). Artificial reefs in the Caribbean: A need for comprehensive monitoring and integration into marine management plans. Ocean and Coastal Management, 209(April), 105672. doi:10.1016/j.ocecoaman.2021.105672

Ilyas, I. S., Astuty, S., Harahap, S. A., & Purba, N. P. (2017). The Diversity of Targeted Reef Fish in Relation to Coral Growth Forms in the Core Zone of the Anambas Archipelago Aquatic Tourism Park. *Journal of Fisheries and Marine Affairs*, *8*(2), 103–111.

IMSA Amsterdam. (2013). *Worldwide rigs-to-reefs experiences: considerations and motivation for the Living North Sea.* Report Feasibility Update Living North Sea Initiative. https://ecoeffective.biz/wp-content/uploads/2016/02/LNSF250_Worldwide-rigs-to-reefs-experiences_LiNSI_D02.pdf

INECC. (2007): Retrieved from: http://www2.inecc.gob.mx/publicaciones2/libros/402/cuencabalsas.html

INEGI. (2016). Retrieved on January 1, 2016, from https://www.inegi.org.mx/

Ishizaka, A., & Labib, A. (2009). Analytic hierarchy process and expert choice: Benefits and limitations. *Or Insight*, *22*(4), 201-220.

Islam, G. M. N., Noh, K. M., Sidique, S. F., & Noh, A. F. M. (2014). Economic impact of artificial reefs: A case study of small scale fishers in Terengganu, Peninsular Malaysia. *Fisheries Research*, *151*, 122–129. doi:10.1016/j.fishres.2013.10.018

Islam, G. M. N., Noh, K. M., Sidique, S. F., Noh, A. F. M., & Ali, A. (2014). Economic Impacts of Artificial Reefs on Small-Scale Fishers in Peninsular Malaysia. *Human Ecology*, *42*(6), 989–998. doi:10.100710745-014-9692-2

Jackson, A., Chapman, M., & Underwood, A. (2008). Ecological Interactions in the Provision of Habitat by Urban Development: Whelks and Engineering by Oysters on Artificial Seawalls. [x.]. *Austral Ecology*, *33*(3), 307–316. doi:10.1111/j.1442-9993.2007.01818.x

Jacob, G., & Schreyer, R. (1980). Conflict in outdoor recreation: A theoretical perspective. *Journal of Leisure Research*, *12*(4), 368–380. doi:10.1080/00222216.1980.11969462

Jakšić, S., Stamenković, I., & Đorđević, J. (2013). Impacts of artificial reefs and diving tourism. *Turizam*, *17*(4), 155–165. doi:10.5937/Turizam1304155J

Jansen, M., Staffell, I., Kitzing, L., Quoilin, S., Wiggelinkhuizen, E., Bulder, B., Riepin, I., & Müsgens, F. (2020). Offshore wind competitiveness in mature markets without subsidy. *Nature Energy*, *5*(8), 614–622. doi:10.103841560-020-0661-2

Jebreen, E. (2005). *An investigation into the effects of artificial reefs on fish stocks*. Queensland Department of Primary Industries. http://www.dpi.qld.gov.au/far/9279.htmS

Jensen, A. (2002). Artificial reefs of Europe: perspective and future. *ICES Journal of Marine Science, 59*, 3-13. Obtenido de: https://www.researchgate.net/publication/228903123Artificial_reefs_of_Europe_Prospective_and_future

Jensen, A. C., Collins, K. J., & Lockwood, P. M. (2000). Introduction and background to 'Artificial Reefs in European Seas'. In A.C. Jensen, K. J. Collins & A. P. M. Lockwood (Ed.), Artificial Reefs in European Seas (pp. ix-xii). Kluwer Academic Publishers.

Jensen, A. (2002). Artificial reefs of Europe: Perspective and future. *ICES Journal of Marine Science*, *59*(1), 3–13. doi:10.1006/jmsc.2002.1298

Jensen, A. C., Collins, K. J., & Lockwood, P. M. (2000). Current Issues Relating to Artificial Reefs in European Seas. In A. C. Jensen, K. J. Collins, & A. P. M. Lockwood (Eds.), *Artificial Reefs in European Seas* (pp. 489–499). Kluwer Academic Publishers. doi:10.1007/978-94-011-4215-1_29

Jimenez, C., Hadjioannou, L., Petrou, A., Andreou, V., & Georojou, A. (2016). Fouling communities of two accidental artificial reefs (modern shipwrecks) in Cyprus (Levantine Sea). *Water (Basel)*, *9*(1), 2–11. doi:10.3390/w9010011

Jiyah. (2016). Study Of Total Suspended Solid (Tss) Distribution In Coastal Waters Of Demak Regency Using Landsat Image. *Jurnal Geodesi Undip Januari*, *6*, 41–47.

Johannes, R.E. (2002). The renaissance of community based marine resource management in oceania. *Annu. Rev. Ecol. Systemat.*, *33*, 317–340. .ecolsys.33.010802.150524 doi:10.1146/annurev

Johansen, K. (2013). Scuba diving education and training. In G. Musa & K. Dimmock (Eds.), *Scuba diving tourism: contemporary geographies of leisure, tourism and mobility* (pp. 81–96). Routledge.

Johansen, K. (2013). Scuba diving education and training. In *Scuba diving tourism* (pp. 99–114). Routledge.

Johns, G. (2004). *Socioeconomic Study of Reefs in Martin County, Florida. Report prepared for Martin County, Florida .*

Johns, G., Leeworthy, V. R., Bell, F. W., & Bonn, M. A. (2001). *Socioeconomic Study of Reefs in Southeast Florida. Report prepared for Miami-Dade County, Florida .*

Jordan, L. K. B., Gilliam, D. S., & Spieler, R. E. (2005). Reef fish assemblage structure affected by small-scale spacing and size variations of artificial patch reefs. *Journal of Experimental Marine Biology and Ecology*, *326*(2), 170–186. doi:10.1016/j.jembe.2005.05.023

Jørgensen, D. (2012). OSPAR's exclusion of rigs-to-reefs in the North Sea. *Ocean and Coastal Management*, *58*, 57–61. doi:10.1016/j.ocecoaman.2011.12.012

Joyce, A., & Paquin, R. L. (2016). The triple-layered business model canvas: A tool to design more sustainable business models. *Journal of Cleaner Production*, *135*, 1474–1486. doi:10.1016/j.jclepro.2016.06.067

Júnior, L. D. O., Garel, E., & Relvas, P. (2021). The structure of incipient coastal counter currents in South Portugal as indicator of their forcing agents. *Journal of Marine Systems*, *214*, 103486. doi:10.1016/j.jmarsys.2020.103486

Kaiser, M. J. (2006). The Louisiana artificial reef program. *Marine Policy*, *30*(6), 605-623. doi:10.1016/j.marpol.2005.04.005

Kaiser, M. J., & Pulsipher, A. G. (2005). Rigs-to-reef programs in the Gulf of Mexico. *Ocean Development and International Law*, *36*(2), 119–134. doi:10.1080/00908320590943990

Kambey, A. D., Bataragoa, N. E., & Wantasen, A. S. (2017). Study Of Placement Of Artificial Reefs From "Bamboreef" Bamboo In Malalayang Waters Two, Malalayang District, Manado City. *Jurnal Ilmiah Platax*, *5*(1), 1–9. doi:10.35800/jip.5.1.2017.14967

Keller, K., Steffe, A. S., Lowry, M. B., Murphy, J. J., Smith, J. A., & Suthers, I. M. (2017). Estimating the recreational harvest of fish from a nearshore designed artificial reef using a pragmatic approach. *Fisheries Research*, *187*, 158–167. doi:10.1016/j.fishres.2016.11.022

Kendall, L. (2008). The conduct of qualitative interview: research questions, methodological issues and researching online. In J. Coiro, M. Knobel, C. Lankshea, & D. Leu (Eds.), *Handbook of Research on New Literacies* (pp. 133–149). Lawrence Erlbaum Associares. doi:10.4324/9781410618894.ch5

Kerlinger, F. N. (1997). Behavior Research. McGraw Hill.

Kheawwongjan, A., & Kim, D. S. (2012). Present status and prospects of artificial reefs in Thailand. *Ocean and Coastal Management*, *57*, 21–33. doi:10.1016/j.ocecoaman.2011.11.001

Kim, C. G., Lee, J. W., & Park, J. S. (1994). Artificial reef designs for Korean coastal waters. *Bulletin of Marine Science*, *55*, 858–866.

Kim, D., Jung, S., & Na, W. B. (2021). Evaluation of turbulence models for estimating the wake region of artificial reefs using particle image velocimetry and computational fluid dynamics. *Ocean Engineering*, *223*, 108673. doi:10.1016/j.oceaneng.2021.108673

Kim, D., Woo, J., Na, W. B., & Yoon, H. S. (2014). Flow and structural response characteristics of a box-type artificial reef. *Journal of Korean Society of Coastal and Ocean Engineers*, *26*(3), 113–119. doi:10.9765/KSCOE.2014.26.3.113

Kim, H. S., Kim, C. G., Na, W. B., & Kim, J. K. (2008). Chemical degradation characteristics of reinforced concrete reefs in South Korea. *Ocean Engineering*, *35*(8-9), 738–748. doi:10.1016/j.oceaneng.2008.02.003

Kim, N., Park, J., & Choi, J. J. (2017). Perceptual differences in core competencies between tourism industry practitioners and students using Analytic Hierarchy Process (AHP). *Journal of Hospitality, Leisure, Sport and Tourism Education*, *20*, 76–86. doi:10.1016/j.jhlste.2017.04.003

Kirkbride-Smith, A. E., Wheeler, P. M., & Johnson, M. L. (2013). The relationship between diver experience levels and perceptions of attractiveness of artificial reefs-examination of a potential management tool. *PLoS One*, *8*(7), e68899. doi:10.1371/journal.pone.0068899 PMID:23894372

Kirkbride-Smith, A. E., Wheeler, P. M., & Johnson, M. L. (2016). Artificial reefs and marine protected areas: A study in willingness to pay to access Folkestone Marine Reserve, Barbados, West Indies. *PeerJ*, *4*, e2175. doi:10.7717/peerj.2175 PMID:27547521

Klöpffer, W. (2003). Lifecycle based methods for sustainable product development. *The International Journal of Life Cycle Assessment*, *8*(3), 157–159. doi:10.1007/BF02978462

Knowlton, N., Brainard, R. E., Fisher, R., Moews, M., Plaisance, L., & Caley, M. J. (2010). Coral reef biodiversity. *Life in the world's oceans: Diversity distribution and abundance*, 65-74.

Kolian, S. R., Sammarco, P. W., & Porter, S. A. (2017). Abundance of corals on offshore oil and gas platforms in the Gulf of Mexico. *Environmental Management*, *60*(2), 357–366. doi:10.100700267-017-0862-z PMID:28526900

Koundouri, P., Giannouli, A., Airoldi, L., Bas, B., Broszeit, S., Elginoz, N., Giannakis, E., Zagonari, F., Krontira, Y., Moussoulides, A., Tsani, S., Troianos, D., Xepapadeas, P., Xepapadeas, A., & Zanuttigh, B. (2017). Socio-economic Analysis of a Selected Multi-use Offshore Site in the Mediterranean Sea. In *The Ocean of Tomorrow: Investment Assessment of Multi-Use Offshore Platforms: Methodology and Applications* (pp. 85–101). Springer International Publishing. doi:10.1007/978-3-319-55772-4_6

Kragt, M. E., Roebeling, P. C., & Ruijs, A. (2009). Effects of Great Barrier Reef degradation on recreational reef-trip demand: A contingent behaviour approach. *The Australian Journal of Agricultural and Resource Economics*, *53*(2), 213–229. doi:10.1111/j.1467-8489.2007.00444.x

Krause-Jensen, D., & Duarte, C. M. (2016). Substantial Role of Macroalgae in Marine Carbon Sequestration. *Nature Geoscience*, *9*(10), 737–742. doi:10.1038/ngeo2790

Krohling, W., & Zalmon, I. R. (2008). Epibenthic colonization on an artificial reef in a stressed environment off the north coast of the Rio de Janeiro State, Brazil. *Brazilian Archives of Biology and Technology*, *51*(1), 213–221. doi:10.1590/S1516-89132008000100025

Kusumawardhani, M., Kumar, R., & Tore, M. (2016). Asset integrity management: Offshore installations challenges. *Journal of Quality in Maintenance Engineering*, *22*(3), 238–251. doi:10.1108/JQME-06-2015-0023

Lakhal, S. Y., Khan, M. I., & Islam, M. R. (2009). An "Olympic" framework for a green decommissioning of an offshore oil platform. *Ocean and Coastal Management*, *52*(2), 113–123. doi:10.1016/j.ocecoaman.2008.10.007

Lam, C. M., Iris, K. M., Medel, F., Tsang, D. C., Hsu, S. C., & Poon, C. S. (2018). Life-cycle cost-benefit analysis on sustainable food waste management: The case of Hong Kong International Airport. *Journal of Cleaner Production*, *187*, 751–762. doi:10.1016/j.jclepro.2018.03.160

Lang, D. J. (2012). Transdisciplinary research in sustainability science: practice, principles, and challenges. *Sustainability Science*, *1*, 25-43.

Lange, C., Ratoi, L., & Co, D. L. (2020). *Reformative Coral Habitats-Rethinking Artificial Reef structures through a robotic 3D clay printing method.* Academic Press.

Langhamer, O. (2012). Artificial reef effect in relation to offshore renewable energy conversion: State of the art. *TheScientificWorldJournal*, *2012*. doi:10.1100/2012/386713 PMID:23326215

Lawal, D. U., Yusof, K. W., Hashim, M. A., & Balogun, A. L. (2014, June). Spatial analytic hierarchy process model for flood forecasting: An integrated approach. [). IOP Publishing.]. *IOP Conference Series. Earth and Environmental Science*, *20*(1), 012029.

Layman, C. A., & Allgeier, J. E. (2020). An ecosystem ecology perspective on artificial reef production. *Journal of Applied Ecology*, *57*(11), 2139–2148. doi:10.1111/1365-2664.13748

Le Diréach, L., Astruch, P., Bonhomme, D., Bonhomme, P., Fourt, M., Rouanet, E., & … . Ruitton, S. (2015). Comparative assessment of the deployment of 6 artificial reef types in Marseille Prado Bay (France) from a five-year seasonal survey of the fish fauna. *Proceedings of the RECIFS Conference on Artificial Reefs: from Materials to Ecosystems*, 27, 29.

Lee, M. O., Otake, S., & Kim, J. K. (2018). Transition of artificial reefs (ARs) research and its prospects. *Ocean and Coastal Management*, *154*(January), 55–65. doi:10.1016/j.ocecoaman.2018.01.010

Leeworthy, V. R., Maher, T., & Stone, E. A. (2006). Can artificial reefs alter user pressure on adjacent natural reefs? *Bulletin of Marine Science*, *78*(1), 29–38.

Leeworthy, V. R., Maher, T., & Stone, E. A. (2006). Can Artificial Reefs Reduce or Alter User Pressure on Adjacent Natural Reefs? *Bulletin of Marine Science*, *78*, 29–37.

Legorburu, I., Johnson, K. R., & Kerr, S. A. (2018). Multi-use maritime platforms-North Sea oil and offshore wind: Opportunity and risk. *Ocean and Coastal Management*, *160*, 75–85. doi:10.1016/j.ocecoaman.2018.03.044

Leitão, F. (2013). Artificial reefs: From ecological processes to fishing enhancement tools. *Brazilian Journal of Oceanography*, *61*(1), 77–81. doi:10.1590/S1679-87592013000100009

Leitao, F., Baptista, V., Teodósio, M. A., Hughes, S. J., Vieira, V., & Chícharo, L. (2016). The role of environmental and fisheries multi-controls in white seabream (Diplodus sargus) artisanal fisheries in Portuguese coast. *Regional Environmental Change*, *16*(1), 163–176. doi:10.100710113-014-0726-5

Leitao, F., Santos, M. N., Erzini, K., & Monteiro, C. C. (2008). Fish assemblages and rapid colonization after enlargement of an artificial reef off the Algarve coast (Southern Portugal). *Marine Ecology (Berlin)*, *29*(4), 435–448. doi:10.1111/j.1439-0485.2008.00253.x

Leitao, F., Santos, M. N., Erzini, K., & Monteiro, C. C. (2009). Diplodus spp. assemblages on artificial reefs: Importance for near shore fisheries. *Fisheries Management and Ecology*, *16*(2), 88–99. doi:10.1111/j.1365-2400.2008.00646.x

Leleu, K., Alban, F., Pelletier, D., Charbonnel, E., Letourneur, Y., & Boudouresque, C. F. (2012). Fishers' perceptions as indicators of the performance of Marine Protected Area (MPAs). *Marine Policy*, *36*(2), 36. doi:10.1016/j.marpol.2011.06.002

Lemke, L., & Olech, L. (2011). Dive tourism. In The Long Tail of Tourism: Holiday Niches and Their Impact on Mainstream Tourism (pp. 105-114). Gabler Research. doi:10.1007/978-3-8349-6231-7_12

Lew, A. A. (2013). A world geography of recreational scuba diving. In G. Musa & K. Dimmock (Eds.), *Scuba Diving Tourism* (pp. 29–51). Routledge.

Lima, J. S., Zappes, C. A., Di Beneditto, A. P. M., & Zalmon, I. R. (2018). Artisanal fisheries and artificial reefs on the southeast coast of Brazil: Contributions to research and management. *Ocean & Coastal Management, 163*(July), 372–382. doi:10.1016/j.ocecoaman.2018.07.018

Lima, J. S., Atalah, J., Sanchez-Jerez, P., & Zalmon, I. R. (2020). Evaluating the performance and management of artificial reefs using artificial reef multimetric index (ARMI). *Ocean and Coastal Management*, *198*, 105350. Advance online publication. doi:10.1016/j.ocecoaman.2020.105350

Lima, J. S., Costa, I. D., & Zalmon, I. R. (2020). Length-weight relationship of fish species captured around an artificial offshore reef (northern Rio de Janeiro, Brazil). *Journal of Applied Ichthyology*, *37*(2), 337–341. doi:10.1111/jai.14109

Lima, J. S., Sanchez-Jerez, P., Santos, L. N., & Zalmon, I. R. (2020). Could artificial reefs increase access to estuarine fishery resources? Insights from a long-term assessment. *Estuarine, Coastal and Shelf Science*, *242*(May), 106858. Advance online publication. doi:10.1016/j.ecss.2020.106858

Lima, J. S., & Zalmon, I. R. (2020). Trends and Obstacles in Artificial Reef Research. In S. A. Bortone & S. Otake (Eds.), *Modern Fisheries Engineering - Realizing a Healthy and Sustainable Marine Ecosystem* (pp. 11–30). CRC Press. doi:10.1201/9780429328039-3

Lima, J. S., Zalmon, I. R., & Love, M. (2019). Overview and trends of ecological and socioeconomic research on artificial reefs. *Marine Environmental Research*, *145*, 81–96. doi:10.1016/j.marenvres.2019.01.010 PMID:30837123

Lima, J. S., Zappes, C. A., Di Beneditto, A. P. M., & Zalmon, I. R. (2019). Ethnoecology and socioeconomic around an artificial reef: The case of artisanal fisheries from southeastern Brazil. *Biota Neotropica*, *19*(2), e20180620. Advance online publication. doi:10.1590/1676-0611-bn-2018-0620

Lindberg, W. J. (1997). Can Science Resolve the Attraction- Production Issue? *Fisheries (Bethesda, Md.), 22*(4), 10–16. doi:10.1577/1548-8446-22-4

Lindberg, W. J., & Seaman, W. (2011). *Guidelines and Management Practices for Artificial Reef Siting, Use, Construction, and Anchoring in Southeast Florida*. Academic Press.

Lino, P. G., Bentes, L., Abecasis, D., dos Santos, M. N., & Erzini, K. (2009). Comparative behavior of wild and hatchery reared white sea bream (Diplodus sargus) released on artificial reefs off the Algarve (Southern Portugal). In Tagging and tracking of marine animals with electronic devices (pp. 23-34). Springer.

Lipcius, R., & Burke, R. P. (2006). *Abundance, biomass and size structure of eastern oyster and hooked mussel on a modular artificial reef in the Rappahannock River*. Chesapeake Bay.

Liu, Y., Zhao, Y., Dong, G., Guan, C., Cui, Y., & Xu, T.-J. (2013). A study of the flow field characteristics around star-shaped artificial reefs. *Journal of Fluids and Structures, 39*, 27–40. doi:10.1016/j.jfluidstructs.2013.02.018

Lizama, P., & Boccardo, G. (2014). *Variable Association Guide* (Pearson and Spearman in SPSS). University of Chile Faculty of Social Sciences (FACSO) Department of Sociology.

Loia, F., Basile, V., Capobianco, N., & Vona, R. (2021a). Value Co-Creation Practices in the Decommissioning of Offshore Platforms: A Case Study Approach. *Journal of Creating Value, 7*(2), 238–251. doi:10.1177/23949643211032040

Loia, F., Capobianco, N., & Vona, R. (2021b). Towards a resilient perspective for the future of offshore platforms. Insights from a data-driven approach, *Transforming Government: People. Process and Policy, 15*(3). Advance online publication. doi:10.1108/TG-04-2021-0067

Lök, A. (1995). *Yapay resiflerin uygulanabilirliği üzerine bir araştırma* [A study on the feasibility of artificial reefs] [Ph.D. dissertation]. Ege University, Science Institute, İzmir, Turkey.

Lök, A. (1997). Last developments on production versus attraction hypothesis in artificial reefs. *Ege Journal of Fisheries and Aquatic Sciences, 14*(1-2), 239-242.

Lök, A. (2012). Artificial reef applications in Turkey. In *GFCM Scientific Advisory Committee (SAC) 12th Session of the Sub-Committee on Marine Environment and Ecosystems (SCMEE)*, 19.

Lök, A., & Tokaç, A. (2000). Turkey: A new region for artificial habitats. In A.C. Jensen, K. J. Collins & A. P. M. Lockwood (Ed.), Artificial Reefs in European Seas (pp. 21-30). Kluwer Academic Publishers.

Lök, A., & Tokaç, A. (2000). Turkey: A new region for artificial habitats. In Artificial Reefs in European Seas. Kluwer Academic Publications.

Lök, A., Düzbastilar, F. O., Gül, B., Özgül, A., & Ulaş, A. (2011). The role of artificial reefs in fisheries management in Turkey. In S. A. Bortone, F. P. Brandini, G. Fabi, & S. Otake (Eds.), *Artificial Reefs in Fisheries Management* (pp. 155–166). CRC Press Taylor & Francis Group.

Lök, A., & Gül, B. (2005). Evaluation of fish fauna associated with experimental artificial reefs in Hekim Island in Izmir Bay (Aegean Sea, Turkey). *Ege Journal of Fishery and Aquatic Sciences, 22*(1-2), 109–114.

Lök, A., Metin, C., Duzbastılar, F. O., Ulaş, A., & Özgül, A. (2013). *Fish Community Structure and Monitoring of Physico-Chemical Parameters in Edremit Bay Artificial Reef Area. Ege University Underwater Research and Application Center Research Report for Turkish Government*. Ministry of Food, Agriculture and Livestock.

Lök, A., Metin, C., Ulaş, A., Düzbastılar, F. O., & Tokaç, A. (2002). Artificial reefs in Turkey. *ICES Journal of Marine Science, 59*, 192–195. doi:10.1006/jmsc.2002.1221

Lök, A., & Özgül, A. (2010). Fish assemblages around fish aggregation devices in Aegean Sea coast of Turkey. *Rapports et Procès Verbaux des Réunions - Commission Internationale pour l'Exploration Scientifique de la Mer Méditerranée, 39*, 573.

London Convention and Protocol/UNEP. (2009). *London Convention and Protocol/UNEP Guidelines for the Placement of Artificial Reefs*. Author.

Loske, A. (2013). Fundamentals of SCUBA-diving physics. *International Journal of Sports Science, 3*(2), 37–45.

Love, M. S., Brothers, E., Schroeder, D. M., & Lenarz, W. H. (2007). Ecological performance of young-of-the-year blue rockfish (*Sebastes mystinus*) associated with oil platforms and natural reefs in California as measured by daily growth rates. *Bulletin of Marine Science, 80*(1), 147–157.

Love, M. S., Schroeder, D. M., Lenarz, W., MacCall, A., Bull, A. S., & Thorsteinson, L. (2006). Potential use of offshore marine structures in rebuilding an overfished rockfish species, bocaccio (*Sebastes paucispinis*). *Fish Bulletin, 104*, 383–390.

Ly, O., Yoris-Nobile, A. I., Sebaibi, N., Blanco-Fernandez, E., Boutouil, M., Castro-Fresno, D., Hall, A. E., Herbert, R. J. H., Deboucha, W., Reis, B., Franco, J. N., Teresa Borges, M., Sousa-Pinto, I., van der Linden, P., & Stafford, R. (2021). Optimisation of 3D printed concrete for artificial reefs: Biofouling and mechanical analysis. *Construction & Building Materials, 272*, 121649. doi:10.1016/j.conbuildmat.2020.121649

MacCarthy, M., O'neill, M., & Williams, P. (2006). Customer satisfaction and scuba-diving: Some insights from the deep. *Service Industries Journal, 26*(5), 537–555. doi:10.1080/02642060600722841

MacDonald, J. M. (1994). Artificial reef debate: Habitat enhancement or waste disposal? *Ocean Development and International Law, 25*(1), 87–118. doi:10.1080/00908329409546027

Macreadie, P. I., Allen, K., Kelaher, B. P., Ralph, P. J., & Skilbeck, C. G. (2012). Paleoreconstruction of estuarine sediments reveal human-induced weakening of coastal carbon sinks. *Global Change Biology, 18*(3), 891–901. doi:10.1111/j.1365-2486.2011.02582.x

Macreadie, P. I., Fowler, A. M., & Booth, D. J. (2011). Rigs-to-reefs: Will the deep-sea benefit from artificial habitat? *Frontiers in Ecology and the Environment, 9*(8), 455–461. doi:10.1890/100112

Macusi, E. D., Katikiro, R. E., & Babaran, R. P. (2017). The influence of economic factors in the change of fishing strategies of anchored FAD fishers in the face of declining catch, General Santos City, Philippines. *Marine Policy, 78*(October), 98–106. doi:10.1016/j.marpol.2017.01.016

Macusi, E. D., Abreo, N. A. S., & Babaran, R. P. (2017). Local Ecological Knowledge (LEK) on Fish Behavior Around Anchored FADs: The Case of Tuna Purse Seine and Ringnet Fishers from Southern Philippines. *Frontiers in Marine Science, 4*(June), 1–13. doi:10.3389/fmars.2017.00188

Maddupa, H. H., Subhanln, B., Bachtiar, R., Ismetl, M. S., Budikartinis, Y., & Bria, D. (2007). Prospects of Biorock Artificial Reefs in Increasing Fish Resources in the Thousand Islands. *Coral Reef National Conference I*, 68–75.

Makatita, J. S., & Mangimbulude, J. (2014). Study of Phosphate and Nitrate Nutrients in Water and Sediment Seagrass Padang Tujuh Seram Island North West Central Maluku. *FMIPA-UT National Seminar 2014, 23*, 54–66. http://repository.ut.ac.id/4959/1/fmipa2014_09.pdf

Malcolm, C. D., Olivas, M. L. B., & Dagostino, R. M. C. (2021). Reported capture, fishery perceptions, and attitudes toward fisheries management of urban and rural artisanal, small-scale fishers along the Bahía de Banderas coast, Mexico. *Environmental Challenges, 4*(April), 100110. doi:10.1016/j.envc.2021.100110

Maleki, H., & Zahir, S. (2013). A comprehensive literature review of the rank reversal phenomenon in the analytic hierarchy process. *Journal of Multi-Criteria Decision Analysis, 20*(3-4), 141–155. doi:10.1002/mcda.1479

Mancero, X. (2001). *The measurement of human development: Elements of a debate*. ECLAC.

Marcus, A., Ikeda, M., & Jones, E. (2016). Buoyant Ecologies. In *ACADIA: Post-Human Frontiers*. Michigan Taubman College of Architecture & Urban Planning. http://acadia.org/projects/QEPKQ4

Margheritini, L., Colaleo, G., Contestabile, P., Bjørgård, T. L., Simonsen, M. E., Lanfredi, C., Dell'Anno, A., & Vicinanza, D. (2020). Development of an eco-sustainable solution for the second life of decommissioned oil and gas platforms: The mineral accretion technology. *Sustainability, 12*(9), 3742. doi:10.3390u12093742

Marin, G. (2004). The end of all the earth. College of Mexico.

Marli, Y. S. (2019). *Development of Marine Tourism Potential in Trenggalek*. https://www.kompasiana.com/yulistianas holiqhahm5879/5db6afc4097f367da76a4812/pengembangan-potensi-wisata-bahari-di-trenggalek

Martens, P. (2006). Sustainability: Science or fiction? *Sustainability Science Practice Policy, 2*(2), 36–41. doi:10.1080 /15487733.2006.11907976

Martin, K. S. (2004). GIS in marine fisheries science and decision-making. *Geographic information systems in fisheries*, 237-258.

Martins, I. D., Moraes, F. F., Távora, G., Soares, H. L. F., Infante, C. E., Arruda, E. F., Bahiense, L., Caprace, J., & Lourenço, M. I. (2020). A review of the multicriteria decision analysis applied to oil and gas decommissioning problems. *Ocean and Coastal Management, 184*, 105000. doi:10.1016/j.ocecoaman.2019.105000

Masera, O., & López-Ridaura, S. (2000). *Sustainability and peasant systems. Five evaluation experiences in rural Mexico*. MundiPrensa, GIRA, UNAM.

Mason, M. (2010). Sample size and saturation in PhD studies using qualitative interviews. Forum Qualitative Sozial for Schung/Forum Qualitative Social Research, 11.

McGinnis, M., Fernandez, L., & Pomeroy, C. (2001). *The Politics, Economics, and Ecology of Decommissioning Offshore Oil and Gas Structures*. DOI Minerals Management Service Document, MMS Publication 2001-006.

McGlennon, D., & Branden, K. L. (1994). Comparison of Catch and Recreational Anglers Fishing on Artificial Reefs and Natural Seabed in Gulf St. Vincent, South Australia. *Bulletin of Marine Science, 55*(2–3), 510–523.

McGurrin, J. M., Stone, R. B., & Sousa, R. J. (1989). Profiling United States artificial reef development. *Bulletin of Marine Science, 44*(2), 1004–1013.

McKean, J. R., & Taylor, R. (2000). *Outdoor Recreation Use and Value: Snake River Basin of Central Idaho. Idaho*. Experiment Station Bulletin.

McLean, D. L., Partridge, J. C., Bond, T., Birt, M. J., Bornt, K. R., & Langlois, T. J. (2017). Using industry ROV videos to assess fish associations with subsea pipelines. *Continental Shelf Research, 141*, 76–97. doi:10.1016/j.csr.2017.05.006

Mendonça, A., Fortes, C. J., Capitão, R., da Graça Neves, M., Moura, T., & Do Carmo, J. S. A. (2012). Wave hydrodynamics around a multi-functional artificial reef at Leirosa. *Journal of Coastal Conservation, 16*(4), 543–553. doi:10.100711852-012-0196-1

Migné, A., Golléty, C., & Davoult, D. (2015). Effect of Canopy Removal on A Rocky Shore Community Metabolism and Structure. *Marine Biology, 162*(2), 449–457. doi:10.100700227-014-2592-6

Miller, R. J., & Etter, R. J. (2008). Shading Facilitates Sessile Invertebrate Dominance in the Rocky Subtidal Gulf of Maine. *Ecology*, *89*(2), 452–462. doi:10.1890/06-1099.1 PMID:18409434

Milon, J. W., Holland, S. M., & Whitmarsh, D. J. (2000). Social and economic evaluation methods. *Artificial reef evaluation: with application to natural marine habitats*, 165-194.

Milon, J. W. (1988a). *The Economic Benefits of Artificial Reefs: An Analysis of the Dade County, Florida Reef System. Sea Grant Extension Program, University of Florida. SGR-90, Florida Sea Grant College Program.* University of Florida.

Milon, J. W. (1988b). A Nested Demand Shares Model of Artificial Marine Habitat Choice By Sport Anglers. *Marine Resource Economics*, *5*(3), 191–213. doi:10.1086/mre.5.3.42628883

Milon, J. W. (1989a). Artificial Marine Habitat Characteristics and Participation Behavior by Sport Anglers and Divers. *Bulletin of Marine Science*, *44*, 853–862.

Milon, J. W. (1989b). Contingent Valuation Experiments for Strategic Behavior. *Journal of Environmental Economics and Management*, *17*(3), 293–308. doi:10.1016/0095-0696(89)90022-3

Milon, J. W., Holland, S. M., & Whitmarsh, D. J. (2000). *Artificial Reef Evaluation with Application to Natural Marine Habitats*. Social and Economic Evaluation Methods.

Ministry of Environment. (2014). *Quality sea water for marine life standarts. Decree of the State Minister of the Environment No. 51 of 2004 concerning Seawater Quality Standards.* Author.

Moira, V. S., Luthf, I. O. M., & Isdianto, A. (2020). Analysis of Relationship between Chemical Oceanography Conditions and Coral Reef Ecosystems in Damas Waters, Trenggalek, East Java. *Journal of Marine and Coastal Science*, *9*(3), 113–126. doi:10.20473/jmcs.v9i3.22294

Mondragon, B. (2014). Use of the Spearman correlation in an intervention study in physiotherapy. *Mov.cient.*, *8*(1), 98-104.

Monteiro, C. C., & Santos, M. N. (2000). Portuguese artificial reefs. In *Artificial Reefs in European Seas* (pp. 249–261). Springer. doi:10.1007/978-94-011-4215-1_15

Monteiro, P. V. (2016). The Role of knowledge-intensive service activities on inducing innovation in co-opetition strategies: Lessons from the maritime cluster of the Algarve region. *International Journal of Management and Enterprise Development*, *15*(1), 78–95. doi:10.1504/IJMED.2016.075876

Montoya, D. (2015). *Pollution in Sydney Harbour: Sewage, Toxic Chemicals and MicroPlastics*. NSW Parliamentary Research Service.

Moral, O., Teresa, M., Ferra, P., & Miguel, Q. L. (2013). Elaboration and validation of a scale of beliefs of students in compulsory secondary education regarding the environment. *Faculty Journal of Curriculum and Teacher Training, 17.*

Moreno, G., Dagorn, L., Sancho, G., García, D., & Itano, D. (2007). Using local ecological knowledge (LEK) to provide insight on the tuna purse seine fleets of the Indian Ocean useful for management. *Aquatic Living Resources*, *20*(4), 367–376. doi:10.1051/alr:2008014

Morgana, O. A., Massey, D. M., Huth, W. L., & Hall, R. (2009). Demand for Diving on Large Ship Artificial Reefs. *Marine Resource Economics*, *24*(1), 43–59. doi:10.1086/mre.24.1.42629644

Mos, B., Dworjanyn, S. A., Mamo, L. T., & Kelaher, B. P. (2019). Building Global Change Resilience: Concrete Has the Potential to Ameliorate the Negative Effects of Climate-Driven Ocean Change on A Newly Settled Calcifying Invertebrate. *The Science of the Total Environment*, *646*, 1349–1358. doi:10.1016/j.scitotenv.2018.07.379 PMID:30235620

Mota, L., & Frausto, O. (2014). The Use of Scuba Diving Tourism for Marine Protected Area Management. *International Journal of Social, Behavioural, Educational, Economic, Business and Industrial Engineering*, *8*(10), 3159–3164.

Moura, A., Boaventura, D., Curdia, J., Carvalho, S., Da Fonseca, L. C., Leitão, F. M., & Monteiro, C. C. (2007). Effect of depth and reef structure on early macrobenthic communities of the Algarve artificial reefs (southern Portugal). In *Biodiversity in Enclosed Seas and Artificial Marine Habitats* (pp. 173–180). Springer. doi:10.1007/978-1-4020-6156-1_15

Moura, A., Boaventura, D., Cúrdia, J., Santos, M. N., & Monteiro, C. C. (2006). Biomass production of early macrobenthic communities at the Faro/Ancão artificial reef (Portugal): Effect of depth and reef layer. *Bulletin of Marine Science*, *78*(1), 83–92.

Moura, A., Da Fonseca, L. C., Cúrdia, J., Carvalho, S., Boaventura, D., Cerqueira, M., Leitão, F., Santos, M. N., & Monteiro, C. C. (2008). Is surface orientation a determinant for colonisation patterns of vagile and sessile macrobenthos on artificial reefs? *Biofouling*, *24*(5), 381–391. doi:10.1080/08927010802256414 PMID:18612908

Moura, A., Fonseca, L. C. D., Boaventura, D., Santos, M. N., & Monteiro, C. C. (2011). Estimation of secondary production of the Faro/Ancão artificial reefs. *Brazilian Journal of Oceanography*, *59*(spe1), 91–94. doi:10.1590/S1679-87592011000500011

Murillas-Maza, A., Moreno, G., & Murua, J. (2013). A socio-economic sustainability indicator for the basque tropical tuna purse-seine fleet with a FAD fishing strategy. *Economía Agraria y Recursos Naturales*, *13*(2), 5–31. doi:10.7201/earn.2013.02.01

Murillo, V. C., Silva, C. G., & Fernandez, G. B. (2009). Nearshore Sediments and Coastal Evolution of Paraíba do Sul River Delta, Rio de Janeiro, Brazil. *Journal of Coastal Research*, *56*, 650–654.

Murray, J. D., & Betz, C. J. (1994). User views of artificial reef management in the southeastern US. *Bulletin of Marine Science*, *55*(2–3), 970–981.

Musa, G., & Dimmock, K. (2013). Introduction: scuba diving tourism. In G. Musa & K. Dimmock (Eds.), *Scuba Diving Tourism* (pp. 1–13). Routledge.

Musiello-Fernandes, J., Zappes, C. A., & Hostim-Silva, M. (2017). Small-scale shrimp fisheries on the Brazilian coast: Stakeholders perceptions of the closed season and integrated management. *Ocean and Coastal Management*, *148*, 89–96. doi:10.1016/j.ocecoaman.2017.07.018

Mustafa, S., Estim, A., Tuzan, A. D., Ann, C. C., Seng, L. L., & Shaleh, S. R. M. (2019). Nature-based and technology-based solutions for sustainable blue growth and climate change mitigation in marine biodiversity hotspots. *Environmental Biotechnology*, *15*(1), 15. doi:10.14799/ebms302

Nakamura, M., & Otake, S. (2015). Enhancement of marine habitats. In International Compendium of Coastal Engineering (pp. 375-396). World Scientific Publishing Co. Pte. Ltd. doi:10.1142/9789814449434_0011

Nakamura, M. (1985). Evolution of artificial fishing reef concepts in Japan. *Bulletin of Marine Science*, *37*(1), 271–278.

Nassar, W. M., Anaya-Lara, O., Ahmed, K. H., Campos-Gaona, D., & Elgenedy, M. (2020). Assessment of multi-use offshore platforms: Structure classification and design challenges. *Sustainability*, *12*(5), 1860. doi:10.3390u12051860

National Oceanic and Atmospheric Administration (NOAA). (2019). *Tourism and recreation*. Retrieved from https://coast.noaa.gov/states/fast-facts/tourism-and-recreation.html

Nautilus Consultants. (2003). *Artificial Reefs, Scotland, Benefit costs and risks*. Highlands & Islands Enterprise. http://www.nautilus-consultants.co.uk/sites/default/files/Nautilus%20artificial%20reefs%20report_0.pdf

Na, W. B. (2021). Structural Complexity of Artificial Reefs. *Computational Structural Engineering, 34*(3), 19–26.

Newman, J. (2011). *Scuba Diving and Snorkeling for Dummies.* John Wiley & Sons.

Ng, K., Phillips, M. R., Calado, H., Borges, P., & Veloso-Gomes, F. (2013). Seeking harmony in coastal development for small islands: Exploring multifunctional artificial reefs for São Miguel Island, the Azores. *Applied Geography (Sevenoaks, England), 44*, 99–111. doi:10.1016/j.apgeog.2013.07.013

NHRCP. (n.d.). *Artificial Reefs: What works and what doesn't.* https://newheavenreefconservation.org/marine-blog/147-artificial-reefs-what-works-and-what-doesn-t

NOAA. (n.d.). *Sea Surface Salinity.* NOAA. Retrieved October 10, 2021, from https://coastwatch.noaa.gov/cw/satellite-data-products/sea-surface-salinity.1.html

Nunes, C. H., & Roeder, F. (2014). Tourism and Algarve's underwater heritage valorisation. A case study. *Journal for Maritime Research, 11*(1), 35–44.

Oh, C. O., Ditton, R. B., & Stoll, J. R. (2008). The economic value of scuba-diving use of natural and artificial reef habitats. *Society & Natural Resources, 21*(6), 455–468. doi:10.1080/08941920701681953

Oliveira, M. T., Ramos, J., & Santos, M. N. (2015). An approach to the economic value of diving sites: Artificial versus natural reefs off Sal Island, Cape Verde. *Journal of Applied Ichthyology, 31*, 86–95. doi:10.1111/jai.12953

Oliveira, P. C., Di Beneditto, A. P. M., Bulhões, E. M. R., & Zappes, C. A. (2016). Artisanal fishery versus port activity in southern Brazil. *Ocean and Coastal Management, 129*, 49–57. doi:10.1016/j.ocecoaman.2016.05.005

Onyango, H. O., Ochiewo, J. O., & Karani, N. J. (2021). Socio-economic prospects and problems in under-exploited offshore marine fisheries: The case of Fish Aggregating Devices (FADs) in Kenya coastal fisheries. *Regional Studies in Marine Science, 44*, 101706. doi:10.1016/j.rsma.2021.101706

OSPAR Commission. (2009). Assessment of construction or placement of artificial reefs. London: Biodiversity Series, publ. no. 438/2009.

OSPAR Commission. (2009). *Assessment of construction or placement of artificial reefs.* OSPAR, Biodiversity Series. https://www.ospar.org/documents?v=7143

Ovalle, A. R. C., Silva, C. F., Rezende, C. E., Gatts, C. E. N., Suzuki, M. S., & Figueiredo, R. O. (2013). Long-term trends in hydrochemistry in the Paraíba do Sul River, southeastern Brazil. *Journal of Hydrology (Amsterdam), 481*(1), 191–203. doi:10.1016/j.jhydrol.2012.12.036

Özgül, A., Lök, A., Tanrıkul, T. T., & Alos, J. (2019). Home range and residency of Scorpaena porcus and Scorpaena scrofa in artificial reefs revealed by fine-scale acoustic tracking. *Fisheries Research, 210*, 22–30. doi:10.1016/j.fishres.2018.10.008

Padang, B. P. S. P. L. (2021). *Conservation Area.* https://kkp.go.id/djprl/bpsplpadang/page/262-konservasi-kawasan

PADI Worldwide Certification. (2021). *Statistics.* https://www.padi.com/corporate/company-info

Palau-Saumell, R., Forgas-Coll, S., Sánchez-García, J., & Prats, L. (2019). Motivation and attachment to a diving destination: The case of Medes Islands (Catalonia, Spain). *Journal of Vacation Marketing, 25*(3), 301–319. doi:10.1177/1356766718778867

Paredes, M. G., Padilla-Rivera, A., & Güereca, L. P. (2019). Life cycle assessment of ocean energy technologies: A systematic review. *Journal of Marine Science and Engineering, 7*(9), 322. doi:10.3390/jmse7090322

Parente, V., Ferreira, D., dos Santos, E. M., & Luczynski, E. (2006). Offshore decommissioning issues: Deductibility and transferability. *Energy Policy, 34*(15), 1992–2001. doi:10.1016/j.enpol.2005.02.008

Parsons, G. R., & Thur, S. M. (2008). Valuing Changes in the Quality of Coral Reef Ecosystems: A Stated Preference Study of SCUBA Diving in the Bonaire National Marine Park. *Environmental and Resource Economics, 40*(4), 593–608. doi:10.100710640-007-9171-y

Pascaline, B., Catherine, S., Charbonnel, E., & Patrice, F. (2011). Monitoring of the artificial reef fish assemblages of golfe juan marine protected area (France, North-Western Mediterranean). *Brazilian Journal of Oceanography, 59*(SPEC. ISSUE 1), 167–176. doi:10.1590/S1679-87592011000500018

Patty, S. I., Arfah, H., & Abdul, M. S. (2015). Nutrients (Phosphate, Nitrate), Dissolved Oxygen and pH Relation to Fertility in Jikumerasa Waters, Buru Island. *Journal of Coastal and Tropical Oceans, 3*(1), 43. doi:10.35800/jplt.3.1.2015.9578

Pears, R. J., & Williams, D. M. (2005). *Potential effects of artificial reefs on the Great Barrier Reef: Background paper.* CRC Reef Research Centre Technical Report No. 60. CRC Reef Research Centre.

Pears, R. J., & Williams, D. M. (2005). *Potential effects of artificial reefs on the Great Barrier Reef: background paper.* CRC Reef Research Centre.

Pears, R., & Williams, D. McB. (2005). *Potential effects of artificial reefs on the GREAT BARRIER reef. Cooperative Research Centre for the great barrier Reef World Heritage Area technical report 60.* James Cook University.

Pendleton, L. H. (2004). Creating Underwater Value: The Economic Value of Artificial Reefs for Recreational Diving, California Artificial Reef Enhancement Program. The San Diego Oceans Foundation.

Pendleton, L. H. (2005). Understanding the potential economic impacts of sinking ships for scuba recreation. *Marine Technology Society Journal, 39*(2), 47–52. doi:10.4031/002533205787443944

Perkol-Finkel, S., Shashar, N., & Benayahu, Y. (2006). Can artificial reefs mimic natural reef communities? The roles of structural features and age. *Marine Environmental Research, 61*(2), 121–135. doi:10.1016/j.marenvres.2005.08.001 PMID:16198411

Pickering, H., & Whitmarsh, D. (1997). Artificial reefs and fisheries exploitation: A review of the 'attraction versus production' debate, the influence of design and its significance for policy. *Fisheries Research, 31*(1-2), 39–59. doi:10.1016/S0165-7836(97)00019-2

Pickering, H., Whitmarsh, D., & Jensen, A. (1998). Artificial reefs as a tool to aid rehabilitation of coastal ecosystems: Investigating the potential. *Marine Pollution Bulletin, 37*(8–12), 505–514.

Pinto, M. F., Mourão, J. S., & Alves, R. R. N. (2015). Use of ichthyofauna by artisanal fishermen at two protected areas along the coast of Northeast Brazil. *Journal of Ethnobiology and Ethnomedicine, 11*(1), 20. doi:10.118613002-015-0007-5 PMID:25888941

Pioch, S., de Monbrison, D., & Simard, F. (2020). Artificial Reefs in France: Current State-of-the-Art and Recent Innovative Projects. In Modern Fisheries Engineering (pp. 31-40). CRC Press.

Pioch, S., Monbrison, D., & Simard, F. (2020). Artificial Reefs in France: Current State-of-the-Art and Recent Innovative Projects. In S. A. Bortone & S. Otake (Eds.), *Modern Fisheries Engineering* (1st ed., pp. 31–40). CRC Press. doi:10.1201/9780429328039-4

Polak, O., & Shashar, N. (2012). Can a small artificial reef reduce diving pressure from a natural coral reef? Lessons learned from Eilat, Red Sea. *Ocean and Coastal Management, 55*, 94–100. doi:10.1016/j.ocecoaman.2011.10.006

Polovina, J. J. (1991). *Fisheries applications and biological impacts of artificial habitats*. Academic Press.

Polovina, J. J., & Sakai, I. (1989). Impacts of artificial reefs on fishery production in Shimamaki, Japan. *Bulletin of Marine Science*, *44*(2), 997–1003.

Poore, A. G., & Lowry, J. K. (1997). New Ampithoid Amphipods from Port Jackson, New South Wales, Australia (Crustacea: Amphipoda: Ampithoidae). *Invertebrate Systematics*, *11*(6), 897–941. doi:10.1071/IT95045

Pordata. (n.d.). *PORDATA: Base de dados Portugal contemporâneo*. Available online from: pordata.pt

Pradella, N., Fowler, A. M., Booth, D. J., & Macreadie, P. I. (2014). Fish assemblages associated with oil industry structures on the continental shelf of north-western Australia. *Journal of Fish Biology*, *84*(1), 247–255. doi:10.1111/jfb.12274 PMID:24344929

Pratiwi, R. A., Widayati, W., & Astuti, P. (2014). Evaluation of Coral Reef Conservation Policy at Polewali Mandar 2012. *Journal of Politic and Government Studies*, *3*(4), 376–385.

Prescott-Allen, R. (1997). *Sustainability barometer. Measurement and communication of well-being and sustainable development* (No. 333.716 P929E). IUCN.

Przedrzymirska, J., & Zaucha, J. (2018). *Multi-use concept in European Sea Basins*. MUSES Project. https://muses-project.com/wp-content/uploads/sites/70/2018/06/D2-6-Final-Report.pdf

Purnawati, B. I., & Hartati, S. T. (2007). Condition of Artificial Reefs in the waters of Pramuka Island and Semak Daun Island, Thousand Islands. *Proceedings of the VI National Fish Seminar*, 91–96.

Queensland Environment Protection Agency (QEPA). (2005). *Sink the Brisbane. Environmental Protection Agency/ Queensland Parks and Wildlife Service*. http://www.epa.qld.gov.au/about_the_epa/coming_events/sink_the_brisbaneS

Quinn, T. P. (2009). *The influence of artificial reef associated fish assemblages and varying substrates on coral recruitment*. Academic Press.

Radonski, G. C., Martin, R. G., & DuBose, W. P. (1985). Artificial reefs: the sportfishing perspective. In F. M. D'Itri (Ed.), *Artificial reefs: marine and freshwater applications* (pp. 529–536). Lewis Publishers, Inc.

Rael, R., San Fratello, V., Arja, L., & Curth, S. (2018). *Coral Reef Seeding Units*. http://www.emergingobjects.com/project/coral-reef-seeding-units/

Rahayu, S. P. P., Anantanyu, S., & Mulyanto, M. (2019). Participation of Tourism Awareness Groups in Tourism Development in Trenggalek Regency. *Proceedings of the Industrial Research Workshop and National Seminar*, *10*(1), 1348–1356. https://jurnal.polban.ac.id/ojs-3.1.2/proceeding/article/view/1421

Rahim, F. (2012). *Tourism Awareness Group Guidelines*. Director General of Tourism Destination Development, Ministry of Tourism and Creative Economy.

Rahman, M. A. A., Nazri, M. N., Fitriadhy, A., Ahmad, M. F., Kasiman, E. H., Musa, M. A., ... Mohd, M. H. (2020). A Fundamental CFD Investigation of Offshore Structures for Artificial Coral Reef Development. *CFD Letters*, *12*(7), 110–125. doi:10.37934/cfdl.12.7.110125

Ramires, M., Clauzet, M., Barrella, W., Rotundo, M. M., Silvano, R. A., & Begossi, A. (2015). Fishers' knowledge about fish trophic interactions in the southeastern Brazilian coast. *Journal of Ethnobiology and Ethnomedicine*, *11*(1), 19. doi:10.118613002-015-0012-8 PMID:25889226

Ramires, M., Clauzet, M., Rotundo, M. M., & Begossi, A. (2012). Artisanal Fishing and Fishermen of Ilhabela - Sao Paulo State - Brazil. *Boletim do Instituto de Pesca*, *38*(3), 231–246.

Ramm, L. A., Florisson, J. H., Watts, S. L., Becker, A., & Tweedley, J. R. (2021). Artificial reefs in the Anthropocene: A review of geographical and historical trends in their design, purpose, and monitoring. *Bulletin of Marine Science, 97*(4), 699–728. doi:10.5343/bms.2020.0046

Ramos, J. H. P. (2007). *Socio-economic implications of artificial reef deployment: a Portuguese case study* (Doctoral dissertation). University of Portsmouth.

Ramos, J., Oliveira, M. T., & Santos, M. N. (2011). Stakeholder perceptions of decision-making process on marine biodiversity conservation on Sal Island (Cape Verde). *Brazilian Journal of Oceanography, 59*(1), 95–105.

Ramos, J., Santos, M. N., Whitmarsh, D., & Monteiro, C. C. (2011). Stakeholder analysis in the Portuguese artificial reef context: winners and losers. *Brazilian Journal of Oceanography, 59*(SPE1), 133-143.

Ramos, J., Santos, M. N., Whitmarsh, D., & Monteiro, C. C. (2011). Stakeholder analysis in the Portuguese artificial reef context: Winners and losers. *Brazilian Journal of Oceanography, 59*(SPEC. ISSUE 1), 133–143. doi:10.1590/S1679-87592011000500015

Ramos, J., Santos, M.N., Whitmarsh, D., & Monteiro, C.C. (2011). Stakeholder analysis in the Portuguese Artificial Reef Context: Winners and Losers. *Brazilian Journal of Oceanography, 59*, 133-143.

Ramos-Esplá, A. A., Guillén, J. E., Bayle, J. T., & Sánchez-Jérez, P. (2000). Artificial anti-trawling reefs off Alicante, South-Eastern Iberian Peninsula: Evolution of reef block and set designs. In *Artificial reefs in European seas* (pp. 195–218). Springer. doi:10.1007/978-94-011-4215-1_12

Ramos, J. (2021). Online Perception of Artificial Reef Risk and Safety by Stakeholders (Including Residents and Tourists) Via the Analytic Hierarchy Process. *Journal of Spatial and Organizational Dynamics, 9*(3), 189–204.

Ramos, J., Caetano, M., Himes-Cornell, A., & dos Santos, M. N. (2017). Stakeholders' conceptualization of offshore aquaculture and small-scale fisheries interactions using a Bayesian approach. *Ocean and Coastal Management, 138*, 70–82. doi:10.1016/j.ocecoaman.2017.01.008

Ramos, J., Lino, P. G., Himes-Cornell, A., & Santos, M. N. (2019). Local fishermen's perceptions of the usefulness of artificial reef ecosystem services in Portugal. *PeerJ, 6*, e6206. doi:10.7717/peerj.6206 PMID:30647998

Ramos, J., & Santos, M. N. (2015). Facilitating fishing decisions in an artificial reef area off southern Portugal: A case study using generalized additive models. *Journal of Applied Ichthyology, 31*, 24–34. doi:10.1111/jai.12948

Ramos, J., Santos, M. N., Whitmarsh, D., & Monteiro, C. C. (2006a). Patterns of Use in an AR System: A Case Study in Portugal. *Bulletin of Marine Science, 78*, 203–211.

Ramos, J., Santos, M. N., Whitmarsh, D., & Monteiro, C. C. (2006a). Patterns of use in an artificial reef system: A case study in Portugal. *Bulletin of Marine Science, 78*(1), 203–211.

Ramos, J., Santos, M. N., Whitmarsh, D., & Monteiro, C. C. (2006b). The usefulness of the analytic hierarchy process for understanding reef diving choices: A case study. *Bulletin of Marine Science, 78*(1), 213–219.

Ramos, J., Santos, M. N., Whitmarsh, D., & Monteiro, C. C. (2006b). The Usefulness of the Analytic Hierarchy Process for Understanding Reef Diving Choices: A Case Study. *Bulletin of Marine Science, 78*, 213–219.

Ramos, J., Santos, M. N., Whitmarsh, D., & Monteiro, C. C. (2007). Stakeholder perceptions regarding the environmental and socio-economic impacts of the Algarve artificial reefs. *Hydrobiologia, 580*(1), 181–191. doi:10.100710750-006-0454-z

Ramos, J., Santos, M. N., Whitmarsh, D., & Monteiro, C. C. (2007). Stakeholder perceptions regarding the environmental and socio-economic impacts of the Algarve artificial reefs. In *Biodiversity in Enclosed Seas and Artificial Marine Habitats* (pp. 181–191). Springer. doi:10.1007/978-1-4020-6156-1_16

Ramos, J., Tuaty-Guerra, M., Almeida, M., Raposo, A. C., Gaudêncio, M. J., Silva, A. D., Rodrigues, N., Leandro, S. M., & Caetano, M. (2021). An artificial reef at the edge of the deep: An interdisciplinary case study. *Ocean and Coastal Management, 210*(May), 105729. Advance online publication. doi:10.1016/j.ocecoaman.2021.105729

Rangel, M. O., Pita, C. B., Gonçalves, J. M. S., Oliveira, F., Costa, C., & Erzini, K. (2014). Developing self-guided scuba dive routes in the Algarve (Portugal) and analysing visitors' perceptions. *Marine Policy, 45*, 194–203. doi:10.1016/j.marpol.2013.10.011

Reef Design Lab. (n.d.). *MARS*. https://www.reefdesignlab.com/mars1

Reguero, B. G., Beck, M. W., Agostini, V. N., Kramer, P., & Hancock, B. (2018). Coral reefs for coastal protection: A new methodological approach and engineering case study in Grenada. *Journal of Environmental Management, 210*, 146–161. doi:10.1016/j.jenvman.2018.01.024 PMID:29339333

Regulation of the Minister of Maritime Affairs and Fisheries of the Republic of Indonesia. (2020). *The Minister of Maritime Affairs and Fisheries of the Republic of Indonesia Number 31/Permen-Kp/2020 concerning Management of Conservation Areas.* Author.

Relini, G., Relini, M., & Torchia, G. (1995). La Barrier artificiale di Loano. *Biologia Marina Mediterranea*, (2), 21–64.

Remmen, A., Astrup-Jensen, A., & Frydendal, J. (2007). Life cycle management: A business guide to sustainability. *UNEP and Danish Standards, 51.* https://www.lifecycleinitiative.org/wp-content/uploads/2013/09/UNEP_Background_document_LCM_2006_Febr.pdf

Restu, I. W. (2016). *Coral Reef Ecosystem and Its Status (Case Study of Coral Reef Condition) in Bali Province*. Udayana.

Revenga, S., Fernández, F., González, J. L., & Santaella, E. (1997). Artificial reefs in Spain: the regulatory framework. In A. C. Jensen (Ed.), *European Artificial Reef Research* (pp. 161–174). Southampton Oceanography Centre.

Ribeiro, C., & Neves, P. (2020). Habitat mapping of Cabo Girão Marine Park (Madeira island): A tool for conservation and management. *Journal of Coastal Conservation, 24*(2), 1–16. doi:10.100711852-019-00724-9

Rinkevich, B. (2000). Steps towards the evaluation of coral reef restoration by using small branch fragments. *Marine Biology, 136*(5), 807–812. doi:10.1007002270000293

Rizka, R. F., Purnomo, P. W., & Sabdaningsih, A. (2020). Effect of total suspended solid (tss) on zooxhanthellae density on Acropora sp. In laboratory scale. *Pasir Laut, 4*(1), 95–101.

Rizqika, C. N. A., & Latifah, N. (2018). Growth rate of coral reef *Acropora formosa* on Menjangan Small Island, Karimunjawa National Park. *Journal of Maquares, 7*(4), 315–322.

Roa-Ureta, R. H., Santos, M. N., & Leitão, F. (2019). Modelling long-term fisheries data to resolve the attraction versus production dilemma of artificial reefs. *Ecological Modelling, 407*, 108727. doi:10.1016/j.ecolmodel.2019.108727

Roberts, J. K., Thompson, M. E., & Pawlyk, P. W. (1985). Contingent Valuation of Recreational Diving at Petroleum Rigs, Gulf of Mexico. *Transactions of the American Fisheries Society, 114*(2), 214–219. doi:10.1577/1548-8659(1985)114<214:CVORDA>2.0.CO;2

Rocha, D. F., Franco, M. A. L., Gatts, P. V., & Zalmon, I. R. (2014). The effect of an artificial reef system on the transient fish assemblages – south-eastern coast of Brazil. *Journal of the Marine Biological Association of the United Kingdom*, *95*(03), 635–646. doi:10.1017/S0025315414001519

Rogers, C. S. (1990). Responses of coral reefs and reef organisms to sedimentation. *Marine ecology progress series. Oldendorf*, *62*(1), 185–202. doi:10.3354/meps062185

Rojas, S. (1997). Guide to realize social investigations. Plaza and Valdez.

Roseman, E. F., McLean, M., Pritt, J. J., Fischer, J. L., & Kennedy, G. (2017). Artificial reefs and reef restoration in the Laurentian Great Lakes. *Biodiversity, Conservation, and Environmental Management in the Great Lakes Basin*, 33-46.

Rosenberg, L. (2013). Sea-spike detection in high grazing angle X-band sea-clutter. *IEEE Transactions on Geoscience and Remote Sensing*, *51*(8), 4556–4562. doi:10.1109/TGRS.2013.2239112

Rouse, S., Kafas, A., Catarino, R., & Peter, H. (2018). Commercial fisheries interactions with oil and gas pipelines in the North Sea: Considerations for decommissioning. *ICES Journal of Marine Science*, *75*(1), 279–286. doi:10.1093/icesjms/fsx121

Saaty, T. L. (1977). A scaling method for priorities in hierarchical structures. *Journal of Mathematical Psychology*, *15*(3), 234–281. doi:10.1016/0022-2496(77)90033-5

Saaty, T. L. (1990). How to make a decision: The analytic hierarchy process. *European Journal of Operational Research*, *48*(1), 9–26. doi:10.1016/0377-2217(90)90057-I

Saaty, T. L. (2000). *Fundamentals of decision making and priority theory with the analytic hierarchy process* (Vol. 6). RWS publications.

Saaty, T. L. (2004). Decision making—The analytic hierarchy and network processes (AHP/ANP). *Journal of Systems Science and Systems Engineering*, *13*(1), 1–35. doi:10.100711518-006-0151-5

Saaty, T. L. (2008). Decision making with the analytic hierarchy process. *International Journal of Services Sciences*, *1*(1), 83–98. doi:10.1504/IJSSCI.2008.017590

Saaty, T. L., & Erdener, E. R. E. N. (1979). A new approach to performance measurement the analytic hierarchy process. *Design Methods and Theories*, *13*(2), 62–68.

SAGARPA. (2010). *Enrichment of Seagrass Meadows by means of Artificial Reefs, as a Coastal Fishing Management and Conservation Strategy in the Municipalities of Aquila and Coahuayana, Michoacán*. SAGARPA.

Salas-Zapata, W. A. (2011, September). The emerging science of sustainability: From scientific practice towards the constitution of a science. *Interciencia*, *36*(9), 699–706.

Salaün, J., Pioch, S., & Dauvin, J. C. (2020). Description of a theoretical social-ecological approach to manage artificial reefs. *Vie et Milieu/Life & Environment, 70*.

Salcido, R. E. (2005). Enduring optimism: Examining the rig-to-reef bargain. *Ecology Law Quarterly*, *32*, 863. doi:10.15779/Z38HZ7Q

Salim, D., Yuliyanto, Y., & Baharuddin, B. (2017). Physical-Chemistry Oceanographic Parameter Characteristics of Kerumputan Island, Kotabaru Regency, South Kalimantan. *Enggano Journal*, *2*(2), 218–228. doi:10.31186/jenggano.2.2.218-228

Salim, N., Bahauddin, A., & Mohamed, B. (2013). Influence of SCUBA divers' specialization on their underwater behavior. *Worldwide Hospitality and Tourism Themes*, *5*(4), 388–397. doi:10.1108/WHATT-03-2013-0015

Salleh, N. H. M., Aman, A. A., & Hamid, S. A. (2018). Selection of Artificial Reef Deployment Sites by Using Evidential Reasoning. *Marine Artificial Reef Research and Development: Integrating Fisheries Management Objectives*, *86*(January), 251–264.

Samples, K. C. (1989). Assessing recreational and commercial conflicts over artificial fishery habitat use: Theory and practice. *Bulletin of Marine Science*, *44*(2), 844–852.

Santos, M. N., Monteiro, C. C., & Leitão, F. (2011). The role of artificial reefs in the sustainability of artisanal fisheries. *Artificial reefs in Fisheries Management*, 221-237.

Santos, M. N., Monteiro, C. C., & Leitão, F. (2011). The role of artificial reefs in the sustainability of artisanal fisheries. In *Artificial reefs in Fisheries Management.* Taylor & Francis Group.

Santos, L. N., Brotto, D. S., & Zalmon, I. R. (2011). Assessing Artificial Reefs for Fisheries Management A 10-Year Assessment off Northern Coast of Rio de Janeiro. In S. A. Bortone, F. P. Brandini, G. Fabi, & S. Otake (Eds.), *Artificial reefs in fisheries management* (1st ed., pp. 125–140). CRC Press.

Santos, L. N., & Zalmon, I. R. (2015). Long-term changes of fish assemblages associated with artificial reefs off the northern coast of Rio de Janeiro, Brazil. *Journal of Applied Ichthyology*, *31*, 15–23. doi:10.1111/jai.12947

Santos, M. G., Monteiro, C. C., & Leitão, F. (2011). The Role of Artificial Reefs in the Sustainability of Artisanal Fisheries. In S. A. Bortone, F. P. Brandini, G. Fabi, & S. Otake (Eds.), *Artificial Reefs in Fisheries Management* (pp. 221–238). CRC Press Taylor & Francis Group.

Santos, M. N., Lino, P. G., Pousão-Ferreira, P., & Monteiro, C. C. (2006). Preliminary results of hatchery-reared seabreams released at artificial reefs off the Algarve coast (southern Portugal): A pilot study. *Bulletin of Marine Science*, *78*(1), 177–184.

Santos, M. N., & Monteiro, C. C. (1997). The Olhao artificial reef system (south Portugal): Fish assemblages and fishing yield. *Fisheries Research*, *30*(1-2), 33–41. doi:10.1016/S0165-7836(96)00567-X

Santos, M. N., & Monteiro, C. C. (1998). Comparison of the catch and fishing yield from an artificial reef system and neighbouring areas off Faro (Algarve, south Portugal). *Fisheries Research*, *39*(1), 55–65. doi:10.1016/S0165-7836(98)00169-6

Santos, M. N., & Monteiro, C. C. (2001). Modern Aquaculture in the Coastal Zone 281. *Modern Aquaculture in the Coastal Zone: Lessons and Opportunities*, *314*, 281.

Santos, M. N., & Monteiro, C. C. (2007). A fourteen-year overview of the fish assemblages and yield of the two oldest Algarve artificial reefs (southern Portugal). In *Biodiversity in Enclosed Seas and Artificial Marine Habitats* (pp. 225–231). Springer. doi:10.1007/978-1-4020-6156-1_19

Santos, M. N., Monteiro, C. C., & Lasserre, G. (1995). A four-year overview of the fish assemblages and yield on two artificial reef systems off Algarve (south Portugal). *Proceedings of the 30th European Marine Biological Symposium*, 345-352.

Santos, M. N., Monteiro, C. C., & Lasserre, G. (2005). Observations and trends on the intra-annual variation of the fish assemblages on two artificial reefs in Algarve coastal waters (southern Portugal). *Scientia Marina*, *69*(3), 415–426. doi:10.3989cimar.2005.69n3415

Santos, M., Monteiro, C., & Lasserre, G. (1996). Faune ichtyologique comparée de deux récifs artificiels du littoral de la Ria Formosa (lagune Portugal): Résultats préliminaires. *Oceanologica Acta*, *19*(1), 89–97.

Sato, M., Inoue, N., Nambu, R., Furuichi, N., Imaizumi, T., & Ushio, M. (2021). Quantitative assessment of multiple fish species around artificial reefs combining environmental DNA metabarcoding and acoustic survey. *Scientific Reports, 11*(1), 1–14. doi:10.103841598-021-98926-5 PMID:34593907

Sayer, M. D. J., & Wilding, T. A. (2002). Planning, licensing, and stakeholder consultation in an artificial reef development: The Loch Linnhe reef, a case study. *ICES Journal of Marine Science, 59*(suppl), S178–S185. doi:10.1006/jmsc.2002.1270

Scarcella, G., Grati, F., Bolognini, L., Domenichetti, F., Malaspina, S., Manoukian, S., Polidori, P., Spagnolo, A., & Fabi, G. (2015). Time-series analyses of fish abundance from an artificial reef and a reference area in the central-Adriatic Sea. *Journal of Applied Ichthyology, 31*, 74–85. doi:10.1111/jai.12952

Schaefer, N., Dafforn, K. A., Johnston, E. L., & Mayer-Pinto, M. (2018). Size, Depth and Position Affect the Diversity and Structure of Rock Pool Communities in An Urban Estuary. *Marine and Freshwater Research, 70*(7), 1034–1044. doi:10.1071/MF18074

Schroeder, D. M., & Love, M. S. (2002). Recreational fishing and marine fish populations in California. *California Cooperative Oceanic Fisheries Investigations Report, 43*, 182-190. https://lovelab.msi.ucsb.edu/Schroeder_Love2002.pdf

Schroeder, D. M., & Love, M. S. (2004). Ecological and political issues surrounding decommissioning of offshore oil facilities in the Southern California Bight. *Ocean and Coastal Management, 47*(1-2), 21–48. doi:10.1016/j.ocecoaman.2004.03.002

Schuett, M. A., Ding, C., Kyle, G., & Shively, J. D. (2016). Examining the behavior, management preferences, and sociodemographics of artificial reef users in the Gulf of Mexico offshore from Texas. *North American Journal of Fisheries Management, 36*(2), 321–328. doi:10.1080/02755947.2015.1123204

Schug, D. M. (1978). *Political and Economic Aspects of Artificial Reefs in Pinellas County, Florida* (Thesis). Department of Marine Science, University of South Florida, Tampa, FL.

Schulze, A., Erdner, D. L., Grimes, C. J., Holstein, D. M., & Miglietta, M. P. (2020). Artificial Reefs in the Northern Gulf of Mexico: Community Ecology Amid the "Ocean Sprawl". *Frontiers in Marine Science, 7*, 447. doi:10.3389/fmars.2020.00447

Schuschny, A., & Soto, H. (2009). *Methodological Guide. Design of composite indicators of sustainable development. Economic Commission for Latin America and the Caribbean.* ECLAC.

Schutter, M., Ven, R. M., van der, Janse, M., Verreth, J. A. J., & Wijffels, R. H. (2012). Light intensity, photoperiod duration, daily light flux and coral growth of Galaxea fascicularis in an aquarium setting: A matter of photons? *Journal of the Marine Biological Association of the United Kingdom, 92*(4), 703–712. doi:10.1017/S0025315411000920

SDGS. (2021). *A Gift Or A Curse In The Management Of Coastal Conflicts.* SDGS Center Universitas Hassanudin. https://sdgscenter.unhas.ac.id/anugerah-atau-kutukan-dalam-pengelolaan-konflik-wilayah-pesisir/

Seaman, W., & Lindberg, W. J. (2009). Artificial reefs. In Encyclopedia of Ocean Sciences (2nd ed., pp. 226–233). Elsevier Ltd. doi:10.1016/B978-012374473-9.00668-8

Seaman, W., Jr., & Jensen, A. C. (2000). Purposes and Practices of Artificial Reef Evaluation. In Artificial habitats for marine and freshwater fisheries. Academic Press.

Seaman, J., & Jensen, A. (2000). Purposes and practices of artificial reef evaluation. In *Artificial Reef Evaluation* (pp. 1–19). CRC Press. doi:10.1201/9781420036633-6

Seaman, W. (2000). *Artificial reef evaluation: with application to natural marine habitats.* CRC Press. doi:10.1201/9781420036633

Seaman, W. Jr. (2002). Unifying trends and opportunities in global artificial reef research, including evaluation. *ICES Journal of Marine Science, 59*(suppl), S14–S16. doi:10.1006/jmsc.2002.1277

Seaman, W. Jr, Grove, R., Whitmarsh, D., Santos, M. N., Fabi, G., Kim, C. G., & Pitcher, T. (2011). *Artificial reefs as unifying and energizing factors in future research and management of fisheries and ecosystems.* CRC Press. doi:10.1201/b10910-3

Seaman, W., & Sprague, L. M. (Eds.). (2000). *Artificial habitats for marine and freshwater fisheries.* Academic Press.

Şensurat-Genç, T., Lök, A., Özgül, A., Irmak, E., & Özkan, E. Y. (2021). Determination of Contribution to Diving Tourism and Environmental Effects of Shipwrecks Used as Artificial Reefs. TUBITAK 117Y0383project report.

Şensurat-Genç, T., Özgül, A., & Lök, A. (2017). The Use of Artificial Reefs for Recreational Diving. *Turkish Journal of Maritime and Marine Sciences, 3*(1), 27–33.

Sepulveda, S. (2002). Micro-regional Sustainable Development. In Sustainable development. Agriculture, Natural Resources and Rural Development. IICA.

Shani, A., Polak, O., & Shashar, N. (2012). Artificial reefs and mass marine ecotourism. *Tourism Geographies, 14*(3), 361–382. doi:10.1080/14616688.2011.610350

Sheehan, M. S., & Sickels-Taves, L. B. (2002). Vernacular building materials and the factors conditioning their use: Tabby, a case study. *Material Culture, 34*(2), 16–28.

Sheehy, D. (1979). *Fisheries development.* Warwe Spectrum. Winter.

Sheehy, D. J. (1982). The use of designed and prefabricated artificial reefs in the United States. *Marine Fisheries Review, 44*(6), 4–15.

Sherman, R. (2002). Artificial reef design: Void space, complexity, and attractants. *ICES Journal of Marine Science, 59*, S196–S200. doi:10.1006/jmsc.2001.1163

Siburian, R., Simatupang, L., & Bukit, M. (2017). Analysis Of The Quality Of Sea Waters On Activities In The Port Environment Of Waingapu- Alor. *East Sumba., 23*(1), 225–232.

Silvano, R. A. M., & Begossi, A. (2012). Fishermen's local ecological knowledge on southeastern Brazilian coastal fishes: Contributions to research, conservation, and management. *Neotropical Ichthyology, 10*(1), 133–147. doi:10.1590/S1679-62252012000100013

Silvano, R. A. M., Gasalla, M. A., & Souza, S. P. (2009). Applications of Fishers' local ecological knowledge to better understand and manage tropical fitropica. In P. Lopes & A. Begossi (Eds.), *Current Trends Human Ecology* (pp. 76–100). Cambridge Scholars Publishing. doi:10.5848/CSP.0441.00004

Sirait, A. P. R., Luthfi, O. M. M., & Isdianto, A. (2021). Physical Characteristics of Artificial Reef Post Deployed on 2017 in Trenggalek Damas Beach. *Journal of Marine and Coastal Science, 10*(1), 58. doi:10.20473/jmcs.v10i1.25607

Smith, J. A., Lowry, M. B., & Suthers, I. M. (2015). Fish attraction to artificial reefs not always harmful: A simulation study. *Ecology and Evolution. Bohnsack, 1989*(20), 4590–4602. doi:10.1002/ece3.1730 PMID:26668725

Smith, J. T., & Tighe, S. L. (2006). Analytic hierarchy process as a tool for infrastructure management. *Transportation Research Record: Journal of the Transportation Research Board, 1974*(1), 2–9. doi:10.1177/0361198106197400101

Soldal, A. V., Svellingen, I., Jørgensen, T., & Løkkeborg, S. (2002). Rigs-to-reefs in the North Sea: Hydroacoustic quantification of fish in the vicinity of a "semi-cold" platform. *ICES Journal of Marine Science, 59*(suppl), S281–S287. doi:10.1006/jmsc.2002.1279

Solomon, A.L. (2016). *Sampling techniques*. EAFIT University.

Sommer, B., Fowler, A. M., Macreadie, P. I., Palandro, D. A., Aziz, A. C., & Booth, D. J. (2019). Decommissioning of offshore oil and gas structures–Environmental opportunities and challenges. *The Science of the Total Environment*, *658*, 973–981. doi:10.1016/j.scitotenv.2018.12.193 PMID:30583191

Soong, K., & Chen, T. A. (2003). Coral transplantation: Regeneration and growth Acrophore Fragments in a nursery. *Restoration Ecology*, *11*(1), 62–71. doi:10.1046/j.1526-100X.2003.00100.x

Spalding, M., Burkeb, L., Wood, S. A., Ashpolee, J., Hutchisone, J., & Zu Ermgassene, P. (2017). Mapping the global value and distribution of coral reef tourism. *Marine Policy*, *82*, 104–113. doi:10.1016/j.marpol.2017.05.014

Spanier, E., Tom, M., & Pisanty, S. (1985). Enhancement of fish recruitment by artificial enrichment of man-made reefs in the southeastern Mediterranean. *Bulletin of Marine Science*, *37*(1), 356–363.

Stolk, P., Markwell, K., & Jenkins, J. (2005). Perceptions of artificial reefs as scuba diving resources: A study of Australian recreational scuba divers. *Annals of Leisure Research*, *8*(2&3), 153–173. doi:10.1080/11745398.2005.10600968

Stolk, P., Markwell, K., & Jenkins, J. M. (2007). Artificial reefs as recreational scuba diving resources: A critical review of research. *Journal of Sustainable Tourism*, *15*(4), 331–350. doi:10.2167/jost651.0

Stone, R. B., McGurrin, J. M., Sprague, L. M., & Seaman, W. Jr. (1991). Artificial Habitats of the World: Synopsis and Major Trends. In W. Seaman Jr & L. M. Sprague (Eds.), *Artificial Habitats for Marine and Freshwater Fisheries* (pp. 31–60). Academic Press Inc. doi:10.1016/B978-0-08-057117-1.50008-1

Streich, M. K., Ajemian, M. J., Wetz, J. J., & Stunz, G. W. (2017). A comparison of fish community structure at mesophotic artificial reefs and natural banks in the western Gulf of Mexico. *Marine and Coastal Fisheries*, *9*(1), 170–189. doi:10.1080/19425120.2017.1282897

Sui, J., Zheng, Z., Chang, Z., & Yu, Z. (2016). Dynamics analysis on the deployment of a frame-type artificial reef. *Fishery Modernization*, *43*, 74–79.

Sumarno, D., & Muryanto, T. (2014). Salinity Levels, Dissolved Oxygen, And Water Temperature In The Artificial Coral Reef Unit (Tkb) Kotok Kecil Island and Harapan Island. *Btl*, *12*(2), 121–126.

Sun, P., Liu, X., Tang, Y., Cheng, W., Sun, R., Wang, X., Wan, R., & Heino, M. (2017). The bio-economic effects of artificial reefs: Mixed evidence from shandong, China. *ICES Journal of Marine Science*, *74*(8), 2239–2248. doi:10.1093/icesjms/fsx058

Suparni. (2019). *Develop Tourism Village, Trenggalek Tourism Office Holds Training for Pokdarwis*. https://faktualnews.co/2019/03/12/kembangkan-desa-wisata-dinas-pariwisata-trenggalek-gelar-pelatihan-untuk-pokdarwis/128534/

Suryanti, Y., & Supriharyono. (2011). Effect Of Depth On Coral Morphology On Cemara Kecil Island, Karimunjawa National Park. *Journal of Fisheries Science*, *7*(1), 63–69.

Susilo, E., Purwanti, P., & Lestariadi, R. A. (2015). Utilization of Mangrove Resources at Pancer Cengkrong and Damas, Prigi Beach. *Trenggalek.*, *5*(2), 19–25.

Sutton, S. G. (2005). *An assessment of the social characteristics of Queensland's recreational fishers. Cooperative Research Centre for the great barrier Reef World Heritage Area technical report 65*. James Cook University.

Sutton, S. G., & Bushnell, S. L. (2007). Socio-economic aspects of artificial reefs: Considerations for the Great Barrier Reef Marine Park. *Ocean and Coastal Management*, *50*(10), 829–846. doi:10.1016/j.ocecoaman.2007.01.003

Sydney Institute Of Marine Science. (n.d.). *Mission. Living Seawalls.* https://www.livingseawalls.com.au/mission

Tabata, R. S. (1992). Scuba diving holidays. In B. Weiler & C. M. Hall (Eds.), *Special Interest Tourism* (pp. 171–184). Belhaven Press.

Tanamal, Y., Tuhumury, S. F., & Sangaji, M. (2019). Analysis of Suitability and Carrying Capacity of the Great Lagoon Rehabilitation Area and Slope Reef of Kipuo Lagoon, Negeri Ihamahu. Triton. *Journal of Aquatic Resource Management, 15*(1), 21–29.

Tang, Y., Zhang, Z., & Wang, X. (2021). Estimation of the Scale of Artificial Reef Sets on the Basis of Underwater 3D Reconstruction. *Journal of Ocean University of China, 20*(5), 1195–1206. doi:10.100711802-021-4751-y

Techera, E. J., & Chandler, J. (2015). Offshore installations, decommissioning and artificial reefs: Do current legal frameworks best serve the marine environment? *Marine Policy, 59*, 53–60. doi:10.1016/j.marpol.2015.04.021

Tessier, Francour, Charbonnel, Dalias, Bodilis, Seaman, & Lenfant. (2015). Assessment of French artificial reefs: due to limitations of research, trends may be misleading. *Hydrobiologia, 753*, 1–29. doi:10.1007/s10750-015-2213-5

Tessier, A., Dalias, N., & Lenfant, P. (2015). Expectations of professional and recreational users of artificial reefs in the Gulf of Lion, France. *Journal of Applied Ichthyology, 31*, 60–73. doi:10.1111/jai.12951

Texas Parks and Wildlife Department Artificial Reef Program. (2016). https://tpwd.texas.gov/landwater/water/habitats/artificial_reef/ 25/10/2017

Thaeraniza, S. T., Muzaky, O., & Andik, L. (2020). *Macroinvertebrata Recruitments in Artificial Reef After Two Years of Sinking*. Academic Press.

Thierry, J. M. (1988). Artificial reefs in Japan—A general outline. *Aquacultural Engineering, 7*(5), 321–348. doi:10.1016/0144-8609(88)90014-3

Thirumoorthi, T., Wong, K. M., & Musa, G. (2013). Scuba diving satisfaction. In *Scuba Diving Tourism* (pp. 170–176). Routledge.

Tiryakioğlu, F. Ö. (2008). *Socio-Economic Evaluation of Artificial Reefs In Aegean Sea, Turkey* (Master's dissertation). University of Portsmouth Business School.

Todd, V. L., Williamson, L. D., Cox, S. E., Todd, I. B., & Macreadie, P. I. (2020). Characterizing the first wave of fish and invertebrate colonization on a new offshore petroleum platform. *ICES Journal of Marine Science, 77*(3), 1127–1136. doi:10.1093/icesjms/fsz077

Tokaç, A., Ünal, V., Tosunoğlu, Z., Akyol, O., Özbilgin, H., & Gökçe, G. (2010). *Aegean Sea Fisheries*. IMEAK Deniz Ticaret Odası İzmir Şubesi Yayınları.

Treeck, V. P., & Schuhmacher, H. (1998). Mass Diving Tourism – A New Dimension Calls for New Management Approaches. *Marine Pollution Bulletin, 37*(8-12), 499–504. doi:10.1016/S0025-326X(99)00077-6

Trialfhianty, T. I., & Suadi. (2017). The role of the community in supporting coral reef restoration in Pemuteran, Bali, Indonesia. *Journal of Coastal Conservation, 21*(6), 873–882. doi:10.100711852-017-0553-1

Trilsbeck, M., Gardner, N., Fabbri, A., Haeusler, M. H., Zavoleas, Y., & Page, M. (2019). Meeting in the middle: Hybrid clay three-dimensional fabrication processes for bio-reef structures. *International Journal of Architectural Computing, 17*(2), 148–165. doi:10.1177/1478077119849655

Tunca, S. (2011). *Assessing the Socio-Economic Effects of Artificial Reef Deployments in the Northern Aegean Sea: Altınoluk Case* (MSc dissertation). Ege University Graduate School of Natural and Applied Sciences.

Tunca, S., Ünal, V., & Lizaso, J. L. S. (2019). Stakeholder Analysis of the Artificial Reefs in Edremit Bay (Northern Aegean), Turkey: Contributions to management. *International Biodiverstiy & Ecology Sciences Symposium.*

Tunca, S. (2012). Decisions of Stakeholders for the Proposed Artificial Reef Deployment: Analytic Hierarchy Process Approach. *Ege J Fish Aqua Sci, 29*(1), 21–29.

Tunca, S. (2013). Benefits of artificial reefs in Altınoluk (Northern Aegean), Turkey: Assessment of potential users' Opinions via Fuzzy Pair Wise Comparison Approach. *Ege J Fish Aqua Sci, 30*(3), 99–104.

Tunca, S. (2014). Perception and demand for artificial reefs by relevant local groups in Altınoluk (Turkey). *Ege J Fish Aqua Sci, 31*(1), 5–10.

Tunca, S., Miran, B., & Ünal, V. (2012). Decisions of stakeholders for the proposed artificial reef deployment: Analytic hierarchy process approach. *Ege Journal of Fisheries and Aquatic Sciences, 29*(1), 21–29.

Tunca, S., Miran, B., & Ünal, V. (2014). Perception and demand for artificial reef by relevant local groups in Altınoluk (Turkey). *Ege Journal of Fisheries and Aquatic Sciences, 31*(1), 5–10. doi:10.12714/egejfas.2014.31.1.02

Tunca, S., Miran, B., & Ünal, V. (2016). Non-Market Economic Value of Artificial Reefs in Edremit Bay (Eastern Mediterranean, Turkey). *Ege Journal of Fisheries and Aquatic Sciences, 33*(2), 97–107. doi:10.12714/egejfas.2016.33.2.02

Tynyakov, J., Rousseau, M., Chen, M., Figus, O., Belhassen, Y., & Shashar, N. (2017). Artificial reefs as a means of spreading diving pressure in a coral reef environment. *Ocean and Coastal Management, 149*, 159–164. doi:10.1016/j.ocecoaman.2017.10.008

Ulaş, A. (2007). A pre-study for determining of efficiency of fish sampling methods on artificial reefs. *Ege Journal of Fisheries and Aquatic Sciences, 24*(3-4), 287–293.

UN. (1987). *Report of the world commission on environment and development.* Retrieved from: http://www.ecominga.uqam.ca/PDF/BIBLIOGRAPHIE/GUIDE_LECTURE_1/CMMAD-Informe-Comision-Brundtland-sobre-Medio-Ambiente-Development.pdf

Ünal, V., Tunca, S., & Miran, B. (2013). Socioeconomic Impact of Edremit Bay Artificial Reefs Research Report. Ministry of Food, Agriculture and Livestock, General Directorate of Fisheries and Aquaculture, National Artificial Reef Monitoring Project.

Utami, T. M. R., Maslukah, L., & Yusuf, M. (2016). Distribution of Nitrate (NO3) and Phosphate (PO4) in Karangsong Waters, Indramayu Regency. *Buletin Oseanografi Marina, 5*(1), 31. doi:10.14710/buloma.v5i1.11293

Uyarra, M. C., Watkinson, A. R., & Cote, I. M. (2009). Managing dive tourism for the sustainable use of coral reefs: Validating diver perceptions of attractive site features. *Environmental Management, 43*(1), 1–16. doi:10.100700267-008-9198-z PMID:18810525

Valentin, J. L. (2001). The Cabo Frio Upwelling System, Brazil. In U. Seelinger & B. Kjerfve (Eds.), *Coastal Marine Ecosystems of Latin America* (Vol. 144). Springer. doi:10.1007/978-3-662-04482-7_8

Valentino, V. H., Setiawan, H. S., Saputra, A., Haryanto, Y., & Putra, A. S. (2021). Decision Support System for Thesis Session Pass Recommendation Using AHP (Analytic Hierarchy Process) Method. *International Journal of Educational Research & Social Sciences, 2*(1), 215-221.

Van den Burg, S. W. K., Schupp, M. F., Depellegrin, D., Barbanti, A., & Kerr, S. (2020). Development of multi-use platforms at sea: Barriers to realising Blue Growth. *Ocean Engineering, 217*, 107983. doi:10.1016/j.oceaneng.2020.107983

Vargas, L. G. (1990). An overview of the analytic hierarchy process and its applications. *European Journal of Operational Research, 48*(1), 2–8. doi:10.1016/0377-2217(90)90056-H

Vargas, L. G., & Zoffer, H. J. (2019). Applying AHP in conflict resolution. *International Journal of the Analytic Hierarchy Process, 11*(1), 143–147. doi:10.13033/ijahp.v11i1.649

Venegas, R. (2004). Property sustainability indicators. *Journal of Agroecology and Development,* (11/12).

Verkoeyen, S., & Nepal, S. K. (2019). Understanding scuba divers' response to coral bleaching: An application of Protection Motivation Theory. *Journal of Environmental Management, 231*, 869–877. doi:10.1016/j.jenvman.2018.10.030 PMID:30419442

Vicente, M., Falcão, M., Santos, M. N., Caetano, M., Serpa, D., Vale, C., & Monteiro, C. (2008). Environmental assessment of two artificial reef systems off southern Portugal (Faro and Olhão): A question of location. *Continental Shelf Research, 28*(6), 839–847. doi:10.1016/j.csr.2007.12.009

Victorian Fisheries Authority. (n.d.). *Rebuilding Shellfish Reefs in Port Phillip Bay.* https://vfa.vic.gov.au/recreational-fishing/recreational-fishing-reefs/shellfish-reefs

Vié, J.-C., Hilton-Taylor, C., & Stuart, S. N. (Eds.). (2009). Wildlife in a Changing World – An analysis of the 2008 IUCN Red List of Threatened Species™. IUCN.

Vivier, B., Dauvin, J. C., Navon, M., Rusig, A. M., Mussio, I., Orvain, F., Boutouil, M., & Claquin, P. (2021). Marine artificial reefs, a meta-analysis of their design, objectives and effectiveness. *Global Ecology and Conservation, 27*, e01538. doi:10.1016/j.gecco.2021.e01538

Volvo. (n.d.). *Living Seawall - An Ocean Conservation Project.* Volvo Car Australia. https://www.volvocars.com/au/why-volvo/discover/living-seawall

Walker, S. J., & Schlacher, T. A. (2014). Limited habitat and conservation value of a young artificial reef. *Biodiversity and Conservation, 23*(2), 433–447. doi:10.100710531-013-0611-4

Wallace, D. L. (2012). Simplified beta-approximations to the Kruskal-Wallis H test. *Journal of the American Statistical Association, 54*(285), 225–230.

Wang, Z., Chen, Y., Zhang, S., Wang, K., Zhao, J., & Xu, Q. (2015). A comparative study of fish assemblages near aquaculture, artificial and natural habitats. *Journal of Ocean University of China, 14*(1), 149–160. doi:10.100711802-015-2455-x

Weiers, R. M. (1986). *Investigación de Mercados.* Prentice Hall Interamericana.

White, A. T., Chou, L. M., De Silva, M. W. R. N., & Guarin, F. Y. (1990). Artificial reefs for marine habitat enhancement in Southeast Asia. *ICLARM Education Series, 11*, 45.

Whitmarsh, D., Santos, M. N., Ramos, J., & Monteiro, C. C. (2008). Marine habitat modification through artificial reefs off the Algarve (southern Portugal): An economic analysis of the fisheries and the prospects for management. *Ocean and Coastal Management, 51*(6), 463–468. doi:10.1016/j.ocecoaman.2008.04.004

Wielgus, J., Glassom, D., & Fishelson, L. (2003). Long term persistence of low coral cover and abundance on a disturbed reef flat in the northern Red Sea. *Journal of Experimental Marine Biology and Ecology, 297*(1), 31–41. doi:10.1016/S0022-0981(03)00354-X

Wilkinson, C. R. (2000). Executive summary. In C. R. Wikinson (Ed.), *Status of Coral Reefs of the World* (pp. 7–19). Australian Institute of Marine Science.

Williams, K., Rooper, C., & Towler, R. (2010). Use of stereo camera systems for assessment of rockfish abundance in untrawlable areas and for recording Pollock behavior during midwater trawls. *Fishery Bulletin–National Oceanic and Atmospheric Administration, 108*, 352–362.

Wilson, C. A., & Van Sickle, V. R. (1987). Development of the Louisiana artificial reef pro-gramme. *Bulletin of Marine Science, 44*(2), 1071–1072.

Wirtz, P., Genner, M., & Relvas, C. (2020). *First record of Cobia (Rachycentron canadum) (Pisces: Rachycentridae) at the coast of Madeira Island.* NE Atlantic Ocean.

Wongthong, P., & Harvey, N. (2014). Integrated coastal management and sustainable tourism: A case study of the reef-based SCUBA dive industry from Thailand. *Ocean and Coastal Management, 95*, 138–146. doi:10.1016/j.ocecoaman.2014.04.004

Yale Center for Environmental Law and Policy Yale University, Center for International Earth Science Information Network Columbia University, World Economic Forum, & Joint Research Centre European Commission. (2005). *2005 Environmental Sustainability Index (ESI).* NASA Socioeconomic Data and Applications Center (SEDAC).10.7927/H40V89R6

Yamane, T. (1989). Status and future of artificial reef projects in Japan. *Bulletin of Marine Science, 44*(2), 1038–1040.

Yaşar, O. (2011). Underwater Diving Tourism on The Coasts of The Gulf of Saros. *Zeitschrift für die Welt der Türken/ Journal of World of Turks ZFWT, 3*(1), 33-55.

Yeasmin, S., & Rahman, K. F. (2012). 'Triangulation' research method as the tool of social science research. *Bup J, 1*(1), 154–163.

Yonar, M., Luthfi, O. M., & Isdianto, A. (2021). Dynamics of Total Suspended Solid (TSS) Around Damas Beach Coral Reef, Trenggalek. *Journal of Marine and Coastal Science, 10*(1), 48–57. doi:10.20473/jmcs.v10i1.25606

Yoon, H. S., Kim, D., & Na, W. B. (2016). Estimation of effective usable and burial volumes of artificial reefs and the prediction of cost-effective management. *Ocean and Coastal Management, 120*, 135–147. doi:10.1016/j.ocecoaman.2015.12.007

Yozzo, D. J., Wilber, P., & Will, R. J. (2004). Beneficial use of dredged material for habitat creation, enhancement, and restoration in New York–New Jersey Harbor. *Journal of Environmental Management, 73*(1), 39–52. doi:10.1016/j.jenvman.2004.05.008 PMID:15327845

Yusuf, M., Handoyo, G., & Wulandari, S. Y. (2012). Characteristics of Current Patterns in Relation to Water Quality Conditions and Phytoplankton Abundance in the Waters of the Karimunjawa Marine National Park. *Buletin Oseanografi Marina, 1*(5), 63–74. doi:10.14710/buloma.v1i5.6918

Zadeh, L. (1965). Fuzzy sets. *Information and Control, 8*(3), 338–353. doi:10.1016/S0019-9958(65)90241-X

Zalmon, I. R., Novelli, R., Gomes, M. P., & Faria, V. V. (2002). Experimental results of an artificial reef programme on the Brazilian coast north of Rio de Janeiro. *ICES Journal of Marine Science, 59*(March), S83–S87. doi:10.1006/jmsc.2002.1273

Zalmon, I. R., Boina, C. D., & Almeida, T. C. M. (2012). Artificial reef influence on the surrounding infauna-north coast of Rio de Janeiro State, Brazil. *Journal of the Marine Biological Association of the United Kingdom, 92*(6), 1289–1299. doi:10.1017/S0025315411001147

Zalmon, I. R., Sá, F. S., Dornellas Neto, E., Rezende, C. E., Mota, P. M., & Almeida, T. C. M. (2014). Impacts of artificial reef spatial configuration on infaunal community structure— Southeastern Brazil. *Journal of Experimental Marine Biology and Ecology, 454*, 9–17. doi:10.1016/j.jembe.2014.01.015

Zavoleas, Y., & Haeusler, M. H. (2017). Extended Modelling: Dynamic Approaches Applied to Design Reef Habitats at Sydney Harbour. *Shock! Sharing Computational Knowledge, 35th International Conference on Education and Research in Computer Aided Architectural Design in Europe (eCAADe), 1*, 67-74. http://papers.cumincad.org/cgi-bin/works/Show?ecaade2017_146

Zavoleas, Y., Haeusler, M. H., Dunn, K., Bishop, M., Dafforn, K., Schaefer, N., Sedano, F., & Yu, K. D. (2020). Designing Bio-Shelters: Improving Water Quality and Biodiversity in the Bays Precinct through Dynamic Data-Driven Approaches. *Journal of Digital Landscape Architecture, 2020*(5), 521–532. doi:10.14627/537690053

Zimme-Faust, R. K., & Tamburri, M. N. (1994). Chemical Identity and Ecological Implications of a Waterborne, Larval Settlement Cue. *Limnology and Oceanography, 39*(5), 1075–1087. doi:10.4319/lo.1994.39.5.1075

Zisserman, A., Fitzgibbon, A., & Cross, G. (1999). VHS to VRML: 3D graphical models from video sequences. *Proceedings IEEE International Conference on Multimedia Computing and Systems*, 51-57. 10.1109/MMCS.1999.779119

About the Contributors

Jorge H. P. Ramos is an Integrated Researcher of the Research Centre for Tourism, Sustainability and Well-being (CinTurs) at the University of Algarve, Portugal. He has worked on socioeconomics related to coastal issues, particularly ecotourism, small-scale fisheries and offshore aquaculture. His research interests include aspects related to ecotourism, stakeholder analysis, artificial reefs, among other areas. He holds a PhD in Fisheries Economics and Management, a PgC in Social Research Methods, a MSc in Fisheries Economics (University of Portsmouth, UK) and a graduate degree in Marine Biology and Fisheries (University of Algarve, Portugal).

* * *

Muhammad Asadi is a lecturer at the Marine Sciences Study Program, Universitas Brawijaya. Completed undergraduate studies in Marine Science at Diponegoro University and master studies in Marine Ecology at Wageningen University, The Netherlands. He teaches Chemical Oceanography, Marine Pollution, Ecotoxicology, Tropical Marine Ecology, Widya Diving, Swimming and Marine Zoology. The author's area of expertise is Marine Pollution and Mangrove Ecology. The author is also an active reviewer in several international publishers such as Elsevier (Journal Marine Pollution Bulletin, Environmental Challenges, and Science of the Total Environment), AACL Bioflux, and Journal of Coastal Research (JCR).

Kate Dunn is an Academic and Research Fellow UNSW Art, Design, Architecture B.A. Visual Arts (USyd), Hons (SCU), Ph.D. Architecture (USyd). Kate's research investigates experimental 3D Printing, Digital Fabrication, and Robotics with a focus on the development of new and sustainable materials for digital fabrication. Kate's research builds on traditional material processes and integrates them into emerging digital technologies for a range of industries including the Built Environment, Medicine, and Multi-Modal Data Visualisation. Kate convenes and teaches Design courses at UNSW Built Environment and has over 15 years of undergraduate and postgraduate teaching experience in Australia's leading Universities.

Mochammad Fattah was born in Malang, East Java, on May 13, 1986. Graduated from SD Taman Harjo 2 Singosari in 1998, MTsN 1 Malang in 2001, and SMAN 7 Malang in 2004. Bachelor's degree at the Faculty of Fisheries and Marine Science, University of Brawijaya , completed in 2008. In 2011, completed his master's study at the Postgraduate of Brawijaya University, in the Environmental Resources Development and Management Study Program. Currently, he is a lecturer at the Faculty of Fisheries and Marine Sciences, University of Brawijaya starting in 2012. The research being developed is about the management of fishery resources. Actively participates in seminars held by several universities or the Ministry of Marine Affairs and Fisheries. The publication of a book in 2012 on the Strategy of Management of Fish Sanctuaries and in 2017 with the title Management of the Fisheries Industry was published by UB Press. As well as actively writing journals related to fisheries resource management, including: sustainability of fishery resources, fishery potential, fisheries ecotourism, fisheries area planning, fishery assessment and others.

Charlotte Firth is a computational designer and researcher at the University of New South Wales. Charlotte has received a Bachelor of Computational Design with Distinction and is currently completing a Masters of the Built Environment from the University of New South Wales. Her research focuses on developing end-effectors for collaborative robotics for on-site construction, focused on improving the well-being and safety of trades workers. Charlotte previously was a student designer at Cox Architecture working on commercial developments.

Prasad G. is Assistant Professor (Sr.Gr.), Aerospace Engineering, Dayananda Sagar University, Bangalore, India. Prof. Prasad G, received his Masters degree and Bachelors degree PhD in Aeronautical Engineering from Anna University. He has 7 years of teaching experience. Prior to joining Dayananda Sagar University, he worked at Anna University. He has completed two funded project sponsored by The Institution of Engineers (India) and Tamilnadu State Council for Science and Technology. Awarded Indian National Science Academy (INSA) Visiting Scientist Programme 2019 and Awarded Science Academies' Summer Research Fellowship Programme (SRPF) 2019. He is a Professional Membership of American Institute of Aeronautics and Astronautics, Institution of Engineers, Life Member in Indian Cryogenic Council and Life Member in Shock Wave Society. Reviewer in Aircraft Engineering and Aerospace Technology, International Journal of Engine Research and Journal of The Institution of Engineers (India): Series C.

Matthias Haeusler is Founding Director of Computational Design / Built Environment at ADA UNSW Sydney; Deputy Director ADA UNSW AI Institute; Head of Research, Foresight, and Innovation at Giraffe Technology; and known as a researcher, educator, entrepreneur, and designer in media architecture and computational design through over 160 publications.

Muchamad Fairuz Haykal, born in Malang, February 6, 1999, is the eldest of two children, Mr. Mochamad Ali Ngisom and Mrs. Nila Farista. Elementary school formal education was taken at Wandanpuro 2 Elementary School until 2011. Continuing junior high school at SMPN 1 Bululawang graduated in 2014 and continued high school at SMAN 4 Malang graduated in 2017. Bachelor's Degree education was taken at Brawijaya University Malang from 2017 to July 2021 take the Marine Science Study Program. Many organizational activities are also followed such as Scouts and Student Council (Junior High School), Basketball Extracurricular, and Actively Become an Athlete in Malang Regency for Basketball (High School). During his studies, Muchamad Fairuz Haykal was the administrator of the Universitas Brawijaya (Himalaya) Marine Science Student Association (Himalaya) from 2019 to 2020. The current author is a graduate of Marine Science from Brawijaya University who is active in ecological research. Various writings on the topic of conservation and marine biology were initiated while in college when completing an internship with the topic of identifying seagrass species on Kondang Buntung Beach, then until now many writings from authors have been presented at national, international seminars and scientific publications through various journals. Current research topics are ecology, coastal management, conservation, and ecosystem resilience.

Candra Adi Intyas studied from Elementary School to Strata Two (S2) in Malang. Graduated with a Bachelor of Fisheries (S1) at the Faculty of Fisheries and Marine Sciences, Department of Fisheries Socio-Economics (2003–2007) and a Bachelor's Degree (S2) at the Postgraduate Program in Agriculture, Universitas Brawijaya Malang (2014–2017) with the title "Cumlaude". After graduating Bachelor (S1), the author works at the feed company PT. Luxindo Internusa, which is headquartered in East Bekasi and has Vannamei Shrimp enlargement projects in several areas in Indonesia such as Sumbawa, Sulawesi, West Kalimantan and East Java with an initial position as a Lab Analyst Advisor and then a supervisor for Human Resources Management (HRD) and data management (2007 –2011). After gaining enough experience, the author decided to resign and settle in Malang with his first job as an administrative staff at the Home Industry for frozen food products "Al Kautsar" (2011–2012), the passion felt by the author in the field of education led the author to become a trainer plus administrative staff in the English Course "Primagama English Malang" and became a teacher at the International Standard Elementary School "Indonesia Interactive Standard School" (2012-2013). In September 2013, the author was accepted to work as an Assistant Lecturer and became the Vice Dean II Staff (2013–2015). Currently, he has become a Permanent Lecturer in the Fisheries Agribusiness Study Program, Department of Fisheries and Marine Socio-Economics, Faculty of Fisheries and Marine Sciences, Universitas Brawijaya, and a staff of the Vice Dean I (2018-2021).

Andik Isdianto is a lecturer at the Marine Science Study Program, Department of Utilization of Fisheries and Marine Resources, Universitas Brawijaya Malang, Indonesia. Graduated with a Bachelor's degree in Urban and Regional Planning, Faculty of Engineering, Universitas Brawijaya Indonesia and a master's degree in Coastal Management Engineering at the Sepuluh Nopember Institute of Technology Surabaya, Indonesia. Teaches Marine and Fishery Resources Conservation, Coastal and Marine Area Management, Coastal Ecosystem Rehabilitation, Marine Protected Areas, Marine Ecotourism, Coastal and Marine Area Management, Marine Fisheries Law and Regulations, Mapping of Marine Biological Resources, Marine Geographic Information Systems, Remote Sensing Marine, Marine Acoustics, Mapping and Environmental Monitoring. Has written various scientific papers on the topics of coastal resilience, coastal management, and fisheries and marine affairs, particularly those related to coral reef ecology.

Sripad Kulkarni S. has post graduated in field of Advanced Material Technology from University Visvesvaraya College of Engineering, Bengaluru and graduated in the field of Mechanical Engineering from APS college of Engineering. His area of interest is manufacturing science, additive manufacturing, Biomechanics. He has completed NPTEL Certification in Structural analysis of Nanomaterials and Introduction to Research.

Juliano Lima is a professor at the Federal Institute of Sergipe and a PhD in Ecology and Natural Resources with an emphasis in Marine Ecology from the State University of Rio de Janeiro (2020). He holds a degree in Biological Sciences from the Federal University of Sergipe (2007). Specialization in Education (2009) and Management in Public Policies (2013) by the Federal University of Sergipe.

Altan Lök completed his bachelor degree in Faculty of Fisheries at the Ege University, Turkey. He focused on fisheries technology and use of artificial reefs in fisheries management. Advised the preparation of the National Artificial Reef Program. He took part in national and international projects on fisheries technology and artificial reefs.

Shankar P. S. is a scholar in Aerospace Department, Dayananda Sagar University, Bangalore, India.

Sylvain Pioch is currently a researcher at LAGAM laboratory and associate professor at University Paul-Valéry Montpellier in environmental/urban planning, dedicated to marine and coastal area. After a PhD in coastal geography partly done in TUMSAT (Japan), he continued as a post-doctoral fellow at Nova University (Florida, USA), and IFREMER (France) working on ecological engineering and scoring methods to value gains and losses of ecological functions. He works on issues linked to the mitigation of anthropogenic impacts, green infrastructure (harbour & maritime infrastructures, eco-designed, eco-friendly materials, artificial habitats), governance of mitigation in France and the US (laws & rules, guideline and ref. frameworks, transactional processes) and ecological engineering tools. He teaches Graduate students at University Paul-Valéry Montpellier, Nova University (Florida) and in 6 other different training programs.

Pudji Purwanti was born in Jember Regency, East Java on February 28, 1964. She studied from basic to undergraduate education in Malang, East Java Indonesia. Bachelor's education at the Faculty of Fisheries, University of Brawijaya was completed in 1988. Masters Education at the Faculty of Agriculture, Gadjah Mada University, Graduated in 1994 in the field of Agricultural Economics. Doctoral study in 2005 at the Doctoral Program of Agricultural Sciences, University of Brawijaya with an interest in Agricultural Economics, Graduated in 2008. Since 1989 until now she has been working as a lecturer at the Faculty of Fisheries and Marine Sciences University of Brawijaya. Currently, she is the Head of the Fisheries and Marine Socio-Economic Laboratory. The author has also been active in managing the Fisheries Research Journal, Faculty of Fisheries, Universitas Brawijaya from 2000 to 2009. During her time as a lecturer, the author is active in research activities and community service. The results of these research and service activities have been published in several reputable national and international journals. In addition, the author also produces several books that have been published, including the Economic Model of Small-Scale Fishermen's Households (UB Press, 2010); Sustainable Mangrove Forest Management: Institutional and Economic Intensive Approach (UB Press, 2017); Human Adaptation, Food Security and Social Security of Resources (UB Press, 2017); Fishery Industry Management (UB Press, 2017).

Berlania Mahardika Putri, born in Malang, October 15th, 1999, has completed her formal education at SDN Bareng 3 in 2011, SMPN 1 Malang in 2014, and SMAN 1 Malang in 2017. The author continues her education at the Bachelor Degrees in the Marine Sciences Study Program, Brawijaya University and completed a final project entitled Differences Growth of Crustose Coralline Algae on Collector Substrates in South Malang Waters in 2021. During college, the author was active in campus non-academic activities such as the Fisheries Diving Club as a general secretary, member and division head of the Marine Science Student Association Universitas Brawijaya (HIMALAYA) and active as a practicum assistant for Marine Zoology and Marine Coralogy. Currently the author is actively writing books and research journals, especially related to coral reef ecosystems and marine biology that published at national and international levels. The current concentration of research topics leads to conservation, especially coral reef ecosystems.

Jose Antonio Santamaria Reyes has a Master's Degree in Public Policy from the Michoacan University of San Nicolás de Hidalgo. This work was carried out during his studies at the Institute of Economic and Business Research of the Universidad Michoacana (ININEE). Currently, a PhD student in Human Ecology at CINVESTAV-Mérida, he works with fishermen and the Mayan octopus (Octopus maya) in Yucatán, México.

Miguel Neves dos Santos has a PhD on Fisheries Science and Technologies. During the first 25 years of his career, he was based in Portugal mainland where he holds a position of Senior Researcher at the (currently) Portuguese Institute for the Sea and Atmosphere. During that period his research activities were related with national and international commercial fisheries, from small-scale to industrial fisheries. In 2016 he moved to the Secretariat of the International commission for the Conservation of Atlantic Tunas (ICCAT), being currently the Assistant Executive Secretariat of the Organization, a position he accumulated with the post of Science Coordinator. He has a large experience of leadership and coordination of collaborative work, having worked closely and cooperatively with scientists from the Academia, Public Laboratories and tuna Regional Fisheries Management Organizations from many; fisheries managers and technical and administrative staff; fishermen; and, NGOs. The results of his scientific work have been widely spread, being author of more than 130 scientific papers and book chapters published in peer review journals, and more than 100 articles and scientific working documents published in other journals. In additional, he has also a large experience as coordinator of over 20 national and international research projects and programs, related to coastal and highly migratory fish biology, ecology and conservation, fishing gear impacts and artificial reefs ecology.

Edi Susilo was born December 5, 1959 in Trenggalek Regency, East Java, Indonesia. Elementary Education in 1972, Junior High School in 1975, and High School 1979 in the Department of Natural Sciences. S1 education at the Faculty of Fisheries, Brawijaya University, Malang (1984) Intensive course for 6 months in Social Science Research, Airlangga University (1986). Masters education at Bogor Agicultural University, in the field of Rural Sociology (1991) and doctoral studies at the Faculty of Agriculture, Brawijaya University in the field of Rural Sociology. Previously worked as Secretary of the Center for Social Science Research, Universitas Brawijaya (2001-2005), became Secretary of the Social Sciences Program (2003-2004) and is now FISIP. In 2003-2005 he became the Site Adviser of the Cofish Project. Actively participates in seminars organized by several universities (UGM, UNHAS, UIN, UNRI, UB), IMFISERN and ISI. He wrote a chapter in a book published by INTRANS (2004), and GHALIA (2015), and in 2010 wrote a book on Social Structure Dynamics in Coastal Ecosystems published by UB Press. The books he has written are Fish Sanctuary Area Management Strategy (2012), Human Adaptation, Food Security and Social Security of Resources (2017), Sustainable Mangrove Forest Management (2017), Fisheries Law (2019). Wrote several articles, among others, published in the Journal of Research in Ecology (2017) and the Heliyon Journal (2020).

Vahdet Ünal completed his bachelor degree in Agricultural Economics at the Ege University, Turkey. He focused on fisheries economics and management during his career. He is a member of the European Association of Fisheries Economists and Mediterranean Conservation Society. He coordinated the Sub-

Committee of Economic and Social Science in GFCM-FAO from 2009 to 2013. He is working at the Ege University, managing and consulting national and international projects related to fisheries co-management and the Ecosystem Approach to Fisheries. He also shares his experiences by delivering relevant lectures in International Master Program on Sustainable Fisheries Management at Alicante University, Spain.

Gabriel Undery is a studio-based designer, research practitioner, and sessional academic working in the UNSW Design Futures Lab, a multi-disciplinary teaching and research space. He has more than 10 years of professional experience as an industrial model maker, delivering product design services to designers, entrepreneurs, and start-ups. In his current role as a workshop-based educator, Gabriel is passionate about enhancing hands-on skills as a core part of the student experience and is a strong advocate for the importance of craft and practice-led research in design. It is this ethos which has mostly influenced and come to define his professional practice as a designer, maker, and educator to date.

Ilana Zalmon is Senior Professor at University of North Rio de Janeiro since 1995. Main research is Marine Community Ecology (artificial reefs, rocky shores, and sandy beaches).

Yannis Zavoleas is Associate Professor (tenured) at University of Ioannina, Greece, and Senior Lecturer (Adjunct) at University of New South Wales, Australia. Experienced Researcher, Architect, and Designer with a demonstrated history of working in academia. Skilled in interdisciplinary design thinking and advanced computing. Autodesk Maya, Dynamic Simulation, Digital Fabrication, Interdisciplinary Research, Architecture and Urban Design, Computational Design, Comprehensive Planning, Sketching, and Sustainable Design Thinking. Strong entrepreneurship professional with a Ph.D. focused on Architectural Engineering from the National Technical University of Athens and Master's degrees from University of California at Los Angeles and Massachusetts Institute of Technology in Architectural Design and Comparative Media Studies.

Index

A

B

C

D

E

F

G

H

L

M

N

O

P

R

www.ingramcontent.com/pod-product-compliance
Lightning Source LLC
LaVergne TN
LVHW062144300926
849161LV00010B/2624